Economics of Water Management in Agriculture

Economics of Water Management in Agriculture

Editors

Thomas Bournaris
Department of Agricultural Economics
Aristotle University of Thessaloniki
Thessaloniki
Greece

Julio Berbel
University of Cordoba
Campus de Rabanales
Cordoba
Spain

Basil Manos
Department of Agricultural Economics
Aristotle University of Thessaloniki
Thessaloniki
Greece

Davide Viaggi
Department of Agricultural Economics and Engineering
University of Bologne Viale Fanin
Bologna
Italy

CRC Press
Taylor & Francis Group
Boca Raton London New York

CRC Press is an imprint of the
Taylor & Francis Group, an **informa** business
A SCIENCE PUBLISHERS BOOK

CRC Press
Taylor & Francis Group
6000 Broken Sound Parkway NW, Suite 300
Boca Raton, FL 33487-2742

Printed on acid-free paper
Version Date: 20140707

International Standard Book Number-13: 978-1-4822-3839-6 (Hardback)

Visit the Taylor & Francis Web site at
http://www.taylorandfrancis.com

and the CRC Press Web site at
http://www.crcpress.com

Preface

Sustainable water management is perceived as an issue of growing importance for society and research, including economic research. Agriculture, being the most important water consuming sector worldwide is at the core of research in the field of water economics and management. Water resources are becoming scarcer in arid and semi-arid regions of the world, in which agriculture plays a crucial role. On the other hand, irrigation contributes significantly to agricultural production and it is one of the most viable forms of agricultural activity.

This book outlines the current trends in the economics of water management in agriculture. Employing state of the art methodologies it addresses some of the most relevant current and perspective issues for the water policy debate offering a wide variety of innovative approaches and original and relevant cases in European irrigated agriculture. It reports recent economic research on the theory, practice and policies of water management and investigates problems on the economics of water management in an agricultural context. Specific case studies applying new and innovative technologies, methods and techniques in water management are considered.

This volume is grounded in the experiences of the editors in this field. We identified the need to present in a book the latest results of relevant research, either in its pure theoretical aspects or in applied instances. The chapters are written for professionals looking for enhancing the knowledge base of this subject. Such a book can serve as an excellent dissemination tool for water management qualitative and quantitative issues, water markets, demand analysis, economic analysis and implementation of economic issues in the water management in agriculture.

The target audience of this book is professionals and researchers working in the field of water management and its use in agriculture. This includes agricultural and environmental economists, especially those concerned with the economics and management of water resources, academics, agronomists, farming industry practitioners and policy makers in governance institutions and regional authorities. The book contains papers from distinguished scholars who have examined critical issues in the economics of the relationship between water and agriculture, with a special focus on irrigation and provides insights and a reference for scholars and PhD students concerned with water management, agricultural economics and climate change. Also, stakeholders at different levels can especially gain from practices and experiences applied by specific case studies applying new and innovative technologies, methods and techniques in water management.

The book is divided into four sections addressing the main areas for the economics of water management in agriculture. The first section concerns Water Management, Distribution and Governance issues and incudes three chapters. The second section covers Water Economics using evaluation and scenario analysis tools and methods, also pricing mechanisms and market issues. The third section, from chapter nine to twelve focuses on issues of effects from the implementation of the Water Framework Directive (pricing and disproportionality criteria) and the connection between Common Agricultural Policy Rural Development Programs and irrigation. The fourth section addresses the Water Demand in agriculture.

Thomas Bournaris
Julio Berbel
Basil Manos
Davide Viaggi

Contents

Section 3: Water Management and Effects from Water Framework Directive, Agriculture and Energy Policy

Section 4: Water Demand

SECTION 1

Water Management, Distribution and Governance Issues

CHAPTER

1

Understanding Equity and Equality in Sustainable Irrigation Water Management

*Solveig Kolberg** and *Julio Berbel*

Introduction

> *"The arguments against existence [of equity] take three different forms. The first is that equity is merely a word that hypocritical people use to cloak self-interest—it has no intrinsic meaning so therefore fails to exist. The second—is that even if equity does exist in some notional sense, it is so hopelessly subjective that it cannot be analyzed scientifically—it fails to exist in an objective sense. The third argument that there is no sensible theory about it—thus it fails to exist in an academic sense."*
>
> Young (1994)

Despite almost every water management system in the world having equity as a fundamental policy objective, there are misconceptions and lack of understanding of what equity and equality mean in irrigation water management that make it difficult to measure and monitor its

Department of Economics, Sociology and Agricultural Policy, University of Cordoba, Campus de Rabanales, C5, 3° floor, 14014 Cordoba, Spain.
Email: berbel@uco.es
* Corresponding author: Solveig.Kolberg@gmail.com

implementation at all scales and levels. Despite attention over several decades, the concept of equity has proven difficult to define. Often the concepts of 'distribution', 'equality' and 'equity' are used as if their meanings are obvious, and at times, they are used interchangeably. Occasionally, 'equality' and 'equity' are also applied interchangeably when qualifying some other concept, such as 'access' which represents another unhelpful lack of distinction (Williams and Doessel 2006). Equality and equity are not necessarily the same. Equality can be defined as the state of being equal and can be measured with descriptive inequality statistics. Equity refers to being fair, impartial or right judgment and is characterized by conflicting perceptions. Equity is a complex idea that is strongly shaped by cultural values by precedent, and by the specific types of goods and burdens being distributed (Young 1994).

Distributional Justice and Dimensions of Equity

> *"Justice is the tolerable accommodation of the conflicting interests of society, and I don't believe there is any royal road to attain such accommodation concretely."*
>
> Judge Learned Hand (1872–1961)

Bojer (2003) describes the main theories of distributional justice (distribution of rights and resources), from utilitarianism and welfare economics, moving to Rawls's social contract and Sen/Nussbaum's capability approach; she also describes empirical methods of inequality measurement. Bojer claims that there is a gap between what philosophers write and what is studied in empirical analysis. Some examples of important moral philosophers that seem quite unconcerned with how their concepts can be made operational for empirical analysis are Roland Dworkin, Martha Nussbaum, John Rawls and Amartya Sen. These are philosophers that are not at all concerned with welfare, but with opportunities, resources, rights and capabilities. According to them, achieving individual welfare and happiness is the person's own responsibility, while the state is responsible to further the means to and remove constraints on the pursuit of happiness. Since the end of the 1980s, there have been a number of studies exploring community perceptions of fairness and justice in water management (Tisdell 2003). Syme et al. (1999) and others have developed social and psychological theories of justice, equity and fairness, which again have explored the adequacy of equity and procedural justice in explaining individual water allocation decisions. These approaches, however, enter into *perceptions* of what is fair, and that is beyond the scope of this article.

Rasinski (1987) and Syme et al. (1999) show, in the context of social welfare policy, that equity comprises two components, 'proportionality' and 'egalitarianism'. *Proportionality* implies that resources should be distributed according to efforts or needs (as in the Marxist mantra 'from each according to their abilities, to each according to their need'), while in the case of egalitarianism, the term suggests that everyone should be treated equally. Boelens et al. (1998) distinguish five levels of equity in irrigation and water management at local levels. These comprise:

- Equitable water distribution and allocation among different water users and uses,
- Equitable distribution of services involved in irrigation development,
- Equitable distribution of the added agricultural production and other benefits under irrigation,
- Equitable distribution of burdens and obligations related to functions and positions,
- Equitable distribution of the rights to participate in decision making processes, since this relates to the fundamental issue of whether or not every farmer has rights to speak, vote, claim an entitlement of irrigated land and enjoy equality of status in leadership elections, etc.

Phansalkar (2007) further divides two of the above mentioned levels, namely equity in access to and use of water, and the distribution of the impacts of water resource development intervention, into four categories:

- Social equity: equity between different groups of people living in the same location.
- Spatial equity: equity between people living in different regions (Saleth and Dinar 2004).
- Gender equity: equity between men and women in sharing labor costs, efforts to access and use water, and its benefits.
- Inter-generational equity: equity in the enjoyment of natural resources, including water, across generations of people (Divan and Rosencranz 2005).

There has been an increasing focus on the concept of social equity or distributive justice as one of the guiding principles of contemporary people-centered development paradigm. Social science literature on developmental practices defines social equity as social justice in benefit sharing or the fair distribution of benefits (Uprety 2005). Moreover, Syme and Fenton (1993) affirm that the concepts of equity and procedural justice (fair process) have greater significance as competition for water resources augments. For further reading on equity related concepts and social justice, see Young (1994).

The causes of socioeconomic inequality have been disputed since the time of Plato. Wilkinson (2005) claims that the main practical argument in favor of reducing economic inequality is because economic inequality weakens society, hinders social and economic development, and could affect social and political stability (Box 1).

Box 1. How greater inequality leads to poorer social relations.

Greater income inequality
⇓
Increased social distances between income groups, less sense of common identity
More 'them' and 'us'
⇓
More dominance and subordination, superiority and inferiority, snobbery and downward discrimination, hierarchical and authoritarian values
⇓
Increased status competition, shift to more anti-social values, emphasis on self-interest and material success, carelessness of other's welfare, aggressive exploitation of society for individual gain
⇓
Others as rivals: poorer quality of social relations
Source: Wilkinson (2005).

Molle (2004) refers to formal equity in water management; however, he does not provide any explicit definition. Kolberg (2012) proposes that the term *'formal equity'* could be used to define the distributional criteria that the law and legislation have established as fair, through a public participation process, and thus be measured with the help of inequality measures.

Equity in International Agreements and Commitments

"If water is so fundamental a biological requirement in agriculture, if irrigation water (or other outstream flows) is now widely recognized to be an economic good, and if irrigation water constitutes about 70 per cent of all diversions, then there is a need for an economics of irrigation."

Merrett (2002)

Not only regional and national institutions set the agenda for water management. Centrally or externally mandated multilateral institutions such as the United Nations (UN) and World Trade Organization (WTO) build on, homogenize and reproduce standard expectations worldwide, stabilizing international order (Bandaragoda 2000). Thus, new paradigms and management approaches have emerged. The 1987 report of the World Commission on Environment and Development, also referred to as The Brundtland Commission, defined 'sustainable development' as 'development which meets the needs of the present without compromising the ability of future generations to meet their own needs' (WCED 1987). Five years later, the Rio Earth Summit concluded that: 'the right to development must be fulfilled so as to equitably meet developmental and environmental needs of present and future generations' (UN 1992). Meanwhile, 'sustainable development' has become one of the most prominent catchwords on the world political agenda.

The majority of governments and multinational firms have committed themselves to the overall concept of sustainable development. Hitherto, sustainable development, which is not just about the environment, but about the economy and the society, has proven hard to define (Böhringer 2004). One reason for this is that sustainable development explicitly incorporates a (normative) equity dimension, which is 'so hopelessly subjective that it cannot be analyzed scientifically' (Young 1994). Another reason is that the scope of the concept seems prohibitively comprehensive to make it operational in concrete practice (Böhringer 2004). Nonetheless, societal policy is being challenged to come up with pragmatic approaches to sustainable development and to this end, requires practical advice from the scientific community. Inherently, the three dimensions of sustainable development, i.e., environmental quality, economic performance (gross efficiency) and equity concerns, are intertwined and subject to tradeoffs (Fig. 1).

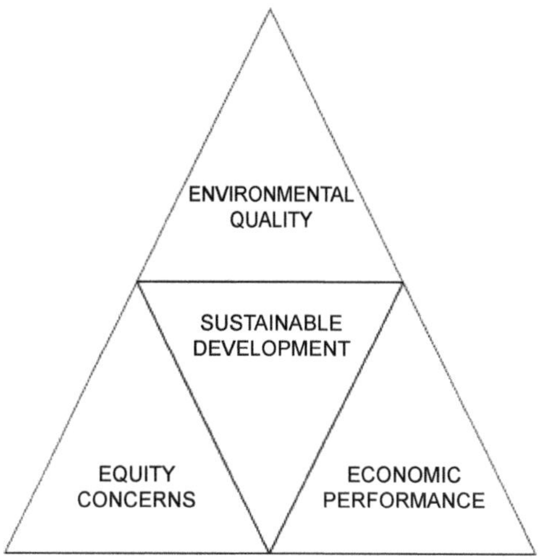

Figure 1. Dimensions of sustainable development. Source: Kolberg 2012.

Lévite and Sally (2002) argue that equity in water allocation involves a fair access for all water users to the water needed for their activities and that attention should also be paid to efficient and beneficial use in order to achieve sustainability. Similarly, Gleick (1998) claims that equity overlaps with sustainability when defining what is to be sustained, for whom, and who should decide. Sustainable water management (SWM) implies managing water in a holistic way, taking into account the various sectors affecting water use, including political, economic, social, technological and environmental considerations. SWM has been high on the international agenda since the Mar del Plata Water Conference, hosted by the UN in 1977 (DAINET n.d.), and it has been redefined several times since then. Current understanding of SWM is founded, above all, upon the principles developed during the International Conference on Water and the Environment in Dublin in 1992 (ICWE 1992) (Box 2).

The interpretation of the concept 'water as an economic good' has taken two directions: i) water should be priced through the market by ensuring it is allocated to the highest valued uses and ii) the process of integrated allocation decision making of scarce resources, which does not necessarily involve financial transactions (see, e.g., McNeill 1998; Perry et al. 1997 cited in Van der Zaag and Savenije 2006). The WFD (EC 2000) and other policy documents acclaim the need of economic analysis to assist in sustainable management of water resources, especially in arid areas where competition and conflicts over water are more prevalent. Most of these economic analyses seem to focus on economic productivity and efficiency as an end in itself and ignore the larger social and equity aspects of water resources.

During the 1990s, water management was extended to include efficient water use, equitable sharing of benefits, and environmental sustainability. This is referred to Integrated Water Resource Management (IWRM). In 2002, at the World Summit on Sustainable Development in Johannesburg, the goal was to develop integrated water resources management plans for all countries by 2005 (WWAP 2009). Equity is the least understood of the 3 E's (equity, economic efficiency and environmental sustainability) in the concept of integrated water resources management (Fig. 2). It remains a nebulous concept, and little efforts have been made to clarify its scope or content within the water context (Peña 2011).

As water scarcity increases and potential conflicts loom, it is crucial to define equity related concepts at different levels of water management to increase transparency and to facilitate dialogue and water negotiations. The research to date, including that which has been done so far for the

Box 2. The Four Dublin Principles.

These principles recommend action at local, national and international levels to reverse the trends of overconsumption, pollution, and rising threats from drought and floods.

Principle No. 1
Fresh water is a finite and vulnerable resource, essential to sustain life, development and the environment. A holistic approach, linking social and economic development with protection of natural ecosystems. Effective management links land and water uses across the whole of a catchment area or groundwater aquifer.

Principle No. 2
Water development and management should be based on a participatory approach, involving users, planners and policy-makers at all levels. Raising awareness of the importance of water among policy-makers and the general public. Decisions to be taken at the lowest appropriate level, with full public consultation and involvement of users in the planning and implementation of water projects.

Principle No. 3
Women play a central part in the provision, management and safeguarding of water. Positive policies required to address women's specific needs and to equip and empower women to participate at all levels in water resources programs, including decision-making and implementation, in ways defined by them.

Principle No. 4
Water has an economic value in all its competing uses and should be recognized as an economic good. Access to clean water and sanitation at an affordable price is a basic right of every person. Past failure to recognize the economic value of water has led to wasteful and environmentally damaging uses of the resource. Managing water as an economic good is an important way of achieving efficient and equitable use, and of encouraging conservation and protection of water resources.

Source: (ICWE 1992).

implementation of The European Water Framework Directive (WFD), has tended to focus on economic efficiency and environment, rather than equity. Expanding and improving irrigation water use provide economic benefits to the society but it may not necessarily imply that the benefits and costs are distributed equally and/or equitably to all sections of society.

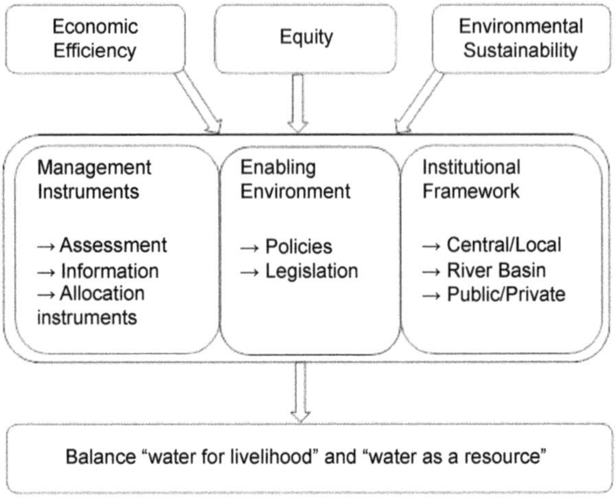

Figure 2. The three pillars of Integrated Water Resources Management.
Source: Kolberg 2012 adopted from UNESCO (2009) cited in the East Asian Seas Congress (2009).

Equitable Water Allocation

"...coping with water scarcity [is] the challenge of the 21st century".

FAO Director-General Dr Jacques Diouf, the World Water Day Celebration 2007

Water can be defined as a good that is homogeneous[1] and divisible (Young 1994) and the supply of it may be fixed or variable. Water is also considered as an economic good (ICWE 1992). In economics, the simplest problem of fairness is that of dividing a homogeneous commodity among a group of agents with equal claims on it. A distinction is made between horizontal and vertical equity, where horizontal equity implies that equals should be treated equally and vertical equity implies that unequals should be treated unequally (Elliot 2009).

In a world of emergent scarcity and growing inequality between water 'haves' and 'have-nots', the issues of equitable water allocation and appropriate water management are likely to become two of the most pressing issues in the 21st century (Boelens et al. 1998). Water use is a frequently studied topic that has gained increasing importance in the period since 1990, as some regions, economies and communities ran out of water

[1] For agriculture the quality of water is probably of less relevance than for domestic water use.

permanently or temporarily, at least for some uses (Allan 1996). In the last 50 years, the world's population has doubled while the water extraction has tripled. Until the 1990s, and continuing in some countries, there was very little interaction between water use sectors. Instead, the sectors worked independently, with specialists in water supply and sanitation, hydropower, irrigation, flood control and so on (WWAP 2009).

Equity in water management appears to be important at all levels. However, the interpretation of the term is often ambiguous, and its impact on water management is not discussed in professional debate (Wegerich 2007). Perceptions of basic liberties and procedural and distributive justice are frequently at the core of numerous water conflicts throughout the world (Tisdell 2003). During the past 15 years there have been a number of studies on community perceptions of fairness and justice in water management and the development of fairness principles (Ibid.). However, there is currently no system or standard methodology in place to measure water allocation related inequality in terms of inputs and outputs, especially in irrigation water management, and above all at basin level. At the same time, efficient water use is increasingly central to the economic well-being of individual regions and countries facing water scarcity (Livingston 1995). Equitable and economic rational uses of water are key objectives of most water policies. A rational use of irrigation water—the world's largest use of water—becomes increasingly important as irrigation water becomes scarce and competition increases.

Equitable Allocation of Irrigation Water

The natural and renewable water resources in the world are by nature unequally distributed within and between countries. For example, in the Mediterranean there is unequal distribution between, the 'rich' north and the 'poor' to 'extremely poor' south and east. As much as 81% of the water resources in Spain are located in the northern half of the country (Kayamanidou 1998). In addition to water resources being naturally distributed unequally within a country, human intervention also unequally distributes water between and within sectors. Irrigated agriculture accounts for a large share of total water withdrawals in the world. In the Mediterranean countries it represents 83% in Greece, 68% in Spain, 57% in Italy, and 52% in Portugal, while it represents less than 10% in Northern European countries (Berbel et al. 2007). Widespread water resource withdrawal and droughts could exacerbate the water supply variability as a result of the drier warmer climate due to the impacts of climate change. States and the local stakeholders' adaptation to growing scarcity implies (Molle et al. 2010):

- Supply responses; by augmenting the supply from existing sources, as well as tapping additional sources;
- Conservation responses; or 'efficiency in use' by making better use of existing resources, without increasing the supply or the number of sources of water; and
- Allocation responses; by reallocating water from one user to another, either within the same sector (e.g., agriculture) or across sectors.

The increase of supply based on building water control structures (dams, polders, drainage ditches, etc.) often changes water regimes, with consequences for the distribution and allocation of water resources among different stakeholders (Chowdhury et al. 1997). The technical, economic and environmental costs related with continued development of new sources during scarcity are high, and makes this approach undesirable, for meeting future demand. Conversely, the allocation of water for irrigation is, in many countries, considered as a low priority (Gorantiwar and Smout 2006). Accordingly, more recently, irrigation has received a reduced share of the total supply due to increased demand from higher valued uses, such as industrial, domestic and recreational ones (Ibid.).

Irrigation is losing out to other sectors in the competition for water (Molle et al. 2010). The world's food production needs immense amounts of water and land and is, by far, the largest consumer of water worldwide. Crops consume about 7,130 km^3 of water annually to meet global food demand. This corresponds to more than 3,000 L per person per day, where 78% comes directly from rainfall and 22% from irrigation (De Fraiture and Perry 2007). Lack of water constrains food production for hundreds of millions of people (Comprehensive Assessment of Water Management in Agriculture 2007). When water becomes a major constraint to agricultural production, farmers are likely to respond by intensifying agricultural production, changing cropping patterns and/or introducing more efficient crops, or irrigating crops that previously only were rain fed. The intra-sectorial allocation criteria for irrigation become crucial, as they eventually define who gets what, and consequently, if distribution is equitable and economically efficient. Allocation of deficit water resources is a complex issue that normally increases the potential for conflicts among farmers, between rural areas and cities, and between upstream and downstream. As a result, the pressures for fair allocation criteria from both outside and within the sector are increasing. The perceived inefficient use of irrigation water has become less tolerable, and so has its adverse impact on water quality. Many consider that the agricultural sector could contribute more to combat both the water quantity and quality challenges in arid RBs. Water restrictions are overwhelmingly imposed on irrigation, while other activities and domestic supply are only affected in cases of very severe shortage. In closing basins, irrigators have to respond to the challenge posed by both short and long term declining water allocations (Molle 2010a).

To date there has been little agreement on the intrinsic meaning of equity and equality in the context of irrigation, and the concepts are sometimes used interchangeably in the literature. Water shortage in arid RBs demands achieving 'fair' sharing of available water resources in order to avoid social tensions. Several authors have proposed measures and attempted to measure inequality in irrigation management (see, e.g., Sampath 1988). However, to date there are no standard methods to measure equity in water management. Most of these studies are irrigation scheme level analyses different from the current study that takes a basin level approach. Sivramakrishna and Jyotishi (2006) stress the importance of addressing both the distribution of inputs and of outputs. Cullis and van Koppen (2007) argue that there is a need for more case studies on specific basins to develop a better understanding of the relationship between equality in the use of water, the benefits of water use [economic output] and equity under different RB conditions.

In irrigation, equity does not necessarily mean that every irrigator receives the same, equal amount of water. It rather implies that each irrigator gets an amount that is fair (Laycock 2007). The question arises: what is a fair water allocation? Though equality can be a key component of equity, the relationship between equity and equality depends very much upon how the concepts are defined (Cullis and van Koppen 2007). In practice, large-scale irrigation systems' water entitlements are almost always based on equality, rather than equity, because it is difficult to accurately determine what a society considers to be a fair way to share water. Further, many larger irrigation systems are constructed in areas that have previously not been irrigated, covering several different communities which may have varying views on fairness, and where there is ineffective communication between system designers and potential water users. In addition, some irrigators may use a larger share of water than others, either due to prior rights (prior appropriation), in compensation for more input in system construction, or maintenance. The result is that it is much easier for irrigation system designers to develop systems based on an assumed concept of equality, which later is assumed to be equitable.

In irrigation systems, the most frequent form of division is by area, implying that each unit area of land is given the same water allocation (Murray-Rust et al. 2000). In some smaller systems managed entirely by the local community, a water share may be assigned to each person irrespective of the area of land they own or cultivate, and can include landless members of the community. Also, and more difficult to estimate and systematically measure, equality of water distribution may be based on the expected potential productivity of land resources, giving more water to more productive land or that soils with high water holding capacities receive less water than soils with lower water holding capacity (Murray-Rust et al. 2000).

Equity and Water Rights

Water rights and equity are among the most debatable water issues (AbuZeid and Elrawady 2008), especially when water resources gets scarce. In arid climates, problems of water scarcity and levels of rainfall are matters of public interest and concern. There is no universally agreed definition on the term 'water right' (Hodgson 2006). The term is used in different contexts and jurisdictions, and has come to mean somewhat different things. Water law, and therefore water rights, reflects economic, social and cultural perceptions of water. These in turn are formed by factors including geography, climate and the extreme variability in water availability and the uses to which water is put. Figure 3 shows that a distinction can be made between 'basic' water rights (defined in primary legislation), from 'allocated' water-use rights or usufruct rights (decided through a defined administrative process). In addition, reserves retained in the river or aquifer for environmental or other sustainability related downstream purposes may either be legislated as a basic right (ADB 2009) or decided administratively through the water resources planning process.

Water rights are closely linked to land rights, as well as rights to the use of irrigation infrastructure. This could include reservoirs and canal systems, tanks, energized tube wells and mechanized pumps. These play

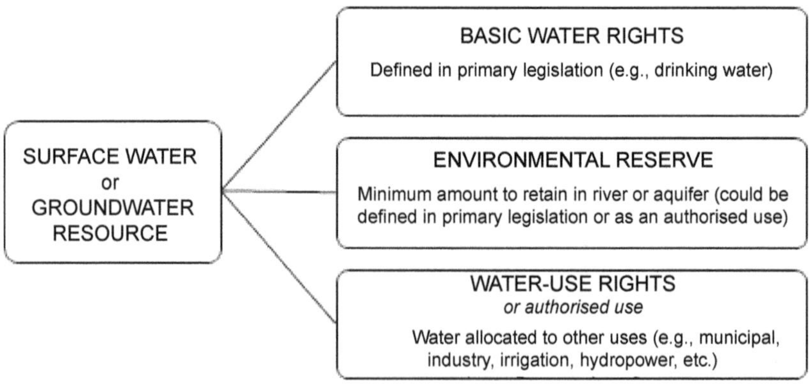

Figure 3. Water Rights, environmental reserve, and water-use rights.
Source: ADB (2009).

a critical role in ensuring access to water. *Access to water* may be defined as the availability of water in the right quantity and quality, at the right moment, and in the right place. Water rights play a critical role in defining who has access to water and who do not (Hodgson 2006). Water rights have been defined as a type of property right that aims, along with other water institutions and 'landed property rights', to assign access, use, liability and control over water from some persons and social groups to others (Wescoat 2002). Uncertainty regarding water quantity and location, in addition to demand for specific amounts of water at specific times and locations, makes water rights a highly complex and controversial issue.

The Need to Link Equity and Efficiency

Tsur and Dinar (1995) define an efficient allocation of water resources as an allocation that maximizes the total net benefit that can be generated by the available quantity of the resource. According to Marsh and Schilling (1994), costs, burdens and amenities, 'efficiency' and 'effectiveness' are the most important criteria in decisions on the allocation of resources. However, generally these criteria are not sufficient for generating acceptable and implementable decisions, and another criterion is required—is the allocation fair? (Ibid.). Water management approaches may be clear on their objectives regarding equality or equity in distribution of input (e.g., water rights and annual allowances of water) without realizing the full implications of such policies on output or outcomes (e.g., economic return, water productivity and employment). In the end, social welfare, however, ultimately depends on the distribution of outcomes, whether equitable or equal. Whether equity of income ought to be a target of irrigation management is uncertain as it goes against the idea that people who work harder and assume higher risk than others deserve more income than others. Also, equity is difficult to ensure because the decisions should be fair and free from bias and should ensure social justice in the distribution of social costs and benefits of water management projects. It is often assumed that the equity objective conflicts with the efficiency objective (e.g., Msangi and Howitt 2007; Molle 2009; Shah et al. 2009). Sampath (1992) argues that this does not necessarily have to be the case as, under certain conditions, the promotion of efficiency can be compatible with improved equity (Sampath 1984, 1988, 1990b, cited in Sampath 1992), while policies introduced to promote equity have sometimes resulted in a simultaneous decrease in efficiency and equity. Dinar and Tsur (1999) investigate efficiency and equity performance of various irrigation water pricing methods, and conclude that the extent to which water pricing methods can affect income redistribution is rather limited. They claim that farm income disparities are due mainly to such factors as farm size and location, and soil quality, but not to water (or other input) prices. Small and

Rimal (1996) analysed several irrigation systems in Asia, and found that efficiency and equity trade-offs become more important with increasing water scarcity.

Measuring Inequality

"All happy families are alike, but every unhappy family is unhappy in its own way."

Opening sentence from the book Anna Karenina by Leo Tolstoy (1828–1910)

Inequality measurement is an attempt to give meaning to comparisons of distributions in terms of criteria which may be derived from ethical principles, mathematical constructs, or simple intuition (Cowell 2009). Inequality measures are most frequently used for dynamic comparisons (comparing inequality measures across time), and for policy analysis (e.g., to compare inequality across regions or by population sub-groups) (Vecchi 2007). The methodology of inequality measurement is not novel, as it has been widely applied in many settings. The empirical application of this methodology to water allocation, however, is relatively novel. However, there is a lack of standard approaches to select relevant variables, the unit of analysis and choice of measurement, not bridging criteria for inequality with, nor what measures are more suitable at different levels and scales. The paucity of studies and agreed upon approaches to apply this methodology justify its further exploration, considering alternative approaches of measurement that allow comparing outcomes of water allocation on not only efficiency, but also equity in water allocation at basin level.

While there is only one way a distribution can be equal, there are infinite ways for it to be unequal, and unequal distributions, like Tolstoy's unhappy family (see quote above), are all unequal in their own way (Bojer 2003). Frequencies, mean, and variance, are well-known statistical measures to describe a distribution. In addition, explicit methods have been developed to describe and measure the inequality of a distribution. There are several established methodologies on how to measure productivity and efficiency; however, currently there are neither standard methods, nor monitoring systems in place to reliably measure the impact of a water allocation on, e.g., social, temporal and territorial equity in water use at basin level.

Despite vast and rapidly expanding empirical research on inequality measurement, to date, few studies have applied inequality measures to quantify how irrigation water is allocated within a RB. The most common, next to standard measures of dispersion, are the coefficient of variation, the Gini coefficient and the Lorenz curve.

Yet there are few empirical studies on water use allocation applying inequality measures and concentration curves. Cullis and van Koppen (2007) use Gini coefficient and Lorenz curve to measure water use inequality and indirect benefit among domestic water users in Olifants water management area in South Africa, and Sun et al. (2010) use environmental Gini coefficient and Lorenz curves to study an allocating wastewater discharge permit in Tianjin, in China. Lorenz curve and Gini coefficient have also been used to assess yield inequality within paddocks (Sadras and Bongiovanni 2003). The coefficient of variation has been used by several authors. For example, Akkuzu et al. (2007) used this measure on water delivery in irrigation systems in irrigation areas in Gediz, Turkey; and Murray-Rust et al. (2000) used it to study water distribution equity in Sindh Province, Pakistan. A descriptive inequality measure can be defined as a statistical measure of the deviation from equality of a distribution and gives a complete ordering over the set of possible distributions of the resource (Bojer 2003). Cowell (2009) defines an inequality measure to be a scalar numerical representation of the interpersonal differences in resources within a given population. The use of scalar indicators implies that all the different features of inequality are compressed into a single number. Coulter (1989) has collected about 50 inequality measures, but probably there exist a few more.

A common inequality measure is the coefficient of variation (CV) that is the ratio of the standard deviation (σ) to the mean ($\bar{y}$) (adapted from Cowell 2009):

$$CV = \frac{\sigma}{\bar{y}} = \frac{\sqrt{v}}{\bar{y}}$$

The CV is independent of measurement unit, and is more relevant than, e.g., the variance (v) as inequality analysis requires comparisons. When all resources are equal then CV = 0, because v = 0 (Bellù and Liberati 2006). There is no upper limit. The CV measures the relative variation independently of the mean resource level.

The Gini coefficient (GI) is one of the most widely used inequality measures and is defined as the area between the line of perfect equality and the observed Lorenz curve. There are various formulas for arithmetic calculation of the Gini coefficient. This is one of them (Bojer 2003):

$$GI = \frac{2\Sigma_j Y_j}{n^2 \bar{y}} - \frac{n+1}{n}$$

Given that resources Y are ranked according to size, and j is the ranking number and $\bar{y}$ is the mean. The advantage of GI is that it is a widely known measure and easy to explain and interpret in a non-technical way. Though it is often claimed that the Gini coefficient tends to give greatest weight to the middle part of the distributions, this is incorrect, as it emphasizes that part of the distribution where the density is greatest (Bojer 2003). Another important inequality measure is the Theil index that has higher resolution for changes to higher resource and is given by (Ibid.):

$$T = \frac{1}{n}\sum_{j}^{n}\frac{Y_j}{\bar{y}}\,\text{In}\left(\frac{Y_j}{\bar{y}}\right)$$

One of the advantages with the Theil index is that it allows for perfect and complete decomposition of the total level of inequality into the inequality within sub-groups of the population (Conceição and Ferreira 2000). Decomposable indexes can provide an analytic and practical way to understand the origin and structure of inequality.[2]

Inequality measures can also be selected on the basis of axioms. The axiomatic approach allows us to obtain a mathematical formula that delivers a class of inequality measures which satisfy a set of elementary properties (axioms) that we think inequality measures ought to have. The most common are (Cowell 2009):

- Anonymity (or symmetry): it does not matter whose the high and low water receiving hectares are.
- Population independence: inequality does not change by changes in the size of the population.
- Scale independence means that if each IU's water allocation changes by the same proportion, then inequality should not change.
- Normalization: if all individual hectares have the same water use, there is no inequality.

Evaluating equity and equality usually involves a comparison of the impact or effect of an action on two or more individuals or groups. Marsh and Schilling (1993) organize groups along four dimensions as provided in the Table 1.

Most studies of irrigation and inequality use a physical irrigator, an irrigation entity or spatial area as a unit of analysis. Gorantiwar and Smout (2005) list equity considerations and indicators for irrigation used by different authors, including statistical measures of dispersion and inequality measures proposed for inequality measurement for irrigation schemes (Table 2).

These studies make use of different indicators to describe irrigation scheme performance in relation to a set of context specific objectives. Moreover, there are few irrigation equity studies and a lack of standard approaches. These are reasons why it is difficult to provide meaningful

[2] Please refer to Cowell (2009) for more details on these and other inequality measures that appear in this chapter.

Table 1. Group dimensions for evaluating inequality.

Group dimension	Description	Examples
Spatial	Jurisdictional boundaries or unit areas that partition a spatial surface into mutually exclusive groups	States, counties, square kilometers & legislative districts
Physical	Geologic, biologic, or geographic features that may divide a spatial surface, or may be distributed throughout the surface	Land use, forest type & habitat
Demographic	Social or human characteristics that are typically distributed over a spatial surface	Population, income, ethnic group & age
Temporal	Time; any category above may also be defined over a period of time	Years, decades, generations

Source: Adopted from Marsh and Schilling (1993).

Table 2. Irrigation performance indicators related to equity by author.

Author	Indicator
Abernethy (1986)	Christianson coefficient (Christianson 1942), standard deviation (Till and Bos 1985), interquartile ratio (Abernethy 1984), Gini coefficient and Shannon-Wiener. However preferred modified interquartile ratio (the average depth of water received by all land in the best quarter, divided by the average depth received in the poorest quarter)
Sampath (1988)	Relative mean deviation, the variance, the coefficient of variation, the standard deviation of logarithms, the Gini coefficient and Theil's information measure (Theil 1967). Preferred Theil's information measure
Molden and Gates (1990); Kalu et al. (1995)	Coefficient of variance (CV) of spatial water distribution to field plots as a measure of inequity and thus (1 - CV) as measure of equity
Steiner (1991)	Relative mean deviation, coefficient of variation, inter-quartile comparison and Gini coefficient
El-Awad et al. (1991)	Absolute average deviation
Bird (1991)	Inter quartile ratio
Goldsmith and Makin (1991)	A normalized equity index called interquartile ratio (Abernethy 1986)
Kaushal et al. (1992)	Christiansen uniformity coefficient, coefficient of variation, modified IQR and Theil index
Bhutta and Van der Velde (1992)	Inter quartile ratio (Abernethy 1986)
Bos et al. (1994)	Modified interquartile ratio (Abernethy 1986) for overall equity and Head: Tail equity ratio (Vander der Velde 1992) for looking at the difference between head and tail of the canal

Source: Gorantiwar and Smout (2005).

comparisons of results. For example Sampath (1988) estimates the ratio of total wetted area (sum of the area wetted by each irrigation over the irrigation season) to cultivable command area, while Bos et al. (1994) applied the flow rate for measuring the performance and El-Awad et al. (1991) used volume. Bos et al. (1994) used discharge in the form of delivery performance ratio for equity measures. Theil index is particularly useful for irrigation related inequality analysis due to its axiomatic properties and decomposition quality (Sampath 1988; Kolberg 2012).

River basin analysis becomes important in many places as basin level water availability becomes the most important constraint to agriculture (Pretorius et al. 2005). Inequality studies at river basin level could provide opportunities to examine the consequences of claims made for basin-level water resources, factors that appear to affect the implementation of integrated water management, and the outcomes of water allocation and hydrological planning. However, some issues are challenging to address at basin level due to the lack of data or poor quality of data, whereas site-specific situations might create the need for investigating other aspects. Kolberg (2012) propose a frequency weighting approach to homogenize management data that has different level and scale for basin level analysis (e.g., mix of individual farms and entire irrigation communities). In order to compare between years and basins, the weighting should be made at the least aggregated level (homogeneous) and with inequality measures that are not scale sensitive.

Inequality measures and concentration curves do not measure equity in water allocation unless equal sharing is the purpose. Charting and measuring inequality could be of assistance to determine if water or related variables are more or less distributed, for example, over time and or between different water planning scenarios. Inequality simply indicates the differences in the resource without regard to the desirability as a system of reward or undesirability as a system running counter to some ideal of equality (Kuznets 1953).

Conclusion

"All science depends on its concepts. These are ideas which receive names. They determine the questions one asks, and the answers one gets. They are more fundamental than the theories which are stated in terms of them."

Sir G. Thompson (1892–1975)

Equitable water allocation is a major policy and management objective, although it is poorly understood, both conceptually and methodologically. How water is shared becomes critical when productive activity becomes constrained. Utilizing different arguments from the public sector, management, and psychology debates, it is argued that the concept of equity is often undefined and ambiguous. Equity in irrigation does not necessarily mean that every irrigator receives the same amount of water; it rather implies that each irrigator gets an amount that is fair. Equality can be an important part of equity, but not necessarily the same. This article proposes that the term 'formal equity' could be useful as the distributional benchmark criteria that the law and legislation have established as fair through a public participation process. Descriptive inequality measures do not measure equity, but have the potential to do so. Currently there are no standard methods to assess or monitor equity and equality in irrigation water allocation, and contextual and measurable definitions must be in place to ensure transparency to mitigate conflict and measurable policies.

The argument put forward in this research is that equity and equality, despite being ambiguous and ill-defined concepts, are highly relevant to rational use of irrigation water and the management of scarce water at all levels, especially when water becomes a constraint to productive activity. Equity or equality is referred to in most water management-related guiding principles, and the terms are prerequisites of hydro-political stability and hydro-solidarity. The nature of the water allocating criteria is at the heart of territorial development, especially where irrigated agriculture gives high added value to crop production. The article shows that these are broad and multifaceted terms that need to be defined depending upon the specific context, the relevant dimensions, levels and scales. That is why it is almost impossible to make a 'one-fits-all' definition for irrigation water management. Moreover, a clear distinction in the use of the terms equity and equality should be made, as they do not necessarily imply the same thing. Equity tends to refer to the state of being fair, impartial, or right judgment and is lacking a mathematical definition (subjective), while equality is concerned with the state of being equal and can be measured (objective). Most literature reviewed considered the terms as management targets in terms of either benefits or/and burdens; however, there are no standard methodologies to measure and monitor their performance, leaving unanswered questions like: What should be equal or equitable? How should we measure and monitor these targets? A confusion of the two terms results in frequent random usage and interchangeability, even though the terms have different connotations and consequences. Still, there seems to be some general conformity in science and public debate that greater equity and equality in irrigation is desirable. Conceivably, it could be that the lack of clear definitions contributes to a general consensus that these objectives are worth striving for (suggesting everyone having their own idea of what the terms mean). It is not likely people would reach a clear consensus of their meanings, and it can be concluded that equity and equality in irrigation water management, though important, is still not well theorized for inputs or outcomes.

Acknowledgements

This chapter is based on parts of literature review of the Ph.D. thesis of Solveig Kolberg with Julio Berbel and Rafaela Dios-Palomares as supervisors.

References

AbuZeid, K.M. and Elrawady, M.H. 2008. Water Rights and Equity in the Arab Region. INWRDAM, Amman, Jordan.

ADB. 2009. Water Rights and Water Allocation: Issues and Challenges for Asia. Asian Development Bank. Mandaluyong City, Philippines.

Akkuzu, E., Ünal, H.B. and Karatas, B.S. 2007. Determination of adequacy and equity of water delivery in irrigation systems: a case study of the Gediz basin. Journal of Applied Sciences 7: 607–611. DOI: 10.3923/jas.2007.607.611.

Allan, J.A. 1996. Policy responses to the closure of water resources: regional and global issues. pp. 3–12. *In*: Howsam, P. and Carter, R.C. (eds.). Water Policy: Allocation and Management in Practice. E&FN Spon, Chapman and Hall, London UK.

Bandaragoda, D.J. 2000. A framework for institutional analysis for water resources management in a river basin context. Working Paper 5. International Water Management Institute, Colombo, Sri Lanka.

Bellù, L.G. and Liberati, P. 2006. Policy impacts on inequality. Simple inequality measures. Analytic tools, Module 080, EASYPol-Resources for policy making, FAO. Available at: www.fao.org/easypol.

Berbel, J., Calatrava, J. and Garrido, A. 2007. Water pricing and irrigation: a review of the European experience. pp. 295–327. *In*: Molle, F., Berkkoff, J.J. and Barker, R. (eds.). Irrigation Water Pricing Policy in Context: Exploring the Gap between Theory and Practice. CABI, Wallingford, UK.

Boelens, R. and Dávila, G. (eds.). 1998. Searching for Equity: Conceptions of Justice and Equity in Peasant Irrigation. VanGorcumand Comp. B.V., Assen, The Netherlands. 472 p.

Böhringer, C. 2004. Sustainability impact assessment: the use of computable general equilibrium models. Économie Internationale 99: 9–26.

Bojer, H. 2003. Distributional Justice: Theory and Measurement. Routledge Frontiers of Political Economy, London and New York, 151 p.

Bos, M.G., Murray-Rust, D.H., Merrey, D.J., Johnson, H.G. and Snellen, W.B. 1994. Methodologies for assessing performance of irrigation and drainage management. Irrigation and Drainage Systems 7: 231–261.

Chowdhury, J.U., Rahman, M. and Salehin, M. 1997. Flood Control in a Floodplain Country: Experiences of Bangladesh. Islamic Educational, Scientific and Cultural Organization (ISESCO), Rabat, Morocco.

Comprehensive Assessment of Water Management in Agriculture. 2007. Water for Food, Water for Life: A Comprehensive Assessment of Water Management in Agriculture. Earthscan (London) and International Water Management Institute (Colombo).

Conceição, P. and Ferreira, P. 2000. The young person's guide to the Theil index: suggesting intuitive interpretations and exploring analytical applications. University of Texas Inequality Project Working Paper 14. Available at: http://utip.gov.utexas.edu/papers/utip_14.pdf.

Coulter, P.B. 1989. Measuring inequality: A Methodological Handbook. Westview Press, Boulder, Colorado, 200 p.

Cowell, F. 2009. Measuring inequality. Part of the series LSE Perspectives in Economic Analysis. Oxford University Press. Available at: http://darp.lse.ac.uk/papersDB/Cowell_measuringinequality3.pdf.

Cullis, J. and van Koppen, B. 2007. Applying the Gini Coefficient to measure inequality of water use in the Olifants River water management area, South Africa. IWMI Research Report 113, Colombo, Sri Lanka. 19 p. DOI: 10.3910/2009.113.

DAINET, n.d. Sustainable water management. How to define the term Sustainable Water Management? Available at: http://www.dainet.org/water/whatisswm.htm.

De Fraiture, C. and Perry, C. 2007. Why is agricultural water demand unresponsive at low price ranges? pp. 94–107. *In*: Molle, F. and Berkoff, J. (eds.). Irrigation Water Pricing: The Gap between Theory and Practice. CABI Publishers, Oxfordshire, UK.

Dinar, A. and Tsur, Y. 1999. Efficiency and equity considerations in pricing and allocating irrigation water. Research Working Papers 1(1): 1–40. DOI: 10.1596/1813-9450-1460.

Divan, S. and Rosencranz, A. 2005. Environmental Law and Policy in India (second edition, seventh imprint). Oxford University Press, New Delhi, India.

EC. 2000. Directive 2000/60/EC of the European Parliament and of the Council of 23rd October 2000 establishing a framework for Community action in the field of water policy. Official Journal 22 December 2000 L 327/1, European Commission, Brussels.

El-Awad, O.M., Gowing, J.W. and Mawdsley, J.A. 1991. An approach to multi-criterion evaluation of irrigation system performance. Irrigation and Drainage Systems 5: 61–75. DOI: 10.1007/BF01102777.

Elliot, S. 2009. Why measure inequality? A discussion of the concept of equality. Oxonomics 4(1): 32–41. Available at: http://pages.uoregon.edu/plambert/SOPHIEinOXON.pdf.

Gleick, P.H. 1998. Water in crisis: paths to sustainable water. Ecological Applications 8(3): 571–579. Available at: http://www.jstor.org/stable/2641249.

Gorantiwar, S.D. and Smout, I.K. 2005. Performance assessment of irrigation water management of heterogeneous irrigation schemes: 1. A framework for evaluation. Irrigation and Drainage Systems 19(1): 1–36. DOI: 10.1007/s10795-005-2970-9.

Gorantiwar, S.D. and Smout, I.K. 2006. Model for performance based land area and water allocation within irrigation schemes. Irrigation and Drainage Systems 20: 345–360. DOI: 10.1007/s10795-006-9012-0.

Hodgson, S. 2006. Modern Water Rights: Theory and Practice. FAO Legislative Study 92, Food and Agriculture Organization of the United Nations, 116 p. Available at: ftp://ftp.fao.org/docrep/fao/010/a0864e/a0864e00.pdf.

ICWE. 1992. The Dublin statement on water and sustainable development. International Conference on Water and the Environment. Dublín, Ireland, January 31, 1992.

Kayamanidou, M.N. 1998. Integrated Water Planning and Management in the Mediterranean. The IPTS Report, Special Euro-Mediterranean Issue, European Commission, DG XII.

Kolberg, S. 2012. Equity and equality in irrigation water allocation under basin closure: concepts and measurement. Doctoral thesis, University of Cordoba.

Kuznets, S. 1953. Economic growth and inequality. American Economic Review 45: 1–28. Available at: http://www.jstor.org/stable/1811581.

Laycock, A. 2007. Irrigation Systems: Design, Planning and Construction. CAB International, Wallingford, UK.

Lévite, H. and Sally, H. 2002. Linkages between productivity and equitable allocation of water. Physics and Chemistry of the Earth 27: 825–830. DOI: 10.1016/S1474-7065(02)00071-2.

Livingston, M.L. 1995. Designing water institutions: market failures and institutional response. Water Resource Management 9(3): 203–220.

Marsh, M.T. and Schilling, D.A. 1994. Equity measurements in facility location analysis: a review and framework. Invited review. European Journal of Operational Research 74: 1–17. Available at: http://fisher.osu.edu/~schilling_1/pub/EquityFacLocn_EJOR1994.pdf.

Merrett, S. 2002. Water for Agriculture: Irrigation Economics in International Perspective. Taylor & Francis, 221 p. DOI: 10.1016/j.bbr.2011.03.031.
Molle, F. 2004. Defining water rights: by prescription or negotiation? Water Policy 6(3): 207–227.
Molle, F. 2009. Water scarcity, prices and quotas: a review of evidence on irrigation volumetric pricing. Irrigation and Drainage Systems 23: 43–58. DOI: 10.1007/s10795-009-9065-y.
Molle, F., Venot, J.-P., Lannerstad, M. and Hoogesteger, J. 2010. Villains or heroes? Farmers' adjustments to water scarcity, irrigation and drainage. Irrigation and Drainage 59: 419–431. DOI: 10.1002/ird.500.
Msangi, S. and Howitt, R.E. 2007. Income distributional effects of using market based instruments for managing common property resources. Agricultural Economics 37: 249–259. DOI: 10.1111/j.1574-0862.2007.00249.x.
Murray-Rust, D.H., Lashari, B. and Memon, Y. 2000. Water distribution equity in Sindh Province, Pakistan. Lahore, Pakistan. Extended Project on Farmer Managed Irrigated Agriculture Under The National Drainage Program (NDP). IWMI Working Paper 9, Pakistan Country Series 1, 31 p. Available at: http://pdf.usaid.gov/pdf_docs/PNACM553.pdf.
Peña, H. 2011. Social Equity and integrated water resources management. TEC Background Paper No. 15. Published by the Global Water Partnership Technical Committee (TEC). Available at: http://www.gwptoolbox.org/images/stories/gwplibrary/background/tec_15_english.pdf.
Phansalkar, S.J. 2007. Water, equity and development. International Journal of Rural Management 3(1): 1–25. DOI: 10.1177/097300520700300101.
Pretorius, E., Woyessa, Y.E., Slabbert, S.W. and Tetsoane, S. 2005. Impact of rainwater harvesting on catchment hydrology: case study of the Modder River basin, South Africa. WIT Press, 7 p.
Rasinski, K.A. 1987. What's fair is fair—or is it? Value differences underlying public views about social justice. Journal of Personality and Social Psychology 53(1): 201–211.
Sadras, V. and Bongiovanni, R. 2003. Use of Lorenz curves and Gini coefficients to assess yield inequality within paddocks. Field Crops Research 90(2): 303–310. DOI: 10.1016/j.fcr.2004.04.003.
Saleth, R.M. and Dinar, A. 2004. The Institutional Economics of Water. A Cross-Country Analysis of Institutions and Performance. A co-publication with the International Bank for Reconstruction and Development/The World Bank 2004.
Sampath, R.K. 1988. Equity measures for irrigation performance evaluation. Water International 13(1): 25–32. DOI: 10.1080/02508068808691984.
Sampath, R.K. 1992. Issues in irrigation pricing in developing countries. World Development 20(7): 967–977. DOI: 10.1016/0305-750X(92)90124-E.
Shah, T., Hassan, M.U., Khattak, M.Z., Banerjee, P.S., Singh, O.P. and Rehman, S.U. 2009. Is irrigation water free? A reality check in the Indo-Gangetic Basin. World Development 37(2): 422–434. DOI: 10.1016/j.worlddev.2008.05.008.
Sivramakrishna, S. and Jyotishi, A. 2006. Hobbes, Coase and Baliraja: Equity and Equality in Surface Water Distribution. Working Paper 169, Gujarat Institute of Development Research. Available at: http://econpapers.repec.org/paper/esswpaper/id_3a1731.htm.
Small, L.E. and Rimal, A. 1996. Effects of alternative water distribution rules on irrigation system performance: a simulation analysis. Irrigation and Drainage Systems 10: 25–45.
Sun, T., Zhang, H., Wang, Y., Meng, X. and Wang, C. 2010. The application of environmental Gini coefficient (EGC) in allocating wastewater discharge permit: the case study of watershed total mass control in Tianjin, China. Resources, Conservation and Recycling 54(9): 601–608. DOI: 10.1016/j.resconrec.2009.10.017.
Syme, G.J. and Fenton, D.M. 1993. Perceptions of equity and procedural preferences for water allocation decisions. Society & Natural Resources 6(4): 347–360. DOI: 10.1080/08941929309380833.

Syme, G.J., Nancarrow, B.E. and McCreddin, J.A. 1999. Defining the components of fairness in the allocation of water to environmental and human uses. Journal of Environmental Management 57(1): 51–70. DOI: 10.1006/jema.1999.0282.

Tisdell, J.G. 2003. Equity and social justice in water doctrines. Social Justice Research 16(4): 401–416. DOI: 10.1023/A:1026365700255.

Tsur, Y. and Dinar, A. 1995. Efficiency and equity considerations in pricing and allocating irrigation water. Policy Research Working paper 1460, The World Bank, May 1995. DOI: 10.1596/1813-9450-1460.

UN. 1992. Report of the United Nations conference on environment and development. Annex 1. Rio de Janeiro, 3–14 June 1992.

Uprety, L.P. 2005. Social equity in farmer-managed irrigation in the Terai of Nepal. Occasional Papers in Sociology and Anthropology 9: 141–175. DOI: 10.3126/opsa.v9i0.1139.

Van der Zaag, P. and Savenije, H.H.G. 2006. Water as an economic good: the value of pricing and the failure of markets. Value of Water Research Series No. 19, UNESCO-IHE.

Vecchi, G. 2007. Measuring inequality. Bosnia and Herzegovina Poverty Analysis Workshop, September 17–21, 2007.

WCED. 1987. Our Common Future. The World Commission on Environment and Development. Oxford University Press, USA, 420 p.

Wegerich, K. 2007. A critical review of the concept of equity to support water allocation at various scales in the Amu Darya basin. Irrigation and Drainage Systems 21(3-4): 185–195. DOI: 10.1007/s10795-007-9035-1.

Wescoat, J.L., Jr. 2002. Water rights in South Asia and the United States: comparative perspectives, 1873–2000. pp. 298–354. *In*: Richards, J.F. (ed.). Land, Property and the Environment. Institute for Contemporary Studies, Oakland, California.

Wilkinson, R.G. 2005. The Impact of Inequality: How to Make Sick Societies Healthier. The New Press, New York, 355 p.

Williams, R.F.G. and Doessel, D.P. 2006. Measuring inequality: tools and an illustration. International Journal of Equity in Health 5(5). DOI: 10.1186/1475-9276-5-5.

WWAP. 2009. The United Nations World Water Development Report 3: Water in a Changing World. UNESCO (Paris) and Earthscan (London).

Young, H.P. 1994. Equity in Theory and Practice. Princeton University Press, Princeton, NJ, 238 p.

CHAPTER

2

Water Management and Institutional Adaptation Strategies in the Irrigation Sector: Two Experiences in Emilia-Romagna

Michele Vollaro,[1,a,*] *Meri Raggi*[2] and *Davide Viaggi*[1,b]

Introduction

Mediterranean Europe is acknowledged as a climate change hot-spot (Giorgi 2006), meaning that the climatic and weather evolution forecast under the different scenarios hypothesized by the Intergovernmental Panel on Climate Change (IPCC 2007) have become a reality. Summer seasons in southern European countries are hotter and drier, characterised by greater uncertainty in terms of temperature and precipitation patterns. The effects of climate

[1] Department of Agricultural Sciences, University of Bologna, viale Fanin 50, Bologna.
[a] Email: michele.vollaro@unibo.it
[b] Email: davide.viaggi@unibo.it
[2] Department of Statistics, University of Bologna, via delle Belle Arti 42, Bologna.
Email: meri.raggi@unibo.it
* Corresponding author

change are evident in the increased risks of extreme events, such as droughts and floods, and a growing reduction in quality water resources. The former are occurring more frequently in territories traditionally endowed with abundant water resources, while the latter is mainly affecting areas with pre-existing systemic water scarcity.

Although in the short run the effects of climate change have repercussions on all economic sectors and the population in general, agriculture is bearing the highest burden of droughts given the high use rates of water resources for irrigation. This is the main reason why adaptation strategies in agriculture are of primary importance, especially for designing and implementing measures for the improvement of water resource management aimed at reducing the effects of weather risks on agricultural production, farmers' income stability and the variability of food prices. Indeed, policies and measures inducing water use efficiency and water saving are particularly important for guaranteeing and securitizing sufficient and competitive food production in the future (Hanjra and Qureshi 2010). In this regard, for the European context, a closer coordination or, preferably, improved integration between water and agricultural policy would be needed to provide a more comprehensive decision-making framework able to balance the trade-off between incentives and constraints in water resource management such to contemporaneously account for agricultural (competitiveness), environmental (safeguard of water resources) and social (food security) objectives (Bartolini et al. 2010).

European institutions have long worked toward diffusing a cultural and normative framework for ecosystem and environmental protection, accounting for several aspects of water resources. In 2000, the European Union (EU) formalised the Water Framework Directive (WFD), providing rules and obligations to Member States (MS) for improving the quality of water resources and establishing the principles of "user pays" and "polluter pays" in the management stages of water resources. The steps forward for a complete framework of water resource management are stated in the European Commission (EC)'s communications on *Addressing the challenge of water scarcity and droughts in the European Union* (EC 2007) and follow-ups, which represent the first efforts in tackling water resource issues from a quantitative perspective.

Despite the efforts by EU institutions to structure directives and frameworks for safeguarding water resources, a comprehensive approach accounting for the qualitative and quantitative aspects of water resource management is still far off. Indeed, EU institutions had to deal with the underlying difficulties in merging multiple and different positions, experiences and constraints (infrastructural and political) in the management of water resources at the national and local level. In this respect, the work of the EC in formalising a policy document followed the process of recognition and evaluation of different management arrangements for water resources and was finalised with the formulation of a set of priority actions, the first two of which are water pricing and metering (which is recognised as a crucial precondition for adequate water pricing mechanisms), regardless of the existing institutional, management and infrastructural arrangements at all levels. Water pricing (and to some extent also water markets) in agriculture has largely been studied through simulation analyses with the aim of exploring the extents of both efficiency improvements and water conservation. Most findings suggest that the adoption or modifications of pricing schemes (or water markets) do indeed incentivize farmers to use irrigation water more efficiently, but also that increased efficiency does not always correspond to improved water conservation (Dono et al. 2010; Balali et al. 2011; Frija et al. 2011). In fact, the adoption of volumetric pricing schemes associated with the adoption of water saving technologies could induce the expansion of crop production with the risk of jeopardising water conservation through the so-called rebound effect (Olmstead 2010). In specific cases of irrigation water spot markets, Janmaat (2011) found that water conservation could be achieved, but could also be relatively costly.

In such context, it is of noteworthy importance to highlight cases of water management that successfully pursued the objectives of efficient allocation and water saving, together with the investigation of the relative management strategies and institutional arrangements adopted.

The chapter reports the water management experiences of two user-based irrigation organisations in Northern Italy and aims at assessing, through a qualitative approach, the relative performances in terms of improvements in water allocation, cost-sharing and water use efficiency. Both cases represent interesting governance developments in irrigation water management that took place in the Italian region of Emilia-Romagna: the first regards the *Tarabina* irrigation district, in which a voluntary change in the tariff system, from a unique area-based payment to a composite tariff accounting for the quantity of water used, set up by the users to resolve distributional issues in the quantity and costs of irrigation water, have resulted in a significant reduction in water use, while the second case regards the creation of *Voluntary Irrigation Boards* (VIB) for overcoming water access issues, in which irrigation water is harvested in small reservoirs and then allocated by quotas, paid for through a volumetric tariff and quotas can be temporally exchanged within the board members.

Although the choices of implementing volumetric tariffs and allowing for the exchange of water rights in the two cases have not been a direct response to the effects of climate change, these experiences *de facto* demonstrate the potential for improvements in water management (water pricing and metering) and the design of new institutions (water rights trading) as effective adaptation strategies aimed at improving water allocation, water use efficiency and water saving.

The next section highlights the recent developments in EU policy orientations for a common framework of quantity water management and the relevant policy motivations. Sections 3 and 4 present the experiences of Tarabina and VIB, respectively.

Background policy on the quantitative aspects of water resources

Nowadays it is commonly recognised that water demand from economic sectors and the population will, in the near future, likely surpass the availability of usable water resources and that such a gap will be significant. One way to limit or arrest the reduction in quality water resources is the environmental safeguard of water bodies, the other is to reduce the inefficiency of water use and to increase the availability of usable water resources.

The Water Framework Directive 2000/60/EC

The Water Framework Directive 60/2000 provides European water authorities with specific guidelines and constraints aimed at improving, in the medium and long-run, the ecological status and the quality of water resources. It corroborates and reinforces previous efforts toward improving the quality of drinking water, reducing the amount of pollutants discharged in water bodies, protecting the aquatic natural habitats of migratory and endangered species and monitoring the qualitative status of water resources. Moreover, the directive represents the first concrete attempt to associate an economic value to water resources by inviting MS to recuperate the costs related to water resources and the respective uses through the application of adequate tariffs. Notwithstanding this, water pricing (basic measure—art. 9) is intended also to work as an economic tool to incentivise a more efficient water use in order to yield a reduction in water abstraction and to meet the objective of improving the ecological status of water bodies.

Although the WFD represents an essential tool to improve the management of water resources, it is intended to address mainly the qualitative issues, omitting future problems concerning quantity, except for inviting MS to apply water pricing. To fill such a political gap, EU countries are working toward assuming a shared position with regard to the problems related to droughts and scarcity and to propose a definitive political framework on water quantity to integrate the other policies related to water management and the use of water resources, such as the WFD and the Common Agricultural Policy (CAP).

Pathway toward a more "quantitative" policy

Since 2006, the EC has been carrying out an assessment of the most applied measures at the EU level devoted to resolving the quantitative issues of water management. This ongoing process has involved stakeholder meetings and collegial evaluations of the measures under examination. The aim of the EC is to elaborate a guideline document, in the form of a collection of recommendations, able to stimulate a better quantitative management of water resources at the EU level, aimed at improving water use efficiency and reducing water losses, without altering the objectives of the WFD.

The first communication document, EC (2007), individuates the options of water supply and water prices as the two main policy instruments to tackle the issues of droughts and water scarcity, while a third is a mix of the previous two and includes other interconnected measures. The water supply option recalls what is proposed in the art. 4(7) of the WFD—new water infrastructures whenever public benefits outweigh the costs—while the water prices option reinforces what is provided in the art. 9 of the WFD—improving the use efficiency of water and cover the costs related to its use. The third option represents a concrete tentative to delineate a comprehensive policy framework for the quantitative management of water resources by identifying a hierarchical set of measures devoted to reducing the exposure to the risks of droughts and water scarcity phenomena. Such a set gives high priority to measures aimed at both preventing droughts, such as mapping and early warning systems, and supporting efficient water allocation and sustainable land use planning. It also proposes to review the existing EU sectorial legislation and funding such to create a common ground policy (harmonized policy) more inclusive with respect to the objectives of increasing water use efficiency and water saving, and giving more importance to leakage reduction and water metering as well.

Policy review on quantitative aspects of water management and the Blueprint

The results of the assessments and evaluations are included in the communication document *Report on the Review of the European Water Scarcity and Droughts Policy* (EC 2012a), which also represents an important cornerstone of the *Blueprint to safeguard Europe's water resources* (EC 2012b). The former is the outcome of the review process, started in 2006, related to the evaluation of measures applied at the national, regional and local level in EU countries. With respect to the first document (EC 2007), the review policy document is limited to the formalisation of a toolkit composed of seven priority actions to tackle the issues of droughts and water scarcity, giving greater attention to water pricing and metering, to the integration of water related compliance measures with CAP subsidies and to more efficient water allocation mechanisms such as water use rights trading. The latter document is broader in scope and is aimed at evaluating the existing EU water policies and analysing the obstacles that likely hamper the implementation of the proposed measures. As regards the issues of droughts and water scarcity in the agricultural sector, the *Blueprint* proposes to enforce the application of art. 9 of the WFD (pricing and metering) and to foster actions for water use reduction as a pre-condition for accessing Rural Development and Cohesion funds, together with the realisation of a guidance document for the development of irrigation water use rights trading schemes.

Motivations underlying the adoption of a "quantitative" policy

Now that the WFD has been acknowledged and implemented in EU countries and the ongoing functioning and efficacy of the applied ecological measures have been verified, the attention of water policies in the EU in the last few years has shifted towards the quantitative aspects of water management. The reasons that pushed the EU to develop a policy document on the safeguard of water quantity can be traced back to two objectives. The first relates to the purpose of reaching the good ecological status of water bodies even through the rise in the quantity of water resources. The second is to induce a more efficient use of quality water in all economic sectors and the population such as to ensure that demand corresponds with the future availability of the resource. Both motivations seem independent, but are in fact closely related because the first relates to the improvement of availability, while the second seeks a rational reduction of the use of water resources. The link between the motivations, which in turn justifies a comprehensive strategy of water management, are the changes in the natural cycle of water, the effect of which translates into the reduced capacity of water retention on and under the ground, the necessity to provide sufficient availability of water at least for the domestic and productive sectors and the duty to guarantee sustainable economic development.

Improvements in irrigation water management are not solely achievable through the adoption of specific directives or the implementation of recommended actions, but also by fostering the active participation of water users in the managerial decision-making process and by establishing water allocation rules that guarantee equitable conditions in the distribution of the related costs. The experiences of Tarabina and the Voluntary irrigation boards, although not entirely inspired by reasons related to water scarcity and drought, are clear cases of anticipation and continuation of the prescriptions suggested by the WFD and by the recent policy orientations in water management, especially regarding the objective of cost recovery, the application of volumetric tariffs and the efficient allocation of irrigation water resources. The unexpected results of such choices, beyond the successful management of cost recovery, are notable in the perception, by water users, of water resources as a good (input) with an economic value and the consequent reduction in water use obtained, not only according to the volumetric tariff, but also by a distribution mechanism of the resource that corresponds better with the marginal values of production, as the theory suggests (Johansson 2000).

Tarabina

Tarabina is an irrigation district located in the province of Ravenna, which is part of the East territory of the region Emilia-Romagna (Fig. 1). Tarabina is defined as a "district" because it is under the authority of and controlled by the Reclamation and Irrigation Board of Romagna Occidentale (RIBRO). Since its design in the late 1970s, the Tarabina district was created to operate as a *cost centre*, meaning that the budgeting procedures of costs and revenues are confined to the served area and relative users, but independently from the general budgeting procedures of the RIBRO. The condition of *cost centre* is a direct managerial outcome (effect) of the infrastructural project of the irrigation network, which is entirely realised with pressure pipes from the water abstraction site to each delivery point. Since the outset of the irrigation operations, the district, being a *cost centre*, allowed also for the creation of

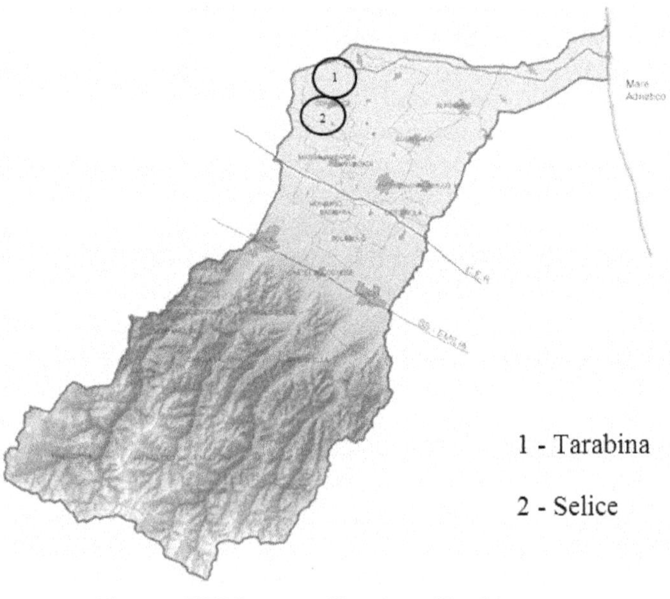

Figure 1. RIBRO map and location of Tarabina area.

a management committee composed of water users, who are responsible for both the technical and administrative procedures.

The Tarabina district receives irrigation water from the Canale Emiliano-Romagnolo (CER), the most important water infrastructure in Northern Italy, which pumps water from the Po River to the Emilia-Romagna region for agricultural and industrial uses. The relationship between RIBRO and the Management Board of the CER is regulated by a long-term contract of water supply which provides a constant water flow of about 13 m^3/s all year long and guarantees the delivery of irrigation water even in periods of drought.

The plain agricultural area served by the pressure pipe system of the Tarabina irrigation district covers 680 Ha, 100 Ha of which are not irrigated, and distributed on 55 farms; about 40% of the irrigated land is managed by a farmhands cooperative. Due to the lack of official statistics on the Tarabina area, only qualitative data have been collected regarding land use and crop mix through unstructured interviews with the technical staff of the RIBRO. Farmers in Tarabina are mainly specialised in growing horticultural crops, in rotation with seed for industrial uses and cereals. Orchards are also common, especially peaches, kiwis, apricots and plums.

Evolution of water management in Tarabina

At the outset of irrigation operations in 1983, farmers agreed to contribute to the costs of the collective irrigation facilities through a proportional tariff based on the agricultural area (€/Ha) owned, as was customary for collectively managed irrigation schemes in Northern Italy. The capital costs were not considered in computing the amount to be paid as the irrigation system was completely financed by national funds and, hence, the tariff was representative of the maintenance and operational (M&O) expenses and the cost of the withdrawn water from CER. The amount of water withdrawn from CER was measured by four meters, each located at the edge of homogeneous agricultural sub-areas, and the cost of water has been stable at around 2€cent per m^3.

Although such a contributive arrangement was not able to cover all the costs (i.e., the capital costs), at least it represented a rational system that guaranteed the financial sustainability of the irrigation facility. However, since the budgeting procedures were based solely on the final balance at the level of cost centre, farmers received the bill for year t in year t+1, implying the unawareness of the current water use.

Since 1983 the Tarabina Management Committee (TMC) has not had its own formal rules for irrigation water management and has based its decisions on the rules established by the neighbouring, and quite similar, Selice irrigation district. For both districts, it was not possible to know the amount of water used by each single farm and there was no rule for regulating or limiting the water drawn at the collective level. In turn, since the *per Ha* tariff in 1983 was as low as 20€, irrigators were induced to use more water than the optimal level implying, in the following years, increases in the management costs of up to 220€. In the first few years, the increase in costs was due to the increment of water used, mainly driven by the extension (about 50%) in 1987 of the area served by the existing infrastructures, while, after the first decade of functioning, costs underwent a second, greater, increase to compensate for the lack of maintenance in the previous years and to cover further increases in water use and energy supply.

The choice of neglecting the maintenance of the collective infrastructure is likely due to the willingness of users to keep the costs of irrigation low. Indeed, the dynamics of cost sharing in user-managed irrigation schemes, according to Levine (1987), involves, *inter alia*, the strategy of neglecting maintenance operations which stems from the users' rational expectations of state intervention for the rehabilitation of infrastructures, especially in cases where the investment costs have not been borne directly by the final users. In Tarabina, the management orientation towards the preservation of low irrigation costs resulted in the noteworthy decline in collective infrastructures such that the ability to deliver water risked being compromised.[1] Moreover, relatively low tariffs incentivised farmers to use more water than they actually needed, implying a generalised attitude to treat water as valueless input. Such behaviour, associated with inadequate, or the lack of, formal management rules, results in an interpretation of water demand that mirrors what Arrojo (1999) defines as "water requirements at quasi-null costs" and is responsible for the establishment of a vicious circle of rising demand—inadequate supply—increased (perceived) scarcity (Dosi and Easter 2000).

By not seeking to establish criteria of efficiency and sustainability in the use of water, the actions of the management of the Tarabina district yielded continuous cost increases, which were equally shared on a *per Ha* basis independently from the amount of water each member used. Although such management could represent a form of mutual economic support between members, its functionality and sustainability were not independent from the amount used by each member. Indeed, despite the costs, each farmer was prone to use as much water as possible, unconsciously triggering a degenerate dynamism which made management costs unsustainable. In fact, some farmers, aware that the costs were shared also by non-irrigators,

[1] The deterioration regarded particularly the pressure pipes that, located for the most part under the ground, suffered from frequent cracks and breaks caused by the clayey-muddy characteristics of the ground, which tends to expand with water and contract during dry periods.

tended to use more water than needed, while others irrigators abused water to keep watersheds (called "chiari"), used for bird hunting, full year round.

The mismanagement of irrigation water implied a condition of contributive dichotomy with respect to the economic value of the resource for each user, since the *big* irrigators[2] gained marginal benefits far greater than the costs, while the *small* or non-irrigators experienced relative losses for the excessive costs. Such a contributive dichotomy represented for *small* and non-irrigators an iniquity too excessive to be tolerated. Therefore, they pushed for the design and implementation of a reformed contributive system by switching from the *monomial* (per Ha) approach to a tariff based on the actual water used by each farmer. The reform of the tariff system took place in 2005, after the institution of a set of formal rules for the management of the water resource based on usage criteria corresponding to the actual irrigation needs of the users. This kind of institutional choice exhibits the characteristics of *congruence* between cost-sharing and resource allocation, which are considered beneficial for the performance and duration of the governance regimes of common resources (Dayton-Johnson 2000).

The tariff reform was implemented in 2006 and irrigators were equipped with at least one mobile meter, purchased at a paltry price (about 180€), to apply at the head-end of each on-farm irrigation system in operation. The new tariff is composed by three parts: a fixed surface contribution of 29€ per Ha, a fixed volumetric contribution of 0.15€ per m^3 and a variable contribution per Ha. The last component is applied whenever the contribution collected through the volumetric tariff is not sufficient to cover the operational costs. In mathematical terms, the new contributive system can be synthetised in the following formula:

$TC = FC + VC$, where FC (fixed costs) = 29 €/Ha and VC (variable costs) = $w \cdot 0{,}15$ €/m^3 + x €/Ha, where w is the amount of water used (m^3), $x > 29$ and $\frac{\partial x}{\partial w} > 0$.

The distribution of the TC between district members is

$$TC = \begin{cases} FC = 29 \text{ €/Ha} & \text{paid for by non-irrigators} \\ VC = w \cdot 0{,}15 \text{ €/m}^3 + x \text{ €/Ha} & \text{paid for by irrigators} \end{cases}$$

In 2005, the last year of the application of the *per-Ha* tariff, the contribution of all farmers, irrigators and non-irrigators, to the M&O costs was 155 €/Ha. In 2006, instead, non-irrigators paid 29 €/Ha, a reduction of almost 81%, while irrigators paid the equivalent average sum of about

[2] Big irrigators can be considered those farmers who exceeded water use, while small irrigators can be considered as those who either needed low amounts of resource or respected the irrigation needs of the crops.

70 €/Ha, a reduction of about 50%. Accordingly, the amount of water distributed in 2005 was about 665,500 m^3, while in 2006 the district used about 332,550 m^3, exactly 50% less, in part, attributable to reduction in use for filling the "chiari".

Economic analysis of water management in Tarabina

A simplified graphical analysis can be of help in understanding the economic rationale driving water users to the choice of abandoning the area-based contributive system and adopting the volumetric tariff. The diagram shows a simplified aggregated demand *D* for irrigation water referred to the irrigation period (summer season).

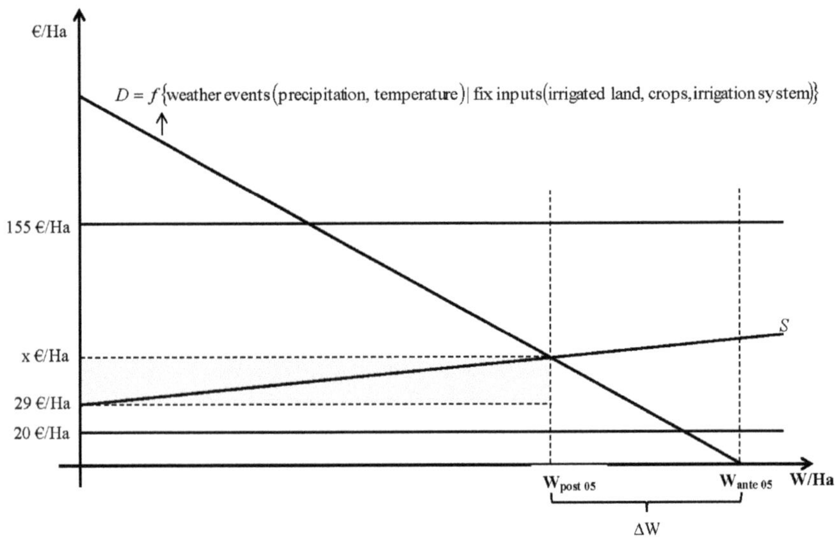

The slope and the curvature of *D* depends upon the structural characteristics of the farms, while the choices of fixed and semi-fixed inputs, such as the type of irrigated crops cultivated, the size of irrigated land and the irrigation system, determine the vertical position (or vertical shifts). At the onset of the irrigation season, farmers have already decided how to allocate their farm inputs and, hence, know how much water their crops need. The amount of water to be used, instead, remains uncertain as it depends upon the frequency and intensity of weather conditions. More specifically, precipitations and temperatures determine the horizontal shifts of *D*. For sake of simplicity, favourable weather conditions, such as abundant rains and low temperatures during irrigation seasons, induce the demand function to shift toward left and vice versa.

Under the regime of areal tariffs, the amount of water used by irrigators tends to reach the maximum of the seasonal demand, indicated by $W_{ante\ 05}$ in the diagram. Accordingly, no matter the level of the tariff, either 20 €/Ha, as in 1983, or 155 €/Ha, as in 2005, the amount of water used will be at least equal to or greater than crop needs. The *break even* condition with respect to water use and crop needs can be reached in those cases where the level of the tariff is equal to or higher than the marginal value of water, while abuse (overuse) of irrigation water can happen when the marginal value of water is higher than the level of the tariff. This is the reason why uncontrolled and unenforced water use limits under regimes of cost-sharing are the potential causes of socially unacceptable water allocations.

In the case of the volumetric tariff, like the one adopted in Tarabina, irrigation water can be identified by a linear upward sloping supply function *S*, characterised by a constant slope, given by the volumetric charge (0,15 €/m^3), and an intercept at 29 €/Ha, the minimum contribution required if no water is used. It follows that under the volumetric tariff regime, water use and water crop needs will likely break-even at the level $W_{post\ 05}$ in the diagram, corresponding to the condition $D = S$. The outcomes of the implementation of a volumetric tariff, highlighted in the diagram in terms of cost distribution and water allocation among farmers, are represented by the stabilisation of the fixed component of the tariff for non-irrigators at a low level (29 €/Ha), the determination of costs for irrigators equal to the shaded area is dependent upon the quantity of water used and a relative structural reduction in water use (ΔW).

Although a reduction in both water use and costs could be directly attributed to the change in the contribution system, simple comparisons between management performances in two consecutive or more years in the same district do not represent a correct method of analysis because of the presence of several influential factors, in particular weather conditions, which are not controlled. For this reason and for the purpose of better understanding the effects of the introduction of a volumetric contributive system on water use, the evolution of water use in Tarabina is assessed through a counterfactual analysis based on water use in Selice.

Counterfactual analysis—Tarabina vs. Selice

The Selice *cost centre* is considered to be the twin of Tarabina since it is identical as regards the agricultural, and infrastructural characteristics. Selice neighbours Tarabina from the South border and its plain agricultural land of about 1300 Ha is shared among 42 farms that receive water from the CER. Since 1983, the contributive system in Selice has been regulated by a *monomial* areal tariff and the district has its own formal set of rules for the management of irrigation infrastructures and water resources. Given the close vicinity to Tarabina, the weather conditions in Selice can be considered as yielding the same effects on water use borne by Tarabina. Indeed, by exploring the linear trends in water use, both Tarabina and Selice record a marginal increase close to 24 m^3/Ha per year until 2005; thereafter Selice continues to marginally increase its water use by 25 m^3/Ha per year while Tarabina shows a null trend. Moreover, the yearly variability of water use also appears similar in both districts and it is possible to suppose that the evolution of water use depends essentially on weather variations and crop patterns, and that it is affected only to a limited extent by the variations in magnitude of the applied tariffs, the relative management of which are independent in both districts. Therefore, the potential factors that can explain the difference in water use between Tarabina and Selice are the changes in the management strategies applied over the years in Tarabina, in particular after the year 2005 when the TMC switched to the tariff system.

The counterfactual analysis is performed by comparing water uses in both irrigation districts according to two measures: standardised (Fig. 2) and unitary (per Ha) (Fig. 3) water use. From both figures it can be noticed that the variations in water use follow similar patterns, but also that the differences in measures are more pronounced in two specific periods.

The first period runs from 1983 to 1991, during which water use in Tarabina grows at higher rates than in Selice for reasons related to low tariffs and a lack of maintenance expenditures, while the second is the period following the tariff change in 2005, in which the use of water in Tarabina diminishes significantly.

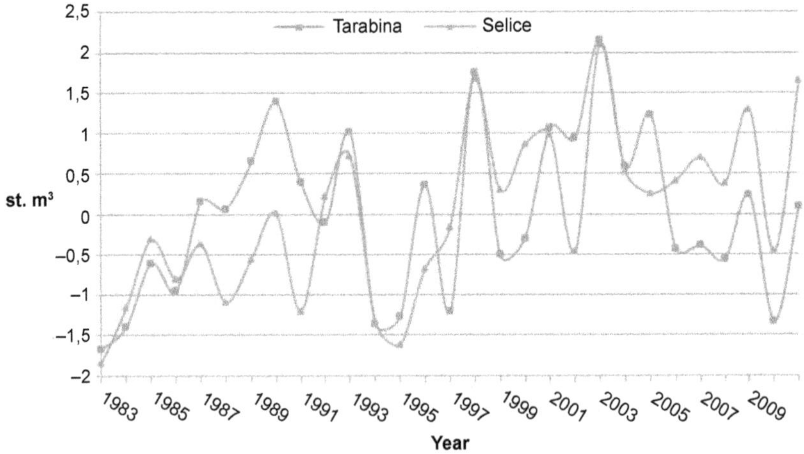

Figure 2. Standardised use of irrigation water in Tarabina and Selice.
Source: own elaboration on RIBRO data.

By taking as reference such periods, the evolution of water use in Tarabina has been analysed with respect to Selice through correlations of standardised values (St. m^3), differences in the coefficient of variation (CV) and means of relative per cent differences in unitary use. The correlation analysis, performed on the standardised water use, could be interpreted as a proxy for the assessment of the different management strategies adopted in Tarabina over the considered periods with respect to the more stable governance in Selice. According to this approach, the first period can be considered as "low tariff—lack of maintenance", the second, "high tariff—maintenance efforts" and the third as "formal rules—tariff change" (Table 1). The evaluation of the difference in CV, computed on the *per Ha* uses, gives the opportunity to track the relative variability of the unitary

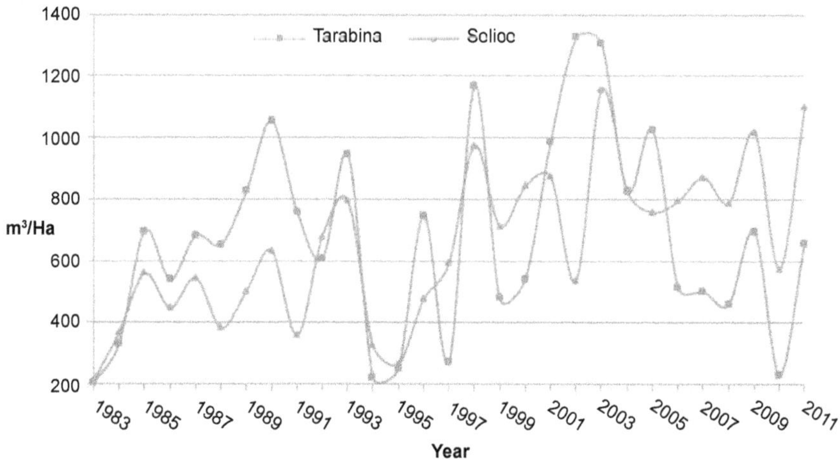

Figure 3. Unitary (per Ha) use of irrigation water in Tarabina and Selice.
Source: own elaboration on RIBRO data.

Table 1. Tarabina vs Selice—Correlation, mean and CV differences in analyses of water use.

Management strategies	Period	Correlation of St. m³	Difference in coefficient of variation	Average of relative % differences in m³/Ha
Low tariff—lack of maintenance	1983–1991	0,67	0,104	+41%
High tariff—maintenance efforts	1992–2005	0,77	0,143	+10%
Formal rules—tariff change	2006–2011	0,96	0,097	–42%

Source: own elaborations on RIBRO data.

water use along the three periods by controlling for the potential effects of exogenous factors, such as weather and crop patterns. The analysis of averages of relative per cent differences in water use was performed on the unitary (per Ha) values of resource use and represents a proxy for measuring the aggregate effects of management strategies on the unitary amount of water used in Tarabina with respect to Selice.

Based on the assumption of equal weather effects and on the observed identical water use trends up until 2005 in both irrigation districts, improvements in water management in Tarabina, as compared to Selice, can be noted in the results, that clearly show a positive evolution of the correlation, especially after the introduction of the volume-based tariff system, and a reduction in the average of relative differences in water used.

More specifically, the improvements experienced in Tarabina after 1991, deduced by the modest increase in correlations (from 0,67 to 0,77) and the significant reduction in mean differences of water use (from +41% to +10%), can be attributable to the choice of strengthening the maintenance efforts, through frequent extraordinary interventions, likely caught by a higher relative variability with respect to Selice (from 0,104 to 0,143), which implied the inevitable increasing trend in tariffs. Although the negative aspects of this type of management actions result in higher costs for users, relative higher tariffs give important "price signals" to irrigators such to be persuaded by the idea of treating water resources as a valuable good, not just for the resource itself, but at least for what is required to work for allowing, first, the user to receive the resource and, second, the resource to provide benefits through quality and quantity. After the institution of the set of rules and the implementation of the tariff reform, further improvements in water management are evident from the results with the correlation close to 1, the difference in CV roughly equal to the first period (0,097) and the average relative difference in water use that turned strongly negative (less than –40%).

The evidence of more rational water management is an outcome that can be referred to as the choice of institutionalising shared rules rather than the implementation of the volumetric tariff itself, because the improved performance of the common-pool water management in Tarabina reflects what is commonly reported in literature, namely that an adequate provision of irrigation water resources is achieved through a good capital endowment (collective irrigation infrastructure), and a better social capital, in the form of a better relationship between users and institutions (Pearce and Atkinson 1998). The switch to the volumetric tariff in Tarabina has its merits in the recorded reduction in the quantity of water used, the zeroing of the water use trend and the relative diffusion among irrigators of the perception of water resources as an economic good. These results are consistent with empirical finding suggesting that communities with appropriate formal rules of water management perform better than societies that are unable to properly allocate water resources (Dosi and Easter 2000), and exactly represents the water quantity sustainability objectives required by the EU for the coming future, clearly stated in EC (2012a).

Voluntary Irrigation Boards

Voluntary Irrigation Boards (VIB) began to spread in the Northern mountain territories of the Emilia-Romagna region in the late 1970s with the purpose of providing access to irrigation water for agricultural areas not reached by the existing irrigation infrastructures and limiting both the spread of small in-farm water reservoirs and the excessive withdrawal of underground water. However, since the institution of the Rural Development Plans (RDP), the realisation of VIB in Emilia-Romagna is limited to mountain-hilly agricultural areas characterised by limited or absent access to water resources infrastructures and allows for the construction of collectively managed reservoirs with a water harvesting capacity of between 50.000 and 250.000 m^3.

The agricultural areas in which VIB were created are characterised by the production of rain-fed fruit crops, especially apricots and kiwis, that require specific water needs in determinate phenological stages. The evolution in climatic patterns has yielded increased production risks to farmers who, operating in conditions of limited access to water infrastructures, are exposed to higher water availability (precipitations) uncertainty. Given the high quality of the fruit production in the mountain-hilly areas of the RIBRO, the stabilisation of irrigation water supply through VIB represents a valid development opportunity, especially for valorising the rural characteristics of agricultural production and to safeguard farm income and local employment.

Evolution and VIB realisation procedures

Up until the year 2000, the realisation of VIB and related infrastructures were totally supported by the financial participation of the Regional administration. Thereafter, with the implementation of the 2000–2006 and 2007–2013 RDPs, VIB and related infrastructures have been partially supported by public funds, through the 3q and 125 measures respectively, providing farmers with the opportunity to select investment choices according to their water needs.

The creation of VIB through the RDP funds begins with a public announcement from the regional authority. The relevant reclamation and irrigation board—RIBRO, as in the cases reported in this chapter—then activates an internal procedure for identifying potential places for the construction of the reservoirs based on a feasibility analysis that considers the number of served farms, their water needs and the relative distance from the artificial basin. Stakeholders, including farmers associations, are invited to a preliminary meeting at which the project is presented. The farms potentially interested in taking part in a collectively managed irrigation infrastructure are invited to assess the project and provide input on its feasibility. Those who agree with the proposal then organise a project committee, the responsibility of which is to initiate the realization of the project. The first step is to undertake necessary verifications for the creation of the reservoir, including inspections of the site and negotiations for evaluating the opportunity either to collectively acquire or rent the area of the artificial basin. The second step is to analyse the realisation costs and to establish participation quotas in terms of individual investments. Farms that decide to invest constitute the VIB and respond to the call.

The rationales underlying the development of VIB through the RDP are perfectly in line with the recommendations expressed by the EU in the water policy documents (EC 2012a), according to which the realisation of new infrastructures, although potentially delaying the achievement of good status for water bodies, need to be legitimated by positive long-run benefit-cost ratios and especially by high economic and social development opportunities. Moreover, the institutions governing the new infrastructures must be inspired by management criteria based on efficient water allocation and economic sustainability of the related costs.

Water management in VIB

VIB are autonomous entities that operate as *cost centres* and are provided with their own statute, a management committee (MC) and a formal set of rules. The statute comprises the institutive principles of the VIB, while the set of rules defines the management thereof. Unlike traditional irrigator associations or boards, VIB are created in order to confer flexibility to the management processes. For example, the statute of the *Rivalta* VIB states at art. 3 that *the board will realise all the financial and economic actions that prove to be necessary, or at least useful, to pursue the objectives of the board,* that is *to optimise through efficiency and affordability criteria the stages of the productive processes of the members;* at art. 10 it is reported that the President of the board has the right *to share the extraordinary costs among members proportionally to their quotas,* while the costs depending upon water use are shared according the measured volume of water used by each member. Art. 5 of the rules states that *members are responsible for the rational use and correct utilisation of the water resources;* art. 12 defines the rule that limits the amount of water each member can use by stating that *unauthorized uses over the established individual quotas yield the immediate stoppage of water allocations and the payment of a fine equal to 3 €/m³ for the water used in excess*; art. 20 defines the rule for transferring the individual quotas which can be exchanged solely through the authorisation of and at a cost determined by the MC. The management flexibility allowed by the statutes and the set of rules would be difficult and costly to implement, in administrative and technical terms, without the realisation of *ad* hoc infrastructures.

The collective irrigation infrastructures are constituted by the reservoir, necessary pumping facilities and pressure pipes, managed in full autonomy by each VIB. The entire infrastructure system is designed to guarantee a continuous water flow to each farmer over the year and the contemporaneity of flows to all farmers. Since 1983, a total of about 20 VIB have been established and five more will commence operation in 2014, involving about 500 farms. All VIB apply the rule of *full cost recovery* and the *user pays principle.*

The distribution of irrigation water among farmers is structured according to fixed quotas which are measured by hydraulic discharge (a water flow of 1 m^3/s) and directly related to a maximum withdrawal amount per year (1 quota = 1000 m^3 per year). This system, therefore, guarantees the full satisfaction of each farmer's water needs, provided that water is harvested at the reservoir's full capacity. Each farmer contributes to the cost recovery of the irrigation system according to a *binomial* tariff system: a fixed amount paid for on a per quota basis (about 30–40€) and a variable amount paid for per m^3 based on actual use of water (volumetric). Given the hilly or mid-mountain location of the small reservoirs, most of the irrigation water is distributed downwards by gravity, remarkably reducing the energy costs such to apply a volumetric tariff of about 10€cent/m^3, on average.

Performance of VIB

The recent experience of the VIB, as compared to traditional irrigation management, shows that, even in years characterised by extreme events of water scarcity, farmers did not suffer from low water availability and did not experience any emergency management situations, such as water rationing or modifications to the tariff system. Another aspect of the virtuous water management of VIB is the possibility for farmers to temporarily exchange their quotas in order to foster efficiencies in the allocation of water within the board. The exchange is regulated by the VIB statute and managed by the MC which every year, at the beginning of the irrigation season, allows farmers who do not intend to use their full water allocation to dispose of part of their quotas in favour of a more beneficial use for other farmers. Farmers who temporarily acquire extra quotas pay the additional fixed per quota part and the variable per m^3 part, augmented by a per cent value to cover the administrative costs of the tariff. Further, the control rules guarantee that each farmer is fined whenever he or she withdraws more than the allowable maximum quota.

The efficient functioning of the quota transferring mechanism is guaranteed by the institutional constraints established at the statute level of the VIB, as imposed by the RDP regulation. In fact, the VIB members are not permitted to apply changes to land use and cropping patterns that generate increments in water needs. However, the statutes provide for some flexibility in changes to crop choices as long as the water balance at farm level remains unchanged. Should members decide to exit the VIB, the available quotas, at first instance, are allocated among the remaining members according to proportions of existing quotas, but farmers, upon the consensus of the VIB, have the opportunity to acquire more quotas and relatively expand their water needs without exceeding the water balance at VIB level. According to information gathered from the RIBRO technical staff, the quota transfer mechanism works quite well in the VIB and transfers are caused mainly by irrigation choices at farm level corresponding to permanent crop water needs which can differ from year to year due to the age of the crop. At VIB level, in the recent years, the yearly amount of transferred quotas represents, on average, about 10% of the available water, while members that relinquish their quotas are about 25%, which roughly corresponds to farmers requesting more water.

Conclusion

The evolution of European policy on water management underscores the necessity of intervening on the supply and demand sides of quality water through strategies promoting a harmonic convergence between the intents of preserving or enhancing the qualitative status and the need for rationalising uses, in particular for agriculture which is the sector most exposed to risks related to climate change, such as water scarcity and drought events.

In line with policy orientations, the study highlights that improvements in water uses are achievable through modifications in water management systems, especially by finding the optimal combination of administrative, economic and infrastructural tools chosen accordingly to local agricultural needs and territorial characteristics.

The experience of the Tarabina irrigation district shows that improvements in water governance and management, i.e., the institution of a formal set of rules and a change in the water tariff system, from area-based to volumetric, has implied a fairer distribution of management costs among farmers and a corresponding more efficient use of water resources determined by actual volumes applied. Such improvements have not borne to farmers any notable charge in transaction costs (only the purchase of a mobile meter) and did not require any change in the infrastructural system. With regard to water quantity, the outcome of the action implemented in Tarabina was a remarkable increase in water savings, computed as the relative difference on water use in the "twin" irrigation district, Selice, in which the tariff system did not change from an area-based system.

The results obtained in the Tarabina irrigation district, after the implementation of the volumetric water pricing system, are in line with the expectations of both WFD and the Blueprint in terms of cost recovery, incentives to reduce water use and safeguarding water resources.

The experience of the VIB reflects the policy indications expressed in the *Blueprint* about the future evolution of water management at the local level, especially by anticipating the intents of integrating sectorial policies for the purpose of fostering rural development conditional on sustainable uses of water resources. The opportunity to have access to irrigation water through investments in reliable infrastructures represents a concrete incentive system for inducing both a self-selection of irrigators that need to reduce the production uncertainties related to weather variability and the institution of water management systems based on efficiency criteria for the allocation and use of water resources. The main outcomes of water management through VIB are evident in the availability of irrigation water despite the critical periods of drought borne by the surrounding areas and in the flexible allocation of water among members through the mechanism of quota transfers, as suggested by the *Blueprint* such to better reflect the economic value of water needs. The VIB also represent the result of a rational integration between the CAP and water policies that, whenever possible, is capable of inducing efficient and sustainable autonomous water resource management and opening up the possibility of developing effective mechanisms of water use rights exchanges in Italy.

From both experiences, the results suggest that an allocation system, considered as a water value-chain from the water sources to the farm delivery, which guarantees a better alignment between quantity and value of water used provides further incentives to treat water as a production input, at the same level as other inputs, and hence improves competitiveness in water use.

The gap between theory and practice in irrigation water management is still a wide one, but this case study has attempted to demonstrate that the trade-off between optimality in water allocation and efficiency in water use can be reduced if measures are devised and implemented according to local needs and specificities. However, the lack of information at farm level has constrained the study to analyse the phenomena from a broader perspective solely. Indeed, a better exploration at farm level of the effects of the change in the tariff system in Tarabina would be desirable in order to shed light on farmers' reactions in terms of technology adoption, cropping patterns and other decisional factors. Equally interesting would be to empirically analyse the mechanism of water quota exchanges within VIBs in order to evaluate the efficacy of the exchange system with the aim of designing and eventually proposing it at a larger scale through a feasibility approach.

Acknowledgements

We would like to thank the technical staff of the Reclamation and Irrigation Board of Romagna Occidentale for the support provided for developing and realising this chapter. Particular gratitude is expressed for the collaboration of Dr. Andrea Fabbri and Mr. Rossano Montuschi, for their important contributions in providing information about the evolution of the irrigation districts and the voluntary irrigation boards, respectively.

References

Arrojo, P. 1999. The impact of irrigation water pricing policy in Spain. pp. 177–193. *In*: European Commission. Directorate-General for Environment (ed.). Pricing Water: Economics, Environment and Society. OOPEC.

Balali, H., Khalilian, S., Viaggi, D., Bartolini, F. and Ahmadian, M. 2011. Groundwater balance and conservation under different water pricing and agricultural policy scenarios: a case study of the Hamadan-Bahar plain. Ecological Economics 70: 863–872.

Bartolini, F., Gallerani, V., Raggi, M. and Viaggi, D. 2010. Water management and irrigated agriculture in Italy: multicriteria analysis of alternative policy scenarios. Water Policy 12: 135–147.

Dayton-Johnson, J. 2000. Choosing rules to govern the commons: a model with evidence from Mexico. Journal of Economic Behavior and Organization 42: 19–41.

Dono, G., Giraldo, L. and Severini, S. 2010. Pricing of irrigation water under alternative charging methods: possible shortcomings of a volumetric approach. Agricultural Water Management 97: 1795–1805.

Dosi, C. and Easter, K.W. 2000. Water scarcity: institutional change, water markets and privatization. pp. 91–116. *In*: Canavari, M., Caggiati, P. and Easter, K.W. (eds.). Economic Studies on Food, Agriculture, and the Environment. Kluwer Academic/Plenum Publisher, New York.

[EC] European Commission. 2007. Addressing the challenge of water scarcity and droughts in the European Union (COM 414). Impact Assessment.

[EC] European Commission. 2012a. Report on the Review of the European Water Scarcity and Droughts Policy (COM 672).

[EC] European Commission. 2012b. A Blueprint to Safeguard Europe's Water Resources (COM 673).

Frija, A., Wossink, A., Buysse, J., Speelman, S. and Van Huylenbroeck, G. 2011. Irrigation pricing policies and its impact on agricultural inputs demand in Tunisia: a DEA-based methodology. Journal of Environmental Management 92: 2109–2118.

Giorgi, F. 2006. Climate change hot-spots. Geophysical Research Letters 33: L08707.

Hanjra, A.M. and Qureshi, M.E. 2010. Global water crisis and future food security in an era of climate change. Food Policy 35(5): 365–377.

[IPCC] Intergovernmental Panel on Climate Change. 2007. Climate Change 2007: Synthesis Report. Contribution of Working Groups I, II and III to the Fourth Assessment Report of the Intergovernmental Panel on Climate Change, Core Writing Team, Pachauri, R.K. and Reisinger, A. (eds.). IPCC, Geneva, Switzerland, 104 p.

Janmaat, J. 2011. Water markets, licenses, and conservation: some implications. Land Economics 87(1): 145–160.

Johansson, R. 2000. Pricing Irrigation Water: A Literature Survey. The World Bank, Washington, DC.

Levine, G. 1987. The challenge of rehabilitation and betterment. *In*: Fowler, D.A. (ed.). Proceedings of International Conference on Irrigation System Rehabilitation and Betterment. Colorado State University, Fort Collins, CO.

Olmstead, M.S. 2010. The economics of managing scarce water resources. Review of Environmental Economics and Policy 4(2): 179–198.

Pearce, D. and Atkinson, G. 1998. The concept of sustainable development: an evaluation of its usefulness ten years after Brundtland. Swiss Journal of Economics and Statistics 134: 251–269.

CHAPTER

3

The Closure of the Guadalquivir River Basin: A DPSIR Framework Approach

Solveig Kolberg,[1,a,*] *Julio Berbel*[1,b] and *Rafaela Dios-Palomares*[2]

Introduction

River basins or sub-basins are said to be 'closed' when their water supply is insufficient to meet consumptive demands and environmental needs (Molle et al. 2010). As demographic, economic and social pressures on water bring more and more basins near closure, the use and transfer of water between and within sectors have become subjects of increasing public attention. At a global scale, about 1.2 billion people live in closed basins, and another 500 million live in basins approaching closure (De Fraiture and Perry 2007).

[1] Department of Agricultural Economics, Sociology and Policy, University of Cordoba, Campus de Rabanales, C5, 3° floor, 14014 Cordoba, Spain.
[a] Email: Solveig.Kolberg@gmail.com
[b] Email: berbel@uco.es
[2] Department of Statistics, Econometrics, Operational Research, Business Organization and Applied Economics, University of Cordoba, Campus de Rabanales, C2, 2° floors, 14014 Cordoba, Spain.
Email: ma1dipar@uco.es
* Corresponding author

There is an increasing literature on basin closure (Molle 2004; Falkenmark and Molden 2008; Venot et al. 2007; Smakhtin 2008; Molle 2008; Molle et al. 2010), but most studies are conceptual and descriptive, focusing on the current situation rather than providing a comprehensive picture of 'how we got here'. The objective of this chapter is to provide a detailed picture of the some of the trajectories that led to the closure of the Guadalquivir River Basin of southern Spain. This river basin (RB) represents a typical water-scarce situation in the Mediterranean, with increasing demand for water resources, especially for irrigation, the main water user. To better understand how the Guadalquivir RB reached closure, we discuss its development using a Driving force-Pressure-State-Impact-Response (DPSIR) framework, applying data from the 'Irrigation Inventory of the Guadalquivir RB' for 2008 (CHG 2010a). We focus on irrigated agriculture because agriculture consumes more than 85% of all water extracted from the RB and it also has the biggest potential for water savings. Furthermore, regions suffering from water scarcity (and the possibility of closure) tend to coincide with regions in which irrigation is a major water user (Berbel et al. 2007).

The DPSIR-framework

There are several approaches for analysing the state and response of a natural system. The Pressure-State-Response (PSR) approach assumes that human activities cause pressure (direct or indirect) on the environment, which affects its state, such as the quality and quantity of water. Society responds to this change in environmental state by adjusting environmental, economic and sector policies. The PSR approach can be applied at the national, sector, community, or individual firm (farm) level. OECD introduced this approach in 1994, and it has since been modified and adjusted. Two examples are the Driving force-State-Response (DSR) model used in the past by the United Nations Conference on Sustainable Development (UNCSD), or the DPSIR-framework currently used by the European Environment Agency (EEA) (WSM 2004). Figure 1 shows a DPSIR-framework for water management at the basin level. It uses a standard EEA approach to describe a dynamic system with various feedback loops (Gabrielsen and Bosch 2003). This framework will be used as a starting point explaining the closure of the Guadalquivir RB.

Guadalquivir River Basin Characteristics

The Guadalquivir RB is located in the southern part of the Iberian Peninsula (Fig. 2). It is born in the Sierra de Cazorla, in south-eastern Spain, flows southwest past Cordoba and Seville, and empties into the Atlantic Ocean's Gulf of Cadiz near Sanlucar de Barrameda.

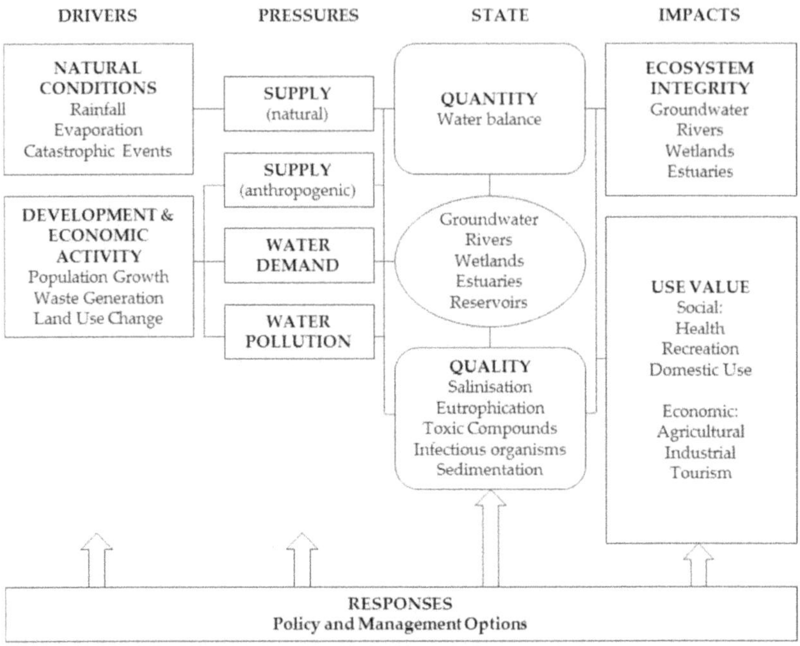

Figure 1. DPSIR-framework for water management at the basin level.
Source: Kolberg (2012) adapted from Walmsley 2004; Kristiansen 2004; Gabrielsen and Bosch 2003.

The basin drains 57,527 km^2 of land, most of which is in the autonomous community (Comunidad Autónoma) of Andalusia (90.2%). Smaller tributaries also drain parts of Castile-La Mancha (7.1%), Extremadura (2.5%) and Murcia (0.2%) (Table 1). From the Gulf of Cadiz to 80 km upstream at Seville, the Guadalquivir River is tidal and navigable. Seville is the only Spanish inland river port although its importance has been decreasing as the size of ocean vessels has been increasing.

The Guadalquivir River is the longest river in southern Spain, with a length of 650 km. The Guadalquivir RB, with its tributaries, comprises 10,700 km of stream channels. The middle reaches of the Guadalquivir River flow through a populous fertile region at the foot of the Sierra Morena where water is primarily used for irrigation. The lower reaches pass through vast marshlands (Las Marismas), which are used for rice cultivation.

The Guadalquivir RB has a typical Mediterranean climate with high intra-annual and inter-annual variation in rainfall and, therefore, in renewable resources. Annual precipitation has an average of 597 mm but ranges from 260 to 983 mm, with a standard deviation of 161 mm. Average annual temperature is 16.81°C, with a strong intra-annual variation in extreme temperatures (CHG 2011). Summer is hot and dry

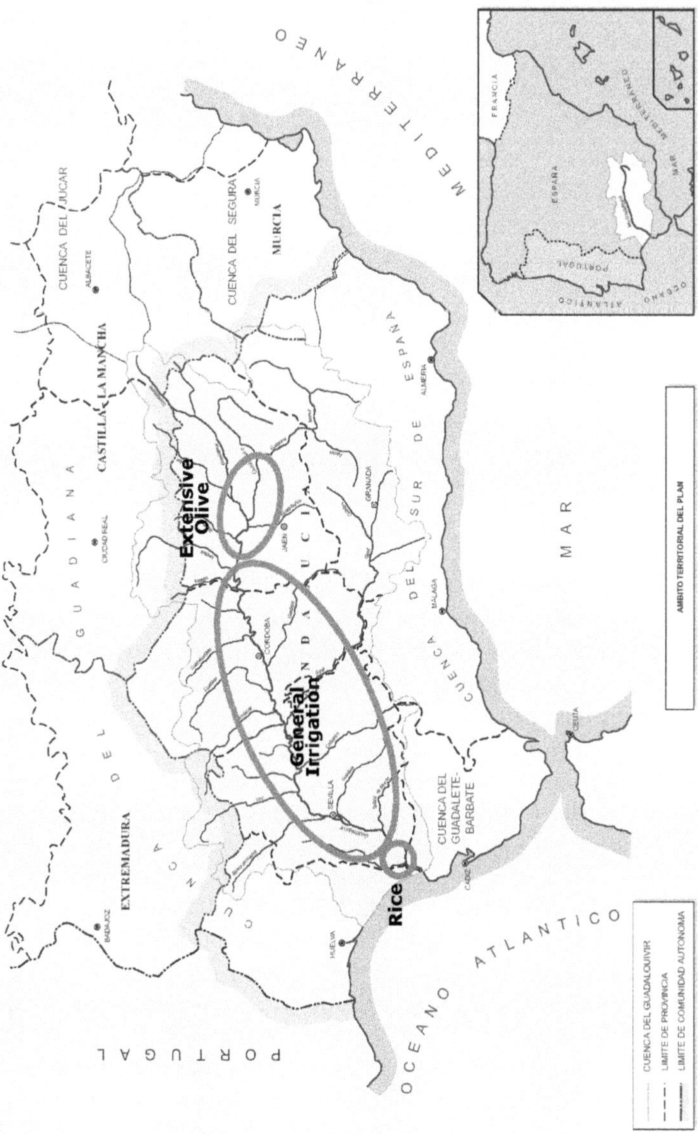

Figure 2. Map of the Guadalquivir River Basin in southern Spain.
Source: CHG (2009).

Color image of this figure appears in the color plate section at the end of the book.

Table 1. Spatial distribution of basin across autonomous communities.

Autonomous communities	Province	Province (km²)	Basin (km²)	Basin/Province (%)	Basin/Total Basin (%)
Andalusia	Almeria	8,774	229	2.6	0.4
	Cadiz	7,385	532	7.2	0.9
	Cordoba	13,718	11,135	81.2	19.4
	Granada	12,531	9,960	79.5	17.3
	Huelva	10,085	2,552	25.3	4.4
	Jaen	13,498	13,002	96.3	22.6
	Malaga	7,276	489	6.7	0.9
	Seville	14,001	14,001	100.0	24.3
Castile-La Mancha	Albacete	14,862	800	5.4	1.4
	Ciudad Real	19,749	3,300	16.7	5.7
Extremadura	Badajoz	21,657	1,411	6.5	2.5
Murcia	Murcia	11,317	116	1.0	0.2
TOTAL		154,853	57,527	37.2	100.0

Source: Kolberg 2012 adapted from CHG (2010b).

(rainfall <10 mm), whereas winters are relatively cold and wet (Sabater et al. 2009). Average annual flow of surface water in the Guadalquivir RB is 7,100 million m³, and average annual recharge of groundwater is 2,576 million m³. Annually, roughly half of this surface water and groundwater is extracted for agriculture (85%), domestic (11%), industry (3%) and tourism (1%). Consumption by tourism, according to the draft of the Guadalquivir Hydrological Basin Plan (GHBP) (CHG 2010b), includes golf courses and winter resorts consumption. The coastal tourism in the basin is negligible.

Land covers in the basin consist of forestry (49%), agriculture (47%), urban areas (2%) and wetlands (2%) (CHG 2010b). Guadalquivir RB contains 25% of Spain's irrigated land (Mesa-Jurado and Berbel 2009). Olive trees cover the largest number of hectares (more than half the RB's surface) while rice cultivation has the highest average water allocation per hectare (Table 2).

Figure 3 shows the spatial distribution of crops as of the division of types of crops in the GHBP: extensive field crops (cotton, maize, wheat, sunflower, etc.), rice, fruits (citrus and peaches), olive and 'others (mainly strawberry). The spatial distribution of net water allocation (m^3 ha^{-1}) is given in Fig. 4 and it can be seen that the average water use goes from less than 2000 m^3 ha^{-1} (olives) to more than 7.000 m^3 ha^{-1} (rice).

Table 2. Land and water allocation for irrigated crops in the Guadalquivir RB.

Crop type (2008)	Irrigated area (ha) (ha)	Net water allocation[1]	
		(m^3 ha^{-1})	(m^3)
Olive (extensive)[2]	393,520	1,500	590,280,000
Cotton	127,031	4,500	571,639,500
Rice	35,530	10,400	369,512,000
Winter cereals	79,598	2,430	193,423,140
Horticulture	34,278	4,500	154,251,000
Olive (intensive)[2]	69,568	2,200	153,049,600
Citrus	27,677	4,000	110,708,000
Sunflower	25,569	3,510	89,747,190
Fruit trees	17,833	4,000	71,332,000
Others	13,612	4,500	61,254,000
Maize	9,300	5,100	47,430,000
Sugar beet	8,072	4,500	36,324,000
Strawberry and raspberry	3,808	3,000	11,424,000
Greenhouse	591	4,500	2,659,500
TOTAL	845,986		

Source: Kolberg 2012 computed from CHG 2010a.

[1] Net water allocation refers to the crop's total irrigation allocation on farm level, without considering the losses in transport.

[2] Extensive olive has a density around 100 trees/ha and have low irrigation demand. Olive orchards with densities higher than 300 tress/ha have been named 'intensive' as they have higher irrigation needs.

The basin has 443 surface water bodies and 60 groundwater bodies. The surface waters consist of 392 rivers, 35 lakes, 13 transitional waters and coastal areas. Among these, 116 are considered heavily modified (reservoirs, and navigation channels).

Over the last 40 years, groundwater has contributed significantly to growth in global irrigated area, and irrigation is the largest user of groundwater globally. The total consumptive groundwater use for irrigation has been estimated to 43% of the total consumptive irrigation water use (Siebert et al. 2010). In Guadalquivir, groundwater constitutes 25% of the total extracted irrigation water, corresponding to an estimated 38% of the irrigated area (211,500 ha). The origin of irrigation water resources is summarized in Table 3.

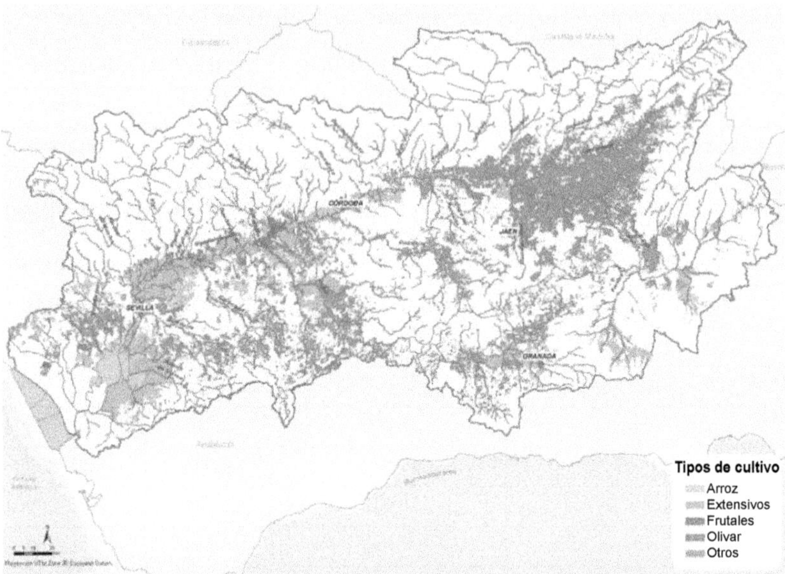

Figure 3. Spatial distribution of crop types in the Guadalquivir RB in 2008.
Source: (CHG 2010b).

Color image of this figure appears in the color plate section at the end of the book.

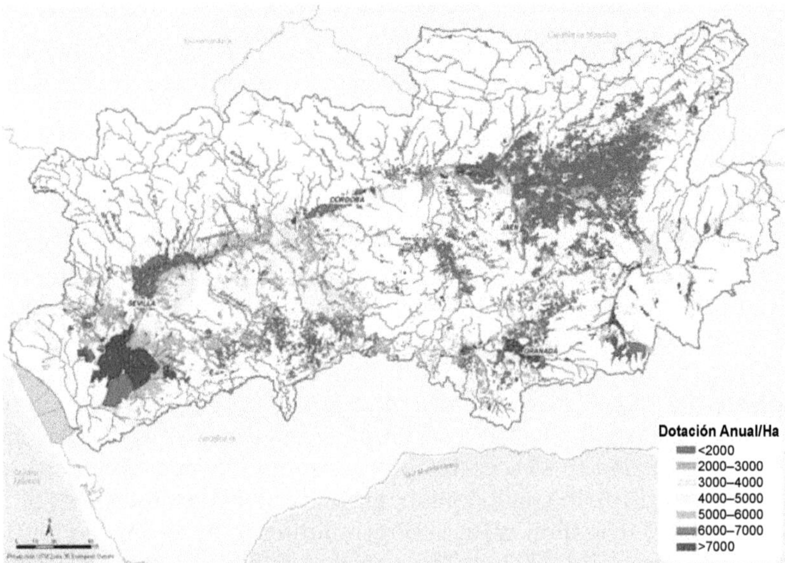

Figure 4. Spatial distribution of annual water allocation (m^3 ha^{-1}).
Source: CHG (2010b).

Color image of this figure appears in the color plate section at the end of the book.

Table 3. Irrigated area and consumption according to water origin in the Guadalquivir River Basin (2008).

Water source	Ha	hm^3	$m^3\ ha^{-1}$
Regulated surface	372,412	2,148	5,666
Non-regulated surface	152,398	574	4,118
Groundwater	308,455	726	2,575
Recycled	11,402	36	3,157
TOTAL	845,000	3,568	4,222

Source: Berbel et al. (2012) adapted from CHG (2010b).

While Northern Spain often has excess water, Southern Spain is often short of water. The process of interfering in this natural imbalance of water by re-allocating between geographical areas started in the aftermath of the Spanish Civil War. The Civil War (1936–1939) left Spain economically and politically isolated. Irrigation was seen as a mean to combat the ailment. Large dams and irrigation channels were constructed, and vast areas of dry land were converted into productive land for irrigated crops (Jiménez Torrecilla and Martínez-Gil 2005). Society adopted the belief that nature and natural hydrological systems were hostile or erroneous, and had to be re-balanced to serve human production. Society then conveyed this belief to civil engineers who, for the first time, had the technology and public funding to change and 'improve' it.

DPSIR and Closure

Drivers & Pressures

Water demand in the Guadalquivir RB is increasing because of the following drivers: (a) population growth; (b) recent irrigated agricultural expansion where rainfed agriculture is converted to irrigation although total cultivated area is relatively constant; (c) economic development including thermosolar plants (there are around 10 under operation); (d) natural conditions: hereunder increasing demand for protection of water quality and quantity for environmental uses, environmental flow has been increased compared with previous hydrological basin plan specification. Recent evolution shows that demand from urban areas and industry remains almost constant and agriculture is the sector that is increasing demand. Irrigated agriculture in Guadalquivir is competitive, that means that crops such as citrus, strawberry, olive oil, vegetables and fruits have found competition mainly from non-European countries but farmers from Guadalquivir have a relative competitive advantage based on proximity to European markets, good farm structure and integrated value chain. Crop profitability has been the driver of growing irrigation demand.

Figure 5 shows the water use scenarios for 2008, 2015 and 2027 as stipulated by the Water Framework Directive (WFD) and GHBP, indicating that urban areas and irrigation will reduce their total water use, but irrigation will remain the largest water user by far.

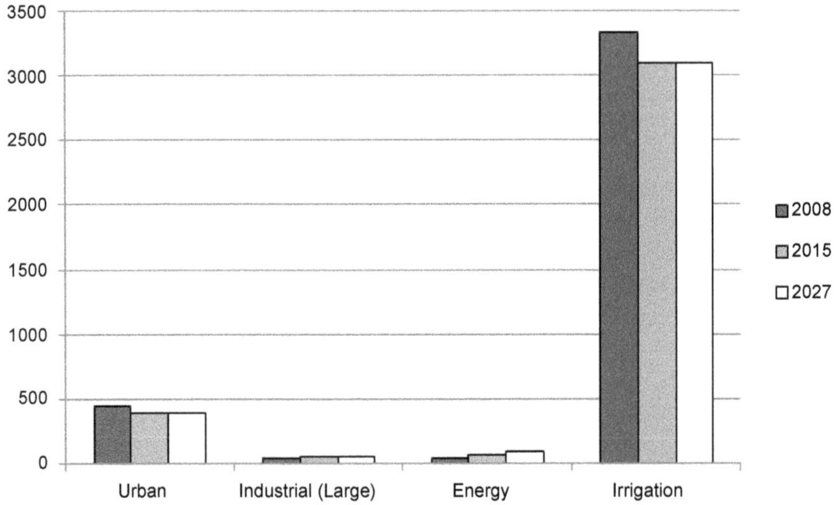

Figure 5. Evolution of water allocation in the Guadalquivir River Basin by sector (2008, 2015 and 2027).
Source: Kolberg 2012 adapted from CHG (2010b).

Population growth

The region's population, roughly 4.2 million people, has been growing, and so have overall levels of production, consumption and trade. Seville, Cordoba, Granada and Jaen are the most populated cities. From 1986 to 1996 the basin experienced 5.51% population growth, compared with 3.1% growth for all of Spain. More rapid growth in the Guadalquivir RB than in the rest of the country is expected to continue (Bhat 2004). Furthermore, urban consumption in the basin has increased from 297 L/person per day in 1992 to 323 L/person per day in 2008 (CHG 2010b).

Agricultural expansion

Irrigated area in the Guadalquivir RB has increased from 142,900 ha in 1904 to 845,000 ha in 2008 according Inventario de Regadios (CHG 2010a). The increase has been particularly rapid in the last few decades (Parias 2007).

High profitability of olive crops in the 1980s, for example, led to significant expansion and intensification of olive cultivation. Comparing the GHBP of 1998 (CHG 1998) with the Draft GHBP of 2010, the total area of land irrigated has expanded from 410,000 ha to more than 845,000 ha. Water demand has increased due to this growth, and was peaking in 2008. An inter-basin transfer ('Negratin Almanzora transfer') of water from the Guadalquivir RB to intensive horticulture in Almeria, located in southeastern Andalusia was approved in 1984.

Economic development

Irrigation in Spain was once considered an engine of economic growth. Regardless, nowadays, irrigated agriculture is subject to the following criticisms by the general public (Fereres and Ceña 1997):

- Irrigation uses too much water;
- Irrigation is inefficient (about 50% of the delivered water is not used by the crop);
- Farmers pay very little if anything, for the water they use;
- Water pollution problems are often caused by agriculture.

In contrast to common perception, the Guadalquivir RB's irrigation water productivity is among the highest in Spain (MAPA 2002). According to Berbel et al. (2011), this high water productivity is a factor that drives demand for irrigation.

In Guadalquivir RB access to irrigation allows farmers to produce high-value crops that would otherwise be impossible to cultivate and helps assure summer production. One irrigated hectare in the Guadalquivir RB contributes 3.5 times more employment than one non-irrigated hectare of farmland (Berbel and Gutiérrez 2004).

Table 4 shows the estimated gross water productivity and water costs for various crops. Though we believe that it could have been more informative to use the net margin per volume, this data was not included in the 'Inventario de Regadios, 2008' however, the high quality data justify the use of this variable.

Water costs depend upon the location, the extraction costs and the quality of the water. The value of the cost of the water is based upon survey data and illustrates the average cost per crop; higher cost for strawberries (0.152 € m^{-3}) and olive (0.140 € m^{-3}) and lower for rice and cotton (0,023 € m^{-3}). Olive has the higher SD (0.100) explained by the variability between locations in the basin.

Gross water productivity (measured as € of crop value generated per m^3 of water applied) is a suitable measure of the value consumers place on a crop, but is not a good measure of producers' economic incentive to grow

Table 4. Gross water productivity (P) and water costs (C) in 2008.

Crop type		P (€ m^{-3})	C (€ m^{3})	C/P (%)	P/C
Cotton	Mean	0.570	0.026	4.5	22.1
	SD	0.560	0.023		
Rice	Mean	0.203	0.026	12.7	7.9
	SD	0.002	0.006		
Citrus	Mean	1.797	0.046	2.6	38.9
	SD	0.332	0.040		
Winter cereals	Mean	0.271	0.042	15.7	6.4
	SD	0.058	0.031		
Strawberry	Mean	8.830	0.152	1.7	58.2
	SD	0.408	0.062		
Fruit trees	Mean	2.485	0.044	1.8	56.1
	SD	0.921	0.039		
Sunflower	Mean	0.177	0.032	18.0	5.5
	SD	0.038	0.023		
Horticulture	Mean	2.260	0.057	2.5	39.8
	SD	0.902	0.051		
Greenhouse	Mean	8.364	0.037	0.4	228.6
	SD	0.610	0.026		
Maize	Mean	0.400	0.044	10.9	9.2
	SD	0.045	0.038		
Olive	Mean	1.757	0.140	8.0	12.5
	SD	0.300	0.102		
Intensive olive	Mean	1.123	0.063	5.6	17.9
	SD	0.262	0.051		
Others	Mean	0.213	0.051	23.9	4.2
	SD	0.148	0.042		
Sugar beet	Mean	0.358	0.041	11.4	8.8
	SD	0.069	0.022		

Source: Kolberg (2012) computed from CHG (2010a). Unit of analysis is 1 ha.

one crop or another. Crops that obtain a high market value are also often expensive to produce, therefore gross value does not necessarily reflect net value, the latter being the one producers actually care about. Net water productivity (i.e., net value per m³ of water applied) is not readily available for every crop but ranges between 0.50 and 0.63 € m^{-3} (Mesa-Jurado et al. 2010). Of course, water is not responsible for all of the gross or net values reported above; other inputs also contribute to these values. The Gross Value Added (GVA) of water, averaged across all crops in the basin, is estimated to 0.50 € m^{-3}. Gross added value productivity is the ratio of the total value of production minus the value of the inputs used divided by water used; meanwhile, residual value is defined (Young 2005) as the ratio of net water value to water when all inputs including opportunity cost of fixed factors such as land and own labour and management are deduced. Berbel et al. (2012) applies this definition to the basin, estimating an average across all crops of 0.31 € m^{-3}. Figure 6 shows the estimated total share of irrigation water allocation (m^3) relative to gross income (€) for all crops in the Guadalquivir RB.

For the Guadalquivir RB as a whole, olives provide more than 35% of the total value of crop production in the basin. They have also become the largest water user, despite their low per-hectare water allocated quota (1,500 $m^3 ha^{-1}$, with an average ratio of irrigation water supply to maximum potential evapotranspiration, of 0.62).

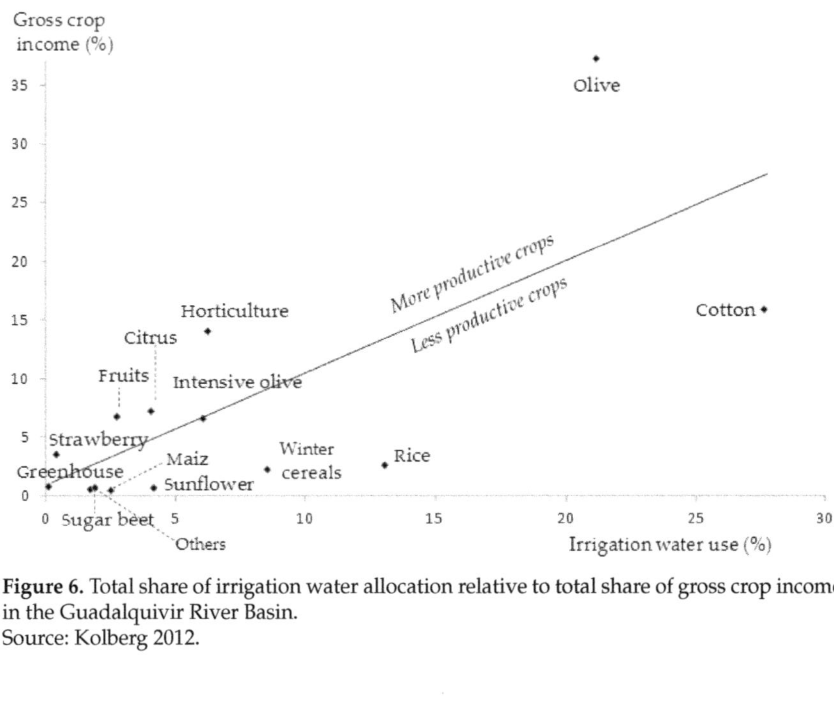

Figure 6. Total share of irrigation water allocation relative to total share of gross crop income in the Guadalquivir River Basin.
Source: Kolberg 2012.

This shows that irrigated agriculture is still an important wealth generator and important for the region's rural based economy. According to regional government (Junta de Andalucía 2010) irrigated agriculture in Guadalquivir produces with 26% of land, 70% of final agricultural product. If we consider that agriculture as a whole contributes 3% to regional GDP, then irrigated agriculture, if considered independently, will represent 2.1% of GDP in the region. Moreover, Berbel and Gutiérrez (2004) argue that agro-industry represents the main sector in industrial production in the region; however the exact contribution from irrigated production is not quantified. As long as the additional value water generates for farmers is higher than the additional cost it generates, farmers seeking increased income will be a main driver of water demand in the basin.

Natural conditions

Natural and human drivers exert pressure on water supply, water demand, and water pollution. Agricultural water demand depends on climate, crop type, soil characteristics, water quality, and cultivation practices. Increased cultivation of high-value irrigated crops such citrus, olive and vegetables has increased demand for water resources. At the same time, however, there have been increases in efficiency of water use per hectare. Not only quantity, but also water quality is a major problem throughout the RB, especially because of pollution from urban and industrial wastewater discharge, erosion, nutrients and pesticide runoff from agricultural land (CHG 2010b). Diffuse pollution from agriculture and urban water use is estimated to cause elevated levels of nitrogen in water bodies. Natural annual flow levels are 7,100 million m^3 for surface water and 2,576 million m^3 for groundwater. About half of these water flows are used for agriculture (80% of total volume extracted). Currently groundwater constitutes 20% of the total water consumed in the basin. Groundwater abstraction has increased over the last few decades due to increasing demands for the irrigation of olive groves in the upper valley. As of 2008, irrigation systems included drip (64%), sprinkler (14%) and surface (27%) techniques (CHG 2010b).

State and impacts

'Filthy water cannot be washed'
West African Proverb

Irrigation improves crop yields, reduces risks during dry spells, and makes it possible to grow more profitable crops. However, irrigation is also the source of various environmental concerns, including excessive extraction of groundwater, irrigation-driven erosion and increased soil salinity. As a consequence of these pressures, the quantitative and qualitative 'states' of the water resource (rivers, lakes, seas, coastal zones, wetlands and groundwater) are affected in terms of physical, chemical and biological environment. Unpredictability in water resource availability, increasing demand from different water sectors, and recurring drought lead to recurrent scarcity events. Scarcity events cause aquifer salinization, higher concentrations of pollutants (lower dilution capacity), and other environmental stresses (CHG 2010b).

Humans have dramatically impacted the landscape, for example, by reducing the natural vegetation to small remnant areas. Large alterations in vegetation and land use can be seen in Cordoba and Jaen, where the natural vegetation of evergreen oaks (Quercus Rotundifolia Lam.) has been replaced with olive tree and other extensive crops (Sabater et al. 2009). Water use in agriculture and other economic sectors is increasingly becoming constrained by concerns about the vulnerability of Mediterranean ecosystems and calls for stricter control of environmental flows and wetlands maintenance. This is true throughout the entire basin, but is especially critical for Doñana National Park, located near the mouth of the river. This park is one of Europe's most important wetland areas and a major site for migrating birds. The impacts of over-extraction of available water include decreases in groundwater levels that in turn can lead to impacts on associated aquatic and terrestrial ecosystems such as wetlands.

The state of water bodies, the fluctuations and availability of the resource were identified as critical during the development of the draft GHBP (Berbel et al. 2012). The impact of agricultural activities (deforestation, use of chemical fertilisers and pesticides, intensification) on the environment (soil erosion, fertility decline, water pollution, salinisation, depletion of the natural water base of ecosystems such as wetlands) is more and more visible and has not been satisfactorily included in calculations of the costs or sustainability of agriculture. If farmers fully accounted for the environmental costs of using water in agriculture, solutions to water scarcity and pollution could be addressed in large part by adjusting farm management practices.

Responses

'By failing to prepare, you are preparing to fail'

Ben Franklin (1706–1790)

Figure 7 shows the current institutional framework of the Guadalquivir river basin management. The institutional framework aims at addressing and responding to the drivers, pressures, state and impacts.

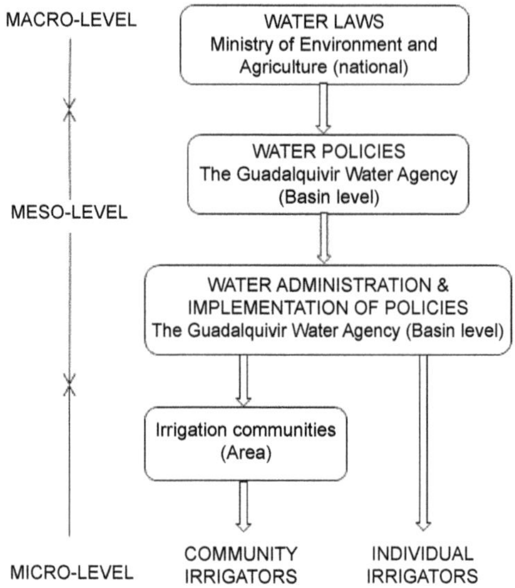

Figure 7. Water Management levels in the Guadalquivir RB.
Source: Kolberg 2012.

The key impacts that forces policy change (response) include: serious environmental degradation, growing water demand, climatic change, agricultural policy, and economic growth (Garrido and Llamas 2008). The response to growing water scarcity and pollution, from the State and the local stakeholders, comprises three distinct categories (Molle et al. 2010): (1) supply-responses, (2) conservation-responses, and (3) allocation-responses. Supply responses consist of increasing supply from existing sources or tapping into additional sources. Conservation responses attempt to improve efficiency by making better use of existing water resources. Allocation responses involve reallocation of water from one user to another, within or between sectors. Irrigation water demand per hectare since 1985 shows a strong tendency to diminish (Camacho 2005). An indepth analysis of a representative sample of 22 irrigation districts in the Guadalquivir RB (30% of the irrigated area in the RB) shows that water consumption per unit of irrigated surface has decreased from an average of 7,000 m^3 ha^{-1} to 5,000 m^3 ha^{-1} in 2004 (Camacho 2005).

Supply responses

Development of additional surface and groundwater resources in the Guadalquivir RB has reached its maximum; almost no additional reservoirs can be built up. The last large dam constructed was the Breña II finalised in

2008. While this dam was being finalised, a public participatory process led to an agreement called 'Acuerdo por el Agua en la Cuenca del Guadalquivir' (CHG 2005), which states there should be no further expansion of irrigated land (only those that are already approved in the GHBP of 1998, but not yet implemented, prior to the agreement, around 35,000 ha). This agreement relieved some of the pressure for additional water rights, especially in the upper basin. The continuous expansion of irrigated area, when paired with relatively constant supply, has typically implied a gradual decrease in total water allocated to each irrigation district. This reduction in annual allocation is described in the Draft GHBP (CHG 2010b). These reductions have been possible because of water saving measures and stricter control by basin authorities. Obviously, during years with heavy rain and accumulated high reserves, water scarcity is not readily perceived. However, in an average or drought year, water is scarce and an obvious limiting factor to yield.

Improved use of available resources

Both the water authorities and the farmers have made great efforts to improve irrigation efficiency during the last few years to adjust to the concerns about long-term water scarcity, and conserving available supplies. This process is often referred to as 'modernisation' of the irrigation sector and expected to continue to year 2015 and beyond, according to the Draft GHBP. Table 5 shows the net and gross water use for 2008 and the anticipated use for 2015.

In 2012, some of the largest irrigation districts are still in the process of system modernisation. Old open-channel networks are being replaced by 'on demand' pressurised networks. The primary aim of these investments is to achieve more efficient conveyance and use of water. As a consequence, nearly half (45%) of the total irrigated area relies on micro (trickle) irrigation, which is now the most common application method in the basin. This is a drastic change from the situation 15 years ago when surface irrigation was the predominant technique (61%) and trickle irrigation (12%) was still regarded as a specialised minority one.

Table 5. Total irrigated area and water use for 2008 and anticipated use for 2015.

Year	Irrigated area (ha)	Gross water allocation		Net water allocation	
		Total (hm^3)	Per hectare ($m^3 ha^{-1}$)	Total (hm^3)	Per hectare ($m^3 ha^{-1}$)
2008	845,986	3,330	3,936	2,463	2,911
2015	881,557*	3,105	3,522	2,524	2,863

Source: Kolberg (2012) computed from CHG (2010a). (*) area included in the GHBP of 1998.

A case study of water saving in Bembezar irrigation district by Rodríguez-Díaz et al. (2011) showed a 40% reduction in water diverted for irrigation. However, total consumptive water use increased substantially, primarily due to new crop patterns, e.g., increased irrigated citrus, a high value crop. Most of the perceived reduction in water use is due to return flow reductions and not to real water savings (Ward and Pulido-Velazquez 2010).

Allocation responses

Today, water resources of the Guadalquivir RB are highly regulated. This is not only to re-balance natural injustice, but also to store water in case of droughts and floods. There are a total of 65 dams in the RB. These regulate 7,145 hm^3 and will amount to over 8.562 hm^3 when the last dam built starts to operate (Argüelles et al. 2012). From the amendment of the 1985 Water Act in 1999, water rights holders are allowed to trade water rights in drought years through basin authorities. The demand for new more productive uses cannot be attended because of the administrative closure in 2005; therefore some market for water right trade has been operating, allowing flexibility to move water rights to most productive uses. An example is water transfer from agriculture to thermosolar plants; for an analysis of water rights trade in Guadalquivir, see Giannoccaro et al. (2013).

Conclusion

This case study adds to a growing body of literature on basin closure, by applying a DPSIR-framework to explain basin closure in a European, or more specifically, Mediterranean, setting. Management of water resources is of vital importance for people's lives and livelihoods, and for society's wealth and economic development. Drivers of change from plenty of water to basin closure include growing population, expanding agriculture, and increasing environmental awareness. Spanish water law is focusing on a firm control of resources and demand. The majority of the water is public and water right allocations are revised every 25 to 40 years. Since 1925, water resources have been managed at basin level. Nevertheless, due to governance problems and stakeholder lobbying (rural municipalities, farmers and others) the demand has increased and 'provisional' allocations have been made. In our opinion there is no need of additional laws; on the contrary, the key issue is to apply the existing one that was revised in the year 2000 due to the approval of the WFD. There is a lack of social acceptance of basin closure by many stakeholders. There is a need to acknowledge the fact that there is no possibility of supply increase and consequently, the resources of the basin should be considered capped.

Acknowledgements

This chapter is based on parts of the background chapter of the PhD thesis of Solveig Kolberg with Julio Berbel and Rafaela Dios-Palomares as supervisors. Thanks are due to the Guadalquivir Basin Authorities for access to their data base.

References

Argüelles, A., Berbel, J. and Gutiérrez-Martín, G. 2012. La evolución de la Cuenca del Guadalquivir (España). Rev. Obras Públicas. Nov-2012, 3537: 51–64.

Berbel, J. and Gutiérrez, C. (eds.). 2004. Sustainability of European Irrigated Agriculture under Water Framework Directive and Agenda 2000. Office for Official Publications of the European Communities, European Commission, Luxembourg.

Berbel, J., Calatrava, J. and Garrido, A. 2007. Water pricing and irrigation: a review of the European experience. pp. 295–327. *In*: Molle, F., Berkkoff, J.J. and Barker, R. (eds.). Irrigation Water Pricing Policy in Context: Exploring the Gap between Theory and Practice. CABI, Wallingford, UK.

Berbel, J., Mesa-Jurado, M.A. and Pistón, J.M. 2011. Value of irrigation water in Guadalquivir basin (Spain) by residual value method. Water Resources Management 25(6): 1565–1579. DOI: 10.1007/s11269-010-9761-2.

Berbel, J., Kolberg, S. and Martin-Ortega, J. 2012. Assessment of the draft Hydrological basin plan of the Guadalquivir river basin (Spain). International Journal of Water Resources Development 28(1): 43–55. DOI: 10.1080/07900627.2012.640875.

Bhat, A. 2004. Policy, politics and water management in the Guadalquivir river basin, Spain. Water Resources Research 40(8): 1–11. DOI: 10.1029/2003WR002726.

Camacho, E. 2005. Análisis de la eficiencia y el ahorro de agua en el regadío de la cuenca del Guadalquivir. Inversiones en la modernización de regadíos. FERAGUA, Spain.

CHG (Confederación Hidrográfica del Guadalquivir). 1998. Plan Hidrológico del Guadalquivir. Url: http://www.chguadalquivir.es/opencms/portalchg/marcoLegal/planHidrologicoCuenca/.

CHG. 2005. Acuerdo por el Agua en la Cuenca del Guadalquivir.

CHG. 2009. La Demarcación. Retrieved November 12, 2009, from url: www.chguadalquivir.es. Retrieved October 15, 2011, url: http://www.chguadalquivir.es.

CHG. 2010a. Inventario de Regadíos 2008. Irrigation Inventory of the Guadalquivir River Basin. Access database, version May 2010.

CHG. 2010b. Propuesta de proyecto de Plan Hidrológico de la Demarcación Hidrográfica del Guadalquivir. Documento para Consulta Pública. Memoria.

CHG. 2011. La Demarcación. Retrieved October 15, 2011, from http://www.chguadalquivir.es.

De Fraiture, C. and Perry, C. 2007. Why is agricultural water demand unresponsive at low price ranges? pp. 94–107. *In*: Molle, F. and Berkoff, J. (eds.). Irrigation Water Pricing: The Gap between Theory and Practice. CABI Publishers, Oxfordshire, UK.

Falkenmark, M. and Molden, D. 2008. Wake up to realities of river basin closure. International Journal of Water Resources Development 24(2): 201–215. DOI: 10.1080/07900620701723570.

Fereres, E. and Ceña, F. 1997. Social benefits and environmental constraints of irrigation in an era of water scarcity. pp. 128–136. *In*: Kay, M., Franks, T. and Smith, L. (eds.). Water: Economics, Management and Demand. E & FN SPON, Chapman and Hall, London, UK.

Gabrielsen, P. and Bosch, P. 2003. Internal working paper Environmental indicators: typology and use in reporting. European Environment Agency, Copenhagen, 20 p.

Garrido, A. and Llamas, M.R. 2008. Water management in Spain: an example of changing paradigms. *In*: Expozaragoza 2008. Exposición Internacional 'Agua y Desarrollo Sostenible'.

Giannoccaro, G., Pedraza, V. and Berbel, J. 2013. Analysis of Stakeholders' Attitudes towards Water Markets in Southern Spain. Water 5(4): 1517–1532.

Jiménez Torrecilla, N. and Martínez-Gil, J. 2005. The new culture of water in Spain: a philosophy towards a sustainable development. *In*: E-WATER, Official Publication of the Water Association (EWA). Url: http://www.ewaonline.de/journal/2005_07.pdf.

Junta de Andalucía. 2010. Agenda del Regadío Andaluz: Horizonte 2015. Consejería de Agricultura, Pesca y Medio Ambiente. Available on: http://www.juntadeandalucia.es/agriculturaypesca/portal/areas-tematicas/infraestructuras-agrarias/regadios-e-infraestructuras-agrarias/agenda-del-regadio.html.

Kolberg, S. 2012. Equity and equality in irrigation water allocation under basin closure: concepts and measurement. PhD dissertation. University of Cordoba, Spain. Url: http://helvia.uco.es/xmlui/handle/10396/7686.

MAPA (Ministerio de Agricultura, Pesca y Alimentación). 2002. Plan Nacional de Regadíos-Horizonte 2008.

Mesa-Jurado, M.A. and Berbel, J. 2009. Water value, water demand and WFD implementation in an olive intensive Guadalquivir sub-basin. Options Méditerranéennes A, 88—Technological Perspectives for Rational Use of Water Resources in the Mediterranean Region, pp. 99–107. Url: http://ressources.ciheam.org/om/pdf/a88/00801184.pdf.

Mesa-Jurado, M.A. and Berbel, J. 2009. Water value, water demand and WFD implementation in an olive intensive Guadalquivir sub-basin. pp. 99–107. *In*: El Moujabber, M., Mandi, L., Trisorio-Liuzzi, G., Martín, I., Rabi, A. and Rodríguez, R. (eds.). Technological perspectives for rational use of water resources in the Mediterranean region. Bari: CIHEAM (Options Méditerranéennes: Série A. Séminaires Méditerranéens; n. 88). 2. MELIA Workshop Technological Perspectives for Rational Use of Water Resources in the Mediterranean Region, 2008/10/29-2008/11/02, Marrakesh (Morocco). http://om.ciheam.org/om/pdf/a88/00801184.pdf.

Mesa-Jurado, M.A., Berbel, J. and Orgaz, F. 2010. Estimating marginal value of water for irrigated olive grove with the production function method. Spanish Journal of Agricultural Research 8(S2): 197–206.

Molle, F. 2004. Technical and institutional responses to basin closure in the Chao Phraya River Basin, Thailand. Water International 29(1): 70–80. DOI: 10.1080/02508060408691750.

Molle, F. 2008. Why enough is never enough: the societal determinants of river basin closure. International Journal of Water Resources Development 24(2): 217–226. DOI: 10.1080/07900620701723646.

Molle, F., Wester, P. and Hirsch, P. 2010. River basin closure: processes, implications and responses. Agricultural Water Management 97(4): 569–577. DOI: 10.1016/j.agwat.2009.01.004.

Parias, P. 2007. Perspectiva del regadío andaluz desde la óptica del regante. Jornadas sobre Recursos Hídricos de Andalucía: Evaluación y Uso Sostenible. Colegio Oficial De Ingenieros Agrónomos de Andalucía. Sevilla. Enero.

Rodríguez-Díaz, J.A., Pérez-Urrestarazu, L., Camacho-Poyato, E. and Montesinos, P. 2011. The paradox of irrigation scheme modernization: more efficient water use linked to higher energy demand. Spanish Journal of Agricultural Research 9(4): 1000–1008. DOI: 10.5424/sjar/20110904-492-10.

Sabater, S., Feio, M.J., Graça, M.A.S., Muñoz, I. and Romaní, A. 2009. The Iberian rivers. pp. 113–149. *In*: Tockner, K., Robinson, Ch. and Uhlinger, U. (eds.). Rivers of Europe. Elsevier Ltd.

Siebert, S., Burke, J., Faurès, J.-M., Frenken, K., Hoogeveen, J., Döll, P. and Portmann, F.T. 2010. Groundwater use for irrigation—a global inventory. Hydrology and Earth System Sciences 14: 1863–1880.

Smakhtin, V. 2008. Basin closure and environmental flow requirements. International Journal of Water Resources Development 24(2): 227–233. DOI: 10.1080/07900620701723729.

Venot, J.P., Turral, H., Samad, M. and Molle, F. 2007. Shifting Waterscapes: Explaining Basin Closure in the Lower Krishna Basin, South India. IWMI Research Report 121, International Water Management Institute (IWMI), Colombo, Sri Lanka.

Ward, F.A. and Pulido-Velazquez, M. 2010. Water conservation in irrigation can increase water use. Proceedings of the National Academy of Sciences of the United States of America 105(47): 18215–18220.

WSM (Water Strategy Man). 2004. Indicators and Indices for decision making in water resources management. Issue 4, January–March 2004. Url: http://environ.chemeng.ntua.gr/WSM/Newsletters/Issue4/Indicators.htm.

Young, R.A. 2005. Determining the economic value of water: Concepts and methods. Resources for the Future, Washington, DC.

SECTION 2

Water Economics, Pricing and Markets

CHAPTER

4

Economics of Rainfed and Irrigated Potato Production in a Humid Environment

J. Morris, K. Ahodo, E.K. Weatherhead, A. Daccache, A. Patel and *J.W. Knox**

Irrigation of Potatoes in the UK

This section reviews the UK potato industry, including recent changes in the nature and composition of production, the underlying trends in crop yield and patterns of water use. The factors influencing the economics of irrigation are briefly highlighted.

The UK potato (*Solanum tuberosum* L.) industry has changed dramatically in recent decades, from a sector comprised of many small individual farms to one with far fewer but much larger agribusinesses, driven by the need to provide high quality product to the major processors and supermarkets (Daccache et al. 2011). Nationally, potatoes represent the most important irrigated crop, accounting for half the total irrigated area and over half the total volume of water abstracted (Knox et al. 2012). For

Environmental Science and Technology Department, Cranfield University, Bedford MK43 0AL, UK.
* Corresponding author: j.knox@cranfield.ac.uk

many growers, the crop is the driving force for long-term investment in irrigation infrastructure. A humid climate means irrigation is supplemental to rainfall but nevertheless important for attaining crop quality, particularly in the drier eastern regions, where fertile soils and flat topography provide ideal conditions for large-scale intensive production. However, there are concerns regarding the future availability and reliability of water supplies for agriculture (Knox et al. 2010b).

Current potato production, yield and underlying trends

The UK is the eleventh largest producer of potatoes globally, with an annual average production of around 6 million tons. The value to the British potato industry is estimated to be around £0.7 billion, making it a significant element of the agricultural sector (Defra 2009). But the industry is undergoing major change and restructuring; potato farming has shifted from being centred around small traditional family farms typically growing less than 5 ha of potatoes per grower to much larger agribusinesses typically growing >50 ha per site (Fig. 1). Over the last 50 years, the total production of potatoes in the UK has remained roughly similar but the area cultivated has almost halved. In 2009, it was reported that 94 000 ha of potatoes were cropped in England and Wales with an average yield of 48 t ha^{-1} (PCL 2010). Improved varieties, better management and uptake of new technology have underpinned productivity increases, with yields increasing from 22 to

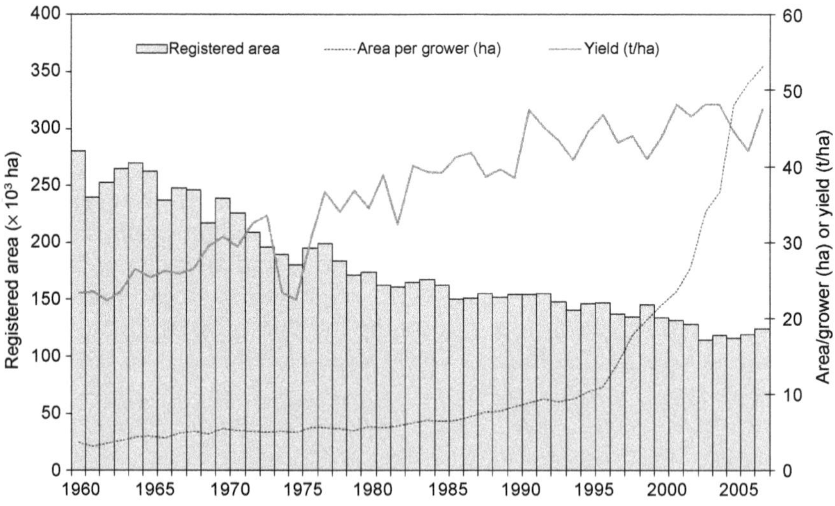

Figure 1. Reported potato cropped area (ha), average cropped area per grower and average yield (t/ha) in the UK between 1960 and 2009 (Source: Daccache et al. 2011).

47 t/ha (Fig. 1); not surprisingly, the UK is now considered to be one of the highest potato yielding areas globally (Daccache et al. 2011).

In most years, national potato consumption exceeds total production, with a small (*c* 10%) proportion being exported. Despite increasing productivity, the UK is still a net importer, equating to approximately a quarter (25%) of total national production. UK produce is destined for either the processing (e.g., crisps, frozen chips) or fresh (supermarket) markets. Nearly half (47%) of total production is sold for fresh consumption with 39% for processing, 10% for seed production and the remainder going into storage. However, in recent years there has been a shift in market share, with a gradual increase in the areas being grown for processing rather than for the fresh market. Similar trends are known to be happening elsewhere in Europe.

There are currently in excess of 160 potato varieties grown commercially in the UK. These can be grouped into two categories depending on their planting and harvest dates. "Earlies" are usually planted between mid-March and early April and then lifted 10 to 13 weeks later. "Maincrop" potatoes are planted between late March and early April and then typically harvested 15 to 20 weeks later. However, these dates vary from year to year depending on weather conditions and summer rainfall. "Earlies" are generally grown as a rainfed crop and concentrated in the wetter regions (e.g., Scotland, Wales and South West England); "maincrop" are concentrated in the drier regions (e.g., Midlands and Eastern England) (Fig. 2). A summary of the main varieties grown and their production statistics is given in Table 1.

Table 1. A summary on the planted area (ha), percentage irrigated (%), average yield (t/ha) and total production (t) of the top 10 varieties grown in England and Wales (Source: Daccache et al. 2012).

Variety	Cropped area (ha)	Proportion Irrigated (%)	Average yield (t/ha)	Total production (t)	Maturity
Maris Piper	19102	63.2	52.9	964652	Maincrop
Estima	9154	46.2	52.7	434758	Maincrop
Lady Rosetta	6491	59.9	52.3	332901	Maincrop
Markies	5867	53.6	51.6	297301	Maincrop
Maris Peer	4577	66.3	34.7	156660	Earlies
Marfona	4277	60.6	52.8	216216	Earlies
Saturna	3364	83.2	45.5	150518	Maincrop
Pentland Dell	2952	17.3	46.5	127195	Maincrop
Hermes	2914	27.8	50.0	144034	Maincrop
Harmony	2759	56.6	54.6	145919	Maincrop

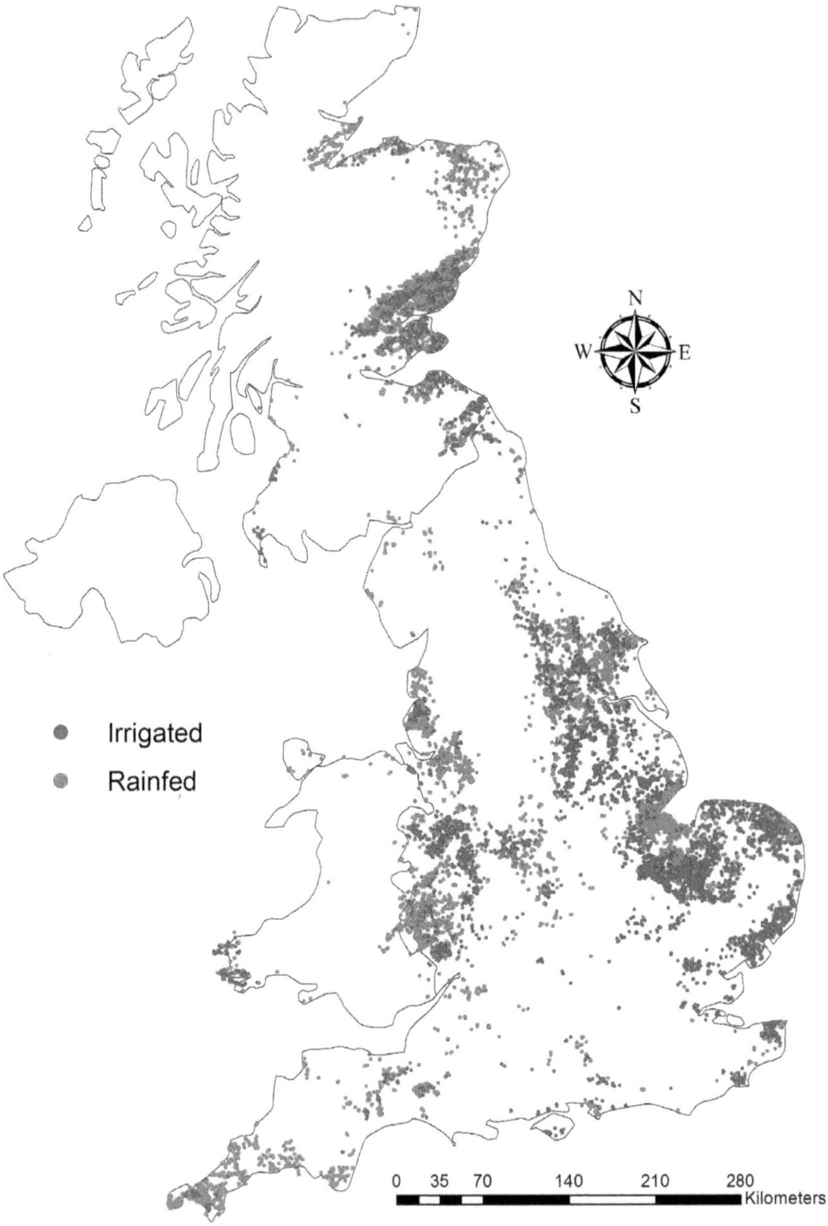

Figure 2. contd....

Figure 2. contd.

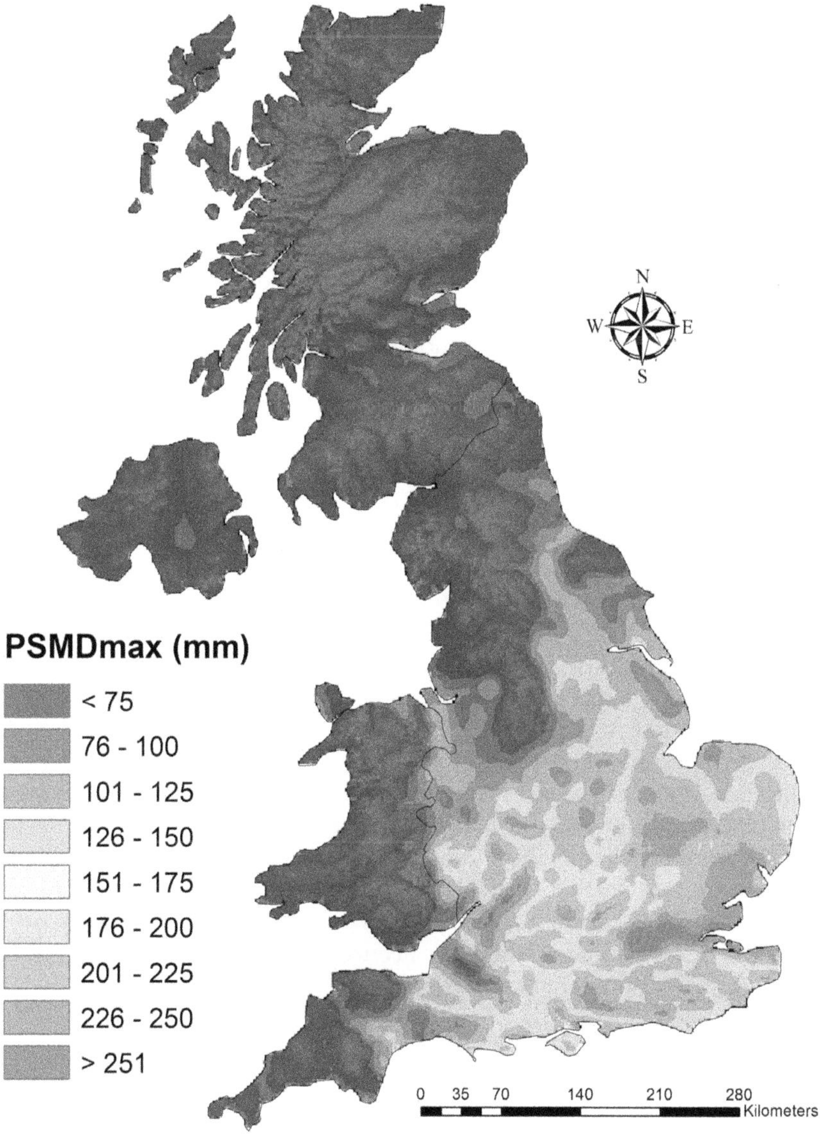

Figure 2. Spatial distribution of irrigated and rainfed potato farms across the UK in 2009 (left panel) and long-term average (1961–1990) spatial variation in agroclimate, using potential soil moisture deficit (PSMD) as an aridity index (right panel).

Color image of this figure appears in the color plate section at the end of the book.

Irrigated areas and volumes applied

Most agricultural cropping in England and Wales is rainfed and, even in a dry year, only a small proportion of land (<1%) is typically irrigated. Over the last 20 years, there have been significant changes in the types of crops irrigated. The proportion of irrigation on grass, sugar beet and cereals has declined steadily. In contrast, there has been a marked increase in irrigation of high value crops, particularly potatoes and field vegetables. This trend is driven by supermarket demands for quality, consistency and continuity of supply, which can only be guaranteed by irrigation (Knox et al. 2000). Agricultural irrigation has therefore developed considerably over the last 40 years, becoming part of a sophisticated production process (Knox et al. 1996). In 1960, only 10% of the national potato cropped area was irrigated (Palutikof and Lister 1999); by 2010, over half (53%) was irrigated (Knox et al. 2012). Irrigation has also encouraged the movement of potato cropping onto lighter soils, where the potential yield losses are highest, but where conditions are more suited to mechanization, providing greater timeliness for planting and mechanical harvesting.

Most irrigation water is abstracted from rivers and streams, and is used direct with relatively little on-farm storage. Over half (54%) of all irrigation abstraction in 2010 was from surface sources (rivers, streams). Groundwater abstraction (boreholes) accounted for 41% and other sources (e.g., water harvesting) was 4%. In field-scale potato irrigation, very little rainwater is harvested and re-used due to the small volumes that can be captured and the requirement for storage.

Field scale irrigation (potatoes and vegetables) can be the largest abstractor in some catchments in dry summers and concerns have been raised over the potential impacts of irrigation water abstraction on the environment, particularly in catchments where irrigation abstractions are concentrated and where water resources are under pressure (Hess et al. 2010). In many catchments, summer water resources are already over-committed and additional summer licences for surface and groundwater irrigation abstraction are unobtainable (EA 2008).

Irrigation equipment

The majority of potato irrigation is dependent on overhead systems, either hose reels fitted with rain guns or booms. Despite their high energy consumption and relatively poor application performance, these systems fit well onto typical UK mechanized arable farms. Booms and guns are popular for their relatively low capital costs and labour requirement but also because they cope particularly well with the flexibility required by rotational cropping patterns, the nature of supplemental irrigation and irregular field shapes. However, potato farming enterprises are reviewing the suitability of these overhead methods given recent rapid increases in energy costs associated with high pressure systems. Micro (drip) irrigation and fixed set mini sprinklers on potatoes are gaining popularity, albeit slowly, for use on high value varieties for pre-pack (fresh) markets where quality assurance is critical.

Irrigation costs and benefits

Irrigation serves mainly to increase crop yield (t/ha) and improve crop quality (£/t) with consequences for revenue (£/ha), over and above that obtained through rain-fed crop production (Knox et al. 2000; Morris et al. 2003). The two effects are multiplicative, rather than additive. Irrigation benefits depend on the crop type being irrigated, the stages in crop development when irrigation is applied, the standard of crop husbandry, market factors which influence commodity prices, and spatially varying environmental factors including agroclimate and soil type. In addition, irrigation may enable a wider range of crops to be grown, support multiple cropping, help in improving seed bed preparation, provide protection against frost damage, enable more effective use of herbicides and fertilisers, and soften tillage pans and clods; these additional benefits are not specifically considered in this assessment. Since irrigation is supplemental to rainfall, in very dry years the benefits can be substantially higher due to the higher commodity prices associated with reduced market supply (Morris et al. 2004).

Historically, most irrigation was applied to attain yield increment over and above rainfed production. However, potato irrigation is now driven principally by the need for timeliness and quality assurance. For most irrigated crops, including potatoes, the quality assurance benefits associated with irrigation can be substantial, as they relate to the whole crop, not just to the extra yield due to irrigation. Quality criteria are increasingly specified as a condition of contract and sale, and failure to meet these quality standards can lead to large price discounting, and sometimes rejection and/or loss of contract. The recent drought (2011) in England highlighted the sensitivity of the market to potential supply shortages and the significant risks growers face in committing to supply contracts. However, establishing links between the amount of irrigation applied and crop quality is complex.

The costs of irrigation vary considerably according to local circumstances, so generalizations are notoriously difficult (Morris et al. 1997). Irrigation costs vary according to the crop requirements for irrigation, the source/s of water used for irrigation (surface or groundwater), the need for any on-farm storage (reservoirs), the type of application system and the size, configuration and topography of the irrigated area, its distance from and height above the water source. In this study, assumptions were based on detailed interviews with key informants and potato farmers regarding all these aspects in order to estimate the benefits and costs of irrigation to a typical 'representative' farming enterprise, recognising of course the inherent limitations such generalization may impose. A brief overview of the methodology used to estimate irrigation costs and benefits (yield and quality) and the added value of irrigation to farmers and to the national economy is summarized below.

Materials and Methods

Developing a framework for assessing irrigation economics

A conceptual framework was developed to assess irrigation yield and quality benefits and irrigation costs at the farm. This considered both physical and climatic contextual factors, farming systems, irrigation systems, and associated inputs and outputs, prices and associated revenues and expenditures. Irrigation benefits and costs were assessed relative to the counterfactual of 'no irrigation', and the resultant incremental benefits and costs were then identified. Hence:

$$R_i = (Y_i.P_r + Y_i.P_i - C_i) - (Y_r.P_r - C_r) - I_c$$

Where

R_i = *annual extra profit from irrigation (£/ha)*
Y_i = *yield with irrigation (t/ha)*
Y_r = *yield for rainfed crop (t/ha)*
P_r = *price for rainfed crop (£/t)*
P_i = *increase in price per unit yield with irrigation relative to price of rainfed crop (£/t)*
C_i = *other crop production costs for irrigated crop (£/ha)*
C_r = *other crop production costs for rainfed crop (£/ha)*
I_c = *average total annual costs of irrigation (£/ha)*

The above values can be expressed as expected annual values to reflect variation year to year in the need for and response to irrigation over the life of the irrigation investment, as explained below.

The counterfactual used here involves a switch from irrigation to reliance on rainfall; that is a rainfed crop, with consequences for crop yield and quality. However, it is noted that in recent decades, potato production has moved onto lighter, more drought prone soils supported by irrigation to suit mechanized harvesting and for crop assurance (quality) reasons.

Thus potato production on many farms may not actually be viable in the absence of supplemental irrigation such that the counterfactual is that of wheat production, the dominant arable crop in England. If this is the case, the returns to irrigation rest on a comparison of expected average annual differences in net margin between the irrigated potato crop and the rainfed wheat crop, less the costs of irrigation. For example, here:

$$R_i = (Y_i(P_r + P_i) - C_i) - (Y_w.P_w - C_w) - I_c$$

Where, additionally:

Y_w = *yield of wheat (t/ha)*
P_w = *price of wheat (£/t)*
C_w = *average total production costs for wheat (£/ha)*

Drawing on earlier work (Knox et al. 2000; Morris et al. 2004) a semi-structured questionnaire was developed and used to support detailed discussions with six growers based in Shropshire, an important region in central England in national terms where potato production is concentrated. Three farms were involved in growing potatoes for the fresh (supermarket) market; the other three for the processing market, with all growing their crops under contract, rather than for the free (open) market. Telephone discussions were also held with a further three potato growers in Eastern England, to capture a broader geographical spread of potato farming and production conditions. The authors also attended a meeting which involved about 25 growers in Shropshire; this provided further opportunity for discussions on irrigation economics with individual participants. Collectively, these interviews provided a comprehensive dataset, albeit limited in sample size, for subsequent analysis.

A separate set of discussions were also made with key informants along the supply chain, including product buyers, consultants and advisors. Secondary information was obtained from published sources (e.g., Nix 2011; ABC 2012) and from irrigation equipment and service suppliers. A parallel study provided estimates of the capital and operating costs of on-farm reservoirs and associated irrigation water supply infrastructure (Morris et al. 2013).

It is noted that the purpose here was to obtain a detailed and updated understanding of irrigation context, practices and outcomes from a representative set of 'cases' that rather than to obtain estimates from a large sample dataset. The data and insights obtained from this exploratory, mixed methods approach must therefore be treated as indicative of the results that might be obtained from a larger sample survey of farmers, although they appear consistent with the findings of earlier work.

Modeling annual yield variability for rainfed and irrigated potatoes

In a humid environment such as England, the irrigation needs for a particular crop vary from year to year depending on summer rainfall. Annual irrigation needs and therefore the costs (and benefits) of irrigation, can vary markedly from year to year. During discussions with famers, the benefits of irrigation were assessed firstly in terms of the expected incremental yield (t/ha) and quality (£/t) attributable to irrigation for a given application depth (mm), for three specified types of 'weather year' based on a 30 year time-series. These were defined as being (i) the top quartile (driest years) which resemble a 'very' dry year, (ii) the lowest quartile (wettest years) and (iii) the remaining two middle quartiles representing 'average' years. Farmers identified, based on their experience, the expected differences in yield between irrigated and non-irrigated potatoes for each of the three types of 'rainfall years'. They also identified the expected difference in potato prices for irrigated (quality assured prices) and non-irrigated potatoes for each type of rainfall year. In this way, an estimate of the incremental revenue attributable to irrigation was determined, which then informed the discussions with farmers about the overall benefits of irrigation, relative to the 'no-irrigation' counterfactual of rainfed potatoes.

The annual variation in irrigation need and yield crop response relative to rainfed yield was also modeled for potatoes using a crop growth model. The SUBSTOR-Potato model embedded within the DSSAT (Decision Support System for Agrotechnology Transfer) program (Jones et al. 2003) has been previously calibrated for UK conditions and used for simulating the impacts of climate variability on UK potato production (Daccache et al. 2011, 2012). Readers interested in a detailed description of the SUBSTOR-Potato model are referred to Griffin et al. (1993). A brief overview is provided here. The SUBSTOR-Potato model simulates the growth and development of the potato crop on a daily time-step using information on climate, soil, management and cultivar. The model is divided into four sub-models simulating simultaneously the phenological development, the biomass formation and partitioning, and the soil water and nitrogen balances to provide a realistic description of the plant-soil-atmosphere system. Phenological development is controlled by cumulative temperature whilst the growth rate is calculated as the product of absorbed radiation, which is a function of leaf area, using a constant ratio of dry matter yield per unit radiation absorbed. Cultivar specific 'genetic coefficients' control the rates of tuber initiation, leaf area development and tuber growth. The soil water balance component embedded within DSSAT is based on Ritchie (1981) where the concept of drained upper and lower limits of the soil are used as the basis for determining available soil water. The nitrogen balance is simulated using the CERES-N model where processes such as mineralization, immobilization, nitrification, denitrification, nitrogen uptake by the plants, distribution and remobilization within the plants are simulated (Godwin and Singh 1998). At each growth stage, deficits in soil water or nitrogen affect the growth of the modeled crop and hence final potato yield.

Using a representative site to reflect typical potato production in a relatively arid part of England (Cambridge), the SUBSTOR-Potato model was first parameterized and then used to simulate rainfed and irrigated yields between 1970 and 2006 (Fig. 3). The ranked data highlight the impacts of extreme dry (1972, 1990, 1975 and 1976) or wet (1993, 1985, 1998 and 1992) years on rainfed potato yields and the buffering effect that irrigation can have on minimizing yield variability between individual climatically contrasting years.

Data on current potato prices for the equivalent period were also obtained. Interestingly, the mean annual potato price and annual irrigation need show a high positive correlation (R = 0.68, P<0.05), and, as might be expected, national average annual crop yield and price are negatively correlated (R = –0.72, P<0.05). This suggests that years of high irrigation yield response also correspond to years of low total production and high potato prices, thus reinforcing the benefits of irrigation. It is noted, however, that farmers sell a large part of their crop on agreed forward prices in order to reduce exposure to market risk.

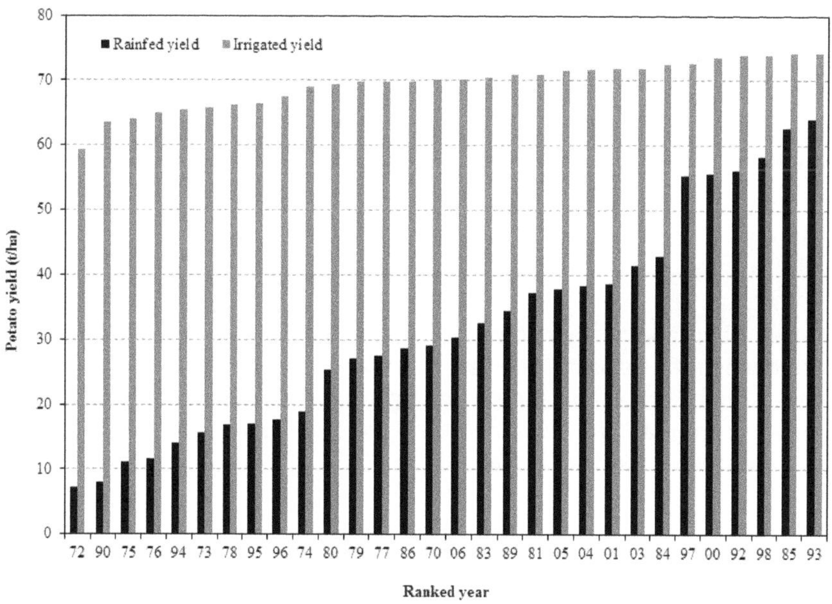

Figure 3. Ranked SUBSTOR-Potato simulated rainfed and irrigated yield (t/ha^{-1}) for maincrop potatoes grown in Cambridgeshire, 1970 to 2006.

The capital, annual fixed and operating costs were derived from the farmer survey data, from secondary sources including irrigation consultants, from a general review of typical irrigation systems, and from first principle estimates informed by observation. Irrigation application rates were based on farm records and other published research sources. A systems diagram was developed and used to guide discussion with key informants regarding the characteristics of their irrigation system, comprising: abstraction and off-take works, reservoir storage (if any), pumping and distribution to the command area and within-field distribution and application. Irrigation water supply costs were then estimated for direct summer abstraction, and for winter abstraction into an on-farm storage reservoir. The in-field application method was assumed to be a hosereel fitted with a raingun, reflecting the dominant overhead irrigation method currently used by potato farming enterprises.

Irrigation costs were estimated for three representative annual irrigated areas, namely 30 ha, 50 ha and 100 ha, intended to reflect a typical small, medium and large irrigated potato enterprise respectively. This broadly covers the observed distribution of irrigation areas by farm at the national scale (Morris et al. 2004) where the mean area of irrigation per farm with irrigation was 85 ha (n = 431; SD = 60 ha), with means of 67 ha of irrigated potatoes for 'potato only' irrigators and 30 ha of irrigated potatoes on farms also irrigating 'other' crops. For the purpose here, we assumed that potato production methods and costs do not vary according to scale, but potential economies of scale are considered in the capital costs of in-field irrigation equipment and water supply infrastructure, notably reservoirs.

Costs were estimated per unit of irrigated area (£/ha) and per unit of water applied (£/m^3) to the crop and then compared with estimated benefits to determine likely return on investment over 20 years at a real discount rate of 6%.

Results and Discussion

As described above, two contrasting approaches were used to estimate potato irrigation benefits. The first was based on the perceived benefits by farmers themselves, and the second approach used the biophysical crop modeling approach for a defined 'dry', 'average' and 'wet' year. The key findings are summarized below.

Farmer assessment of potato irrigation benefits

Estimates of the yield and quality benefits were obtained from the individual farmer interviews for each of the three types of weather year (Table 2). In this approach, rainfall was used as the key variable for classifying years

Table 2. Estimated irrigation benefits for maincrop potatoes grown under three types of 'weather' year based on famer assessments, assuming production conditions (soils and agroclimate) for Shropshire (UK).

Estimated benefit	**Weather year**		
	Wet year	*Average year*	*Dry year*
Extra benefit, yield (% yield increase)	0	25%	50%
Extra benefit, quality (extra £/t)	0	50	30
Yield	**t/ha**	**t/ha**	**t/ha**
With irrigation	50	50	50
Without irrigation	50	40	33
Prices	**£/t**	**£/t**	**£/t**
With irrigation	160	160	160
Without irrigation	160	110	130
Output	**£/ha**	**£/ha**	**£/ha**
With irrigation	8000	8000	8000
Without irrigation	8000	4400	4290
Extra output due to irrigation	0	3600	3710
Relative probability	0.25	0.50	0.25
Expected extra output	0	1800	928
Total expected benefit (£/ha)	-	-	**2728**

by benefit, although it is recognized that total rainfall is not the only factor that determines yield and quality price premia. The temporal distribution of rainfall during the growing season is also critical as well as other factors including temperature and solar radiation, which impact on rate of crop development and growth. The quality price premia are affected largely by general market conditions. The assessment here is based on farmers' perceptions of wet, average and dry years from an irrigation need perspective, grouped by relative frequency at 25%, 50% and 25%, respectively over a run of 30 years or so. Thus, farmers were asked to identify expected irrigation yield and quality benefits in a wet year based on experience of the wettest 7 years out of 30 (2012 was one such year). They were then asked to identify expected benefits for a dry year based on experience of the driest 7 years in 30, and then to assess expected benefits in the remaining 'average' years (Table 2).

Reasonably consistent estimates of extra yield and quality benefits were provided by respondents, although they did not find it easy to separate yield and price benefits. According to the growers and other informants, average potato prices may rise but quality premia may be reduced when markets are short in supply: short markets are often less quality discerning. Most respondents argued that without irrigation, variable potato yields and low prices would render production non-viable in situations where supply and quality assurance was a requisite for market contracts. Farmer estimates produced an average benefit estimate of about £2,730/ha, equivalent to an average benefit of between about £1.60/m^3 and £1.82/m^3 without accounting for the irrigation costs and assuming a total average seasonal volume of irrigation water applied between 150 mm and 170 mm, respectively. It was noted that in wet years, total irrigation application averaged around 50 mm to 60 mm while in dry years, assuming water was available, application rates approached 250 mm or more, thereby averaging about 150 mm overall.

The extra yield attributable to irrigation may also result in some increases in production costs, linked to, for example, harvesting. These were estimated at an overall expected value for the three weather type years of £320 per ha additional costs attributable to irrigation. On this basis, the expected benefits are reduced to about £2,400 per ha (£1.41/m^3 to 1.60/m^3) before accounting for the irrigation costs.

Potato irrigation benefits according crop modeling

Meteorological data from Cambridge NIAB (Latitude 52°22′ N; Longitude 0°10′E) was used to calculate the maximum Potential Soil Moisture Deficit ($PSMD_{max}$) in each year over a 30 year period (1970–1999). The variable, $PSMD_{max}$, is an agroclimatic indicator that has been used previously to reflect the impact of aridity on irrigation need (Rodriguez-Diaz et al. 2007). It is calculated using a simple monthly water balance that takes into account the daily balance between reference evapotranspiration (ETo) and rainfall. The lower the PSMD value, the less arid the year in irrigation terms, and vice versa. The $PSMD_{max}$ in each of the 30 years were ranked to identify the wet (lower quartile), dry (upper quartile) and average years, respectively. For each quartile, the rainfed yields were simulated. This process was repeated based on irrigation need; a similar quartile distribution of years was observed and hence differences in mean wet, mean dry and average yield were similar between the two approaches (Table 3).

Table 3. Mean yield (t/ha) for rainfed potatoes and irrigation yield benefit (%) by type of weather year based on an agroclimate index ($PSMD_{max}$) and irrigation need. Modeled data relate to a farm site at Cambridge for the period 1970 to 1999.

Weather year	Irrigation need (mm)	Based on PSMD (average t/ha)	Based on irrigation need (average t/ha)	Average yield (t/ha)	Yield benefit due to irrigation (% of wet year)	Yield benefit due to irrigation (t/ha)*
Wet	105	58	63	60	0	0
Dry	290	16	15	15	74	37
Average	200	41	37	39	36	18

*assuming 50 t/ha maximum saleable yield.

Assuming an irrigated yield of 50 t/ha saleable crop, applying the estimated yield benefits for mean wet, dry and average years gives a weighted average benefit due to irrigation of around £2,900 per ha for yield benefits (only), excluding any differential price benefit for assured quality. This is broadly similar to the estimates given in Table 2 above. Assuming a crop price of £110 per tonne for a processing grade potato, the estimated benefits will be reduced to an average of about £2,000 per ha over the three types of weather year. If price-quality benefits are also applied to these modeled yield estimates, the benefits rise to about £3,500/ha for pre-pack and about £2,600/ha for processing potatoes.

The implications of these different benefit estimates are considered below.

Potato irrigation costs

Irrigation costs can be grouped into initial capital costs, annual fixed costs and operating costs. Total annual costs can be expressed as £/ha of irrigation area or £/m^3 of water usefully applied.

Capital costs

The size and distribution of capital costs depends on the configuration of a particular scheme, especially the distance (km) and elevation (m) between water source and irrigated area, and whether there is a need to use artificial lining for any on-farm storage (reservoir). In this study, the estimates of the initial capital costs and total average costs were derived for a representative irrigation system to match three different potato production farming units (30 ha, 50 ha and 100 ha), and for two alternative water supply options, (i) direct summer abstraction or (ii) winter abstraction into an on-farm storage reservoir. Farm reservoir costs were determined for 'unlined' (that is lined with natural clay) and an artificial membrane 'lined' reservoir, using a cost function based on data collected from 72 farm reservoirs constructed between 1996 and 2012. A detailed description of the reservoir data and costings is given Morris et al. (2013).

Figure 4 shows the composition of capital costs for a representative 30 ha irrigation scheme, assuming winter abstraction of water into an unlined reservoir provides 50% of annual irrigation water requirements. Capital costs are estimated at about £4,500/ha, of which, in this case, over 50% are attributable to the unlined reservoir, about 30% to supply infrastructure and pumps, and the remainder to infield applicators. If a lined reservoir supplying all irrigation needs is required, total capital costs could rise to over £7,000/ha, of which over two thirds are due to reservoir costs.

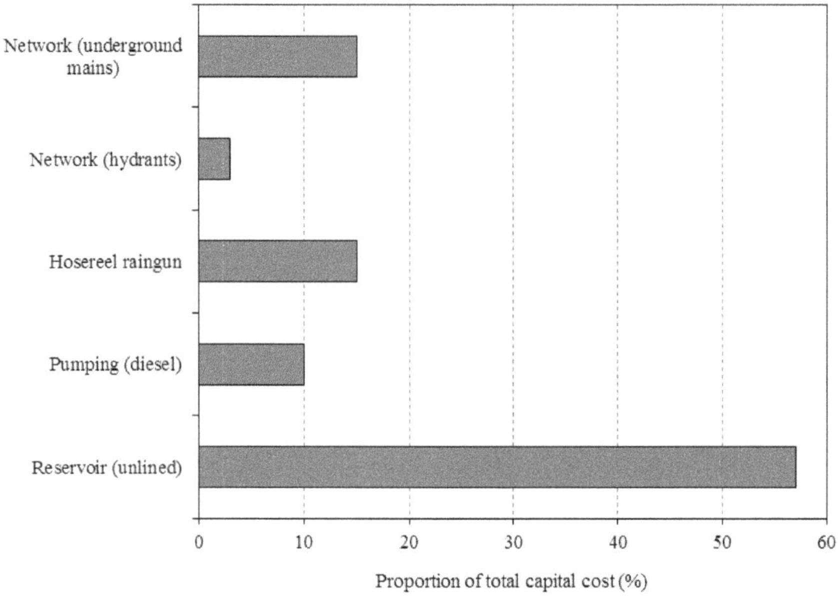

Figure 4. Estimated split in capital cost (%) for a 30 ha irrigation scheme where 50% of the water requirement is met by winter abstraction and stored in an unlined reservoir.

Operating costs

For the same irrigation system, the operating costs differ depending on the volume of water applied and on water charges levied by the Environment Agency (EA), the water regulatory authority. Water charges therefore vary regionally, depending on local hydrological and environmental conditions, and are intended to reflect resource availability and the need for environmental protection. The operating costs for two representative potato production areas located in Shropshire (Midlands) and Cambridgeshire (Eastern England) were estimated in order to account for these differences in water charging regime (Fig. 5).

The annual operating costs for irrigating a 30 ha unit of potatoes (assuming 50% reservoir storage) was estimated to be around £400/ha. Energy for water pumping accounts for between 29% and 32% of the cost, followed by labour and equipment (tractors for moving field applicators) taken together (about 30% repairs and maintenance for infrastructure and equipment account for over a quarter (>25%) of total operating costs. In this study, diesel pumps were assumed. The cost of electrical pumps can be lower but electricity installation costs can be very high.

For Eastern England water rates, water charges range from about £12/ha for 100% winter abstraction and storage (3% of annual operating costs)

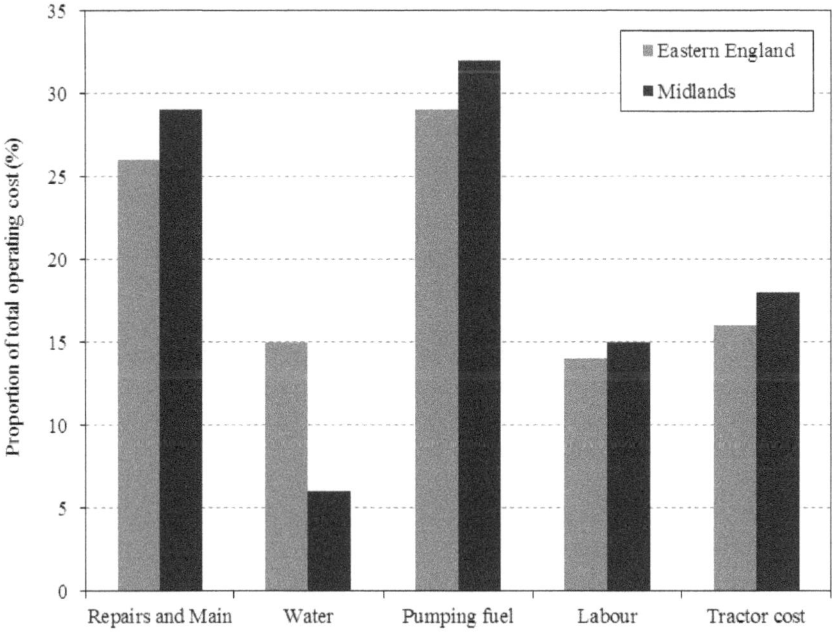

Figure 5. Estimated proportion split in operating costs (%) for a 30 ha irrigation scheme where 50% of the seasonal irrigation water requirement is met by winter abstraction and unlined reservoir storage, for potato production in Eastern England and the Midlands.

through to about £120/ha (18% of operating costs) for 100% direct summer abstraction (assuming water costs £0.06 per m^3 for summer abstraction). However, in the Midlands region, water charges for winter water are about £5/ha (1% of operating costs) and summer water (assuming water costs £0.024 per m^3 for summer abstraction) to about £46/ha (8% of total operating costs) on a 30 ha scheme. In this case, with 50% summer and 50% winter abstracted water, and water charged at rates for Eastern England, water accounts for about 15% of the total average annual operating costs (Fig. 5).

Even though summer abstraction charges are 10 times higher than equivalent winter abstraction charges, the total average costs (£/m^3) are still much lower for direct summer abstraction than winter water once storage costs are added. However, limited availability of new summer abstraction licenses combined with insecurity of future summer supplies are likely to make on-farm reservoirs a necessary investment for future irrigation development, even excluding the potential impacts of climate change on water availability (Knox et al. 2009).

Total average costs

The total costs (capital, fixed and operating costs) for a 30 ha scheme applying an average depth of 150–170 mm/year are presented in Table 4. The average capital costs were estimated to about £4,500/ha (assuming 50% (unlined) reservoir capacity) and total annual costs of about £1,140/ha, made up of about 56% fixed costs and 44% operating costs. This is equivalent to £0.67/m^3 or £6.7/ha mm, and about £69 per acre inch, a cost unit commonly adopted by farmers paying for irrigation services on rented land (where 1 acre (i.e., 4047 m^2) inch (0.0254 m) equals 102.8 m^3, thus £0.67/m^3 x 102.8 m^3 = £69/acre inch). This is higher than estimates of £45 to £50 per acre inch often quoted by farmers during farm surveys, reflecting the use of current rather than historic costs, scheduled repair and maintenance expenditures, and assuming 50% coverage of on-farm reservoirs. Average annual costs are about £0.42/m^3 for installations with no reservoirs, rising about to £0.74/m^3 with 100% unlined reservoir cover. A lined reservoir would increase total average costs to about £0.80/m^3. In future, most irrigation developments are likely to be associated with reservoir investments, such that costs are likely to be around £0.60 to £0.75/m^3 applied (equivalent to £62 to £77 per acre inch).

Table 4. Cost estimates for 30 ha irrigation scheme assuming 50% of water requirements supported by unlined reservoir. Note: Midland water charges give £0.65/m^3. Capital costs and unit costs for other supply options: no reservoir £1.5/m^3 and £0.46/m^3 respectively; 100% reservoir unlined, £4.48/m^3 and £0.74/m^3: 100% reservoir lined £6.9/m^3 and £0.99/m^3.

Capital cost	**£ cap**	**£/year**	**£/ha**	**£/m³**	**% total**
	180628		6021	3.54	
Fixed cost					
Irrigation item					
Unlined reservoir	103507	9542	318	0.19	28%
Diesel pump	18441	2598	87	0.05	8%
Hosereel with raingun	26880	3787	126	0.07	11%
Underground mains	31800	3165	106	0.06	9%
Sub total		**19091**	**636**	**0.37**	**56%**
Operating cost					
Water		1907	64	0.04	6%
Pumping fuel		4339	145	0.09	13%
Labor		2040	68	0.04	6%
Tractor cost		2448	82	0.05	7%
Repairs/maintenance		4464	149	0.09	13%
Sub total		**15198**	**507**	**0.30**	**44%**
Total annual cost		**34289**	**1143**	**0.67**	**100%**

Note: Assumes Anglian Region water charge; 50% winter abstraction with unlined reservoir.

Figure 6 shows the composition of irrigation costs by farm scale, comparing 100% winter abstraction using an unlined (clay) reservoir and 100% summer abstraction without a reservoir. Reservoirs add significantly to capital and average total costs. The average unit costs for irrigation (Fig. 7) on a 50 ha scheme range between £0.33/m^3 applied and £0.61/m^3 for a summer abstraction and a 100% unlined reservoir schemes, respectively. Economies of scale are evident in total average costs per unit of water applied especially for reservoir based systems. Estimated total average costs fall by about 20% over the range 30 ha to 50 ha. Larger equipment, such as infield applicators, have higher work capacities and work rates per unit of investment and per unit of supervision, although beyond a given scale, probably 50 ha for field equipment, systems begin to replicate and scale economies are less evident. Hence, there is little difference in £/m^3 costs between 50 ha and 100 ha installations. Indeed, unit costs for direct abstraction systems tend to rise beyond 50 ha due to increased cost of distribution infrastructure and pumping distances. Additional costs associated with road crossings (typically at £3,500 per crossing) and easements across the land of others can also be substantial for large command areas. In practice, however, much depends on local circumstances such that generalization is difficult.

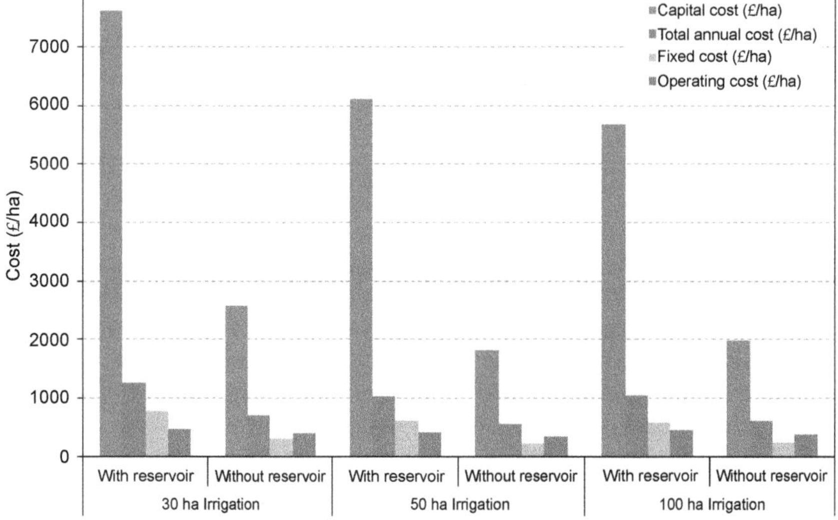

Figure 6. Irrigation costs by area of irrigation and use of unlined on-farm reservoirs providing 100% of water requirements.

Color image of this figure appears in the color plate section at the end of the book.

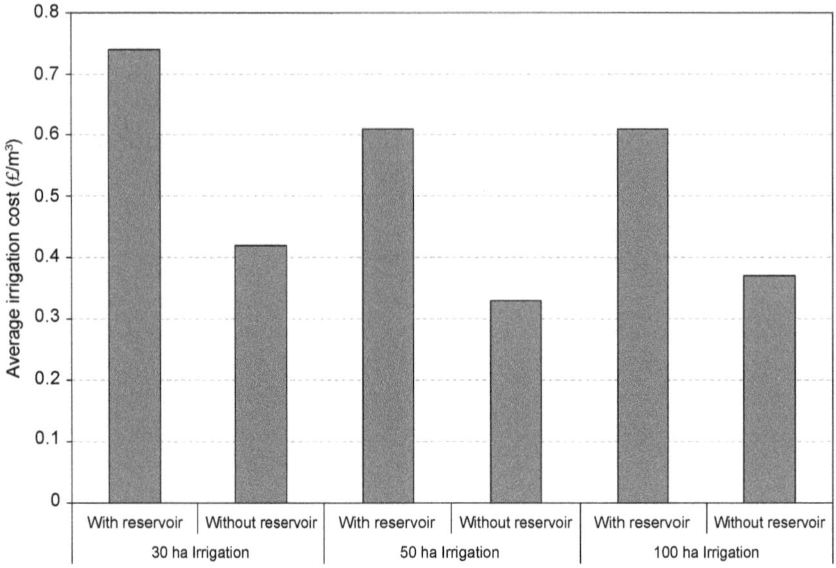

Figure 7. Estimated cost of irrigation water applied (£/m^3) by size of irrigation area, with and without on-farm reservoir storage.

Average value of irrigation water on potatoes

The average value added by irrigation water can be derived by estimating the difference between the extra benefits and extra costs of irrigation relative to the counterfactual 'no–irrigation' situation. A range of benefit estimates were compared with estimated costs (Fig. 8).

Farmer based estimates of incremental yield and quality benefits from the Shropshire case study, assuming a mean irrigation depth of 150 mm was applied, giving an *average* added value by irrigation water of £0.79/m^3 with a reservoir and £1.13/m^3 without a reservoir on 30 ha producing fresh, pre-pack supermarket quality potatoes. Average added value is about 30% lower for process grade potatoes. Added value £/m^3 increases up to about 50 ha due to economies of scale, especially for reservoir options. Modeled estimates of water value on a 30 ha scheme for Cambridgeshire (assuming 170 mm depth applied due to slightly greater aridity) are between £0.82/m^3 and £1.11/m^3 assuming extra yield (only) for fresh grade potatoes. This rises to between £1.34/m^3 and £1.64/m^3 if farmer estimates of quality benefits are combined with modeled yield estimates.

In summary, the average value added by water (£/m^3) on irrigated potatoes is estimated to be between £1.00/m^3 and £1.60/m^3 for fresh potatoes and £0.70/m^3 to £1.20/m^3 for processing potatoes. This is equivalent to a range of about £70/acre inch to £160/acre inch of benefit over and above

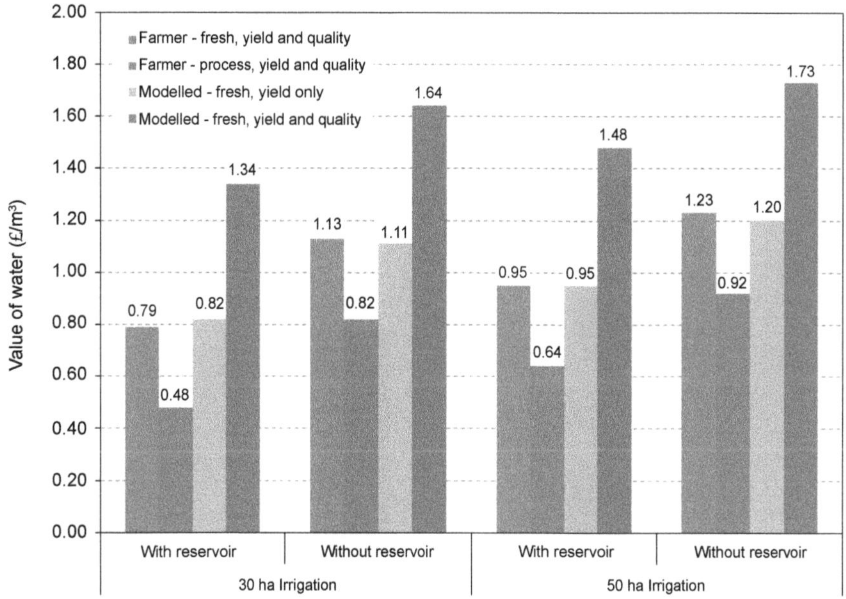

Figure 8. Average value added by irrigation water on potatoes (£/m^3). Note: Farmer estimates based on Shropshire. Modeled estimates based on Cambridgeshire.

Color image of this figure appears in the color plate section at the end of the book.

the costs of water applied depending on assumptions of water supply costs, assuming that all the agronomic crop water requirements are met. If farmers have sufficient water to meet their agronomic needs, then the marginal value of water is zero. Beyond this point, applying more water will add more to costs than revenues. If water availability constrains potato yield and quality, then the marginal value of water will be much higher than the average value, as discussed below. Investment in reservoirs reduces the average value added by irrigation water, possibly by more than 50% where lined reservoirs are required compared with direct summer abstraction (Fig. 9).

The above estimates of the average value of irrigation water are reasonably consistent with previous estimates derived other earlier studies. For example, a review of the impacts of withdrawing water from farmers during periods of water shortage (Morris et al. 1997; Knox et al. 2000), showed average benefits for irrigation of potatoes, field vegetables and salad crops at around £1.75/m^3 to £2.00/m^3 in the Anglian Region, before irrigation costs of about £0.50/m^3, that is £1.25/m^3 to £1.50/m^3 after costs equivalent to about £1.75/m^3 to £2.00/m^3 value added in 2012 prices during the *'design dry year'*.

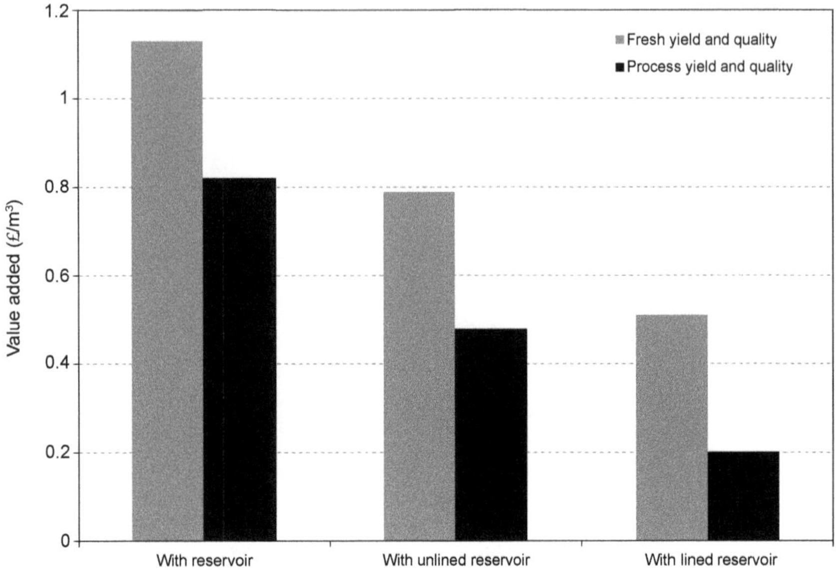

Figure 9. Average value added (£/m^3) by irrigation water supply options on 30 ha farming unit, assuming Shropshire farmer estimates for yield and quality.

Financial investment appraisal

The estimates of benefit and cost of irrigation were used to construct cash flows for an irrigation investment for potato production, discounted over a 20 year period at a real commercial rate of 6%.

Table 5 summarizes the Internal Rates of Return (IRR) for a 30 ha investment for alternative assumptions on water supply for the Shropshire case. Internal rates are very high for direct abstraction, confirming the profitability of irrigation for the assumptions made. IRR remains above the breakeven discount rate (6%) where lined reservoir storage is required.

Table 5. Estimated average value of water (£/m^3) and Internal Rates of Return (%) on investment in irrigation for potato production for different water supply options.

Water supply assumption	**Fresh pre-pack grade**		**Process grade**	
	Ave value of water (£/m³)	*IRR (%)*	*Ave value of water (£/m³)*	*IRR (%)*
Direct abstraction	1.13	78	0.66	60
Winter storage (unlined)	0.96	25	0.65	19
Winter storage (lined)	0.81	15	0.50	10

Sensitivity analysis

Table 6 summarizes a sensitivity analysis of key assumptions of the estimates of water value and the financial return on an investment in irrigation on 30 ha, assuming an unlined reservoir. The most critical assumption here concerns the estimates of extra benefits due to irrigation relative to the alternative rainfed potato crop.

For the assumptions made here, a +/–10% change in revenue due to change in yield and/or quality generates a +/–7% change in the average value of irrigation water applied to the fresh grade potatoes. A fall of 45% in estimated extra crop benefit would make the irrigation investment breakeven at 6% discount rate. For process grade potatoes and a lined reservoir, a 20% fall in assumed revenues from potatoes would make the investment break-even. Other variables are less critical.

Table 6. Sensitivity analysis of estimates of water value and IRR (%) to change in critical variables for a 30 ha irrigation scheme producing pre-pack grade potatoes with an unlined reservoir (Estimates rounded).

Descriptor	% change in costs £/m³ for 10% change in variable	% change in value of water applied for 10% change in variable	Switch value: % change in variable to make value of water = 0
Extra value of potatoes	-	7	–45% (–40% processed)
Capital costs (all)	6	4	+300
Reservoir costs (lined)	5	5	+280
Reservoir costs (unlined)	3	3	+480
Energy	2	2	+800
Water charges	<1	<1	+1000

Marginal value of irrigation water

The preceding analysis estimates the average value of irrigation water over the relevant range of irrigation volume applied, weighted according to types of irrigation response years. From a policy perspective, the concern is not so much with the total value of water applied in the agricultural sector, but rather with the economic implications of making more or less water available. Estimates of the marginal value of water (£/m^3) in England over the relevant range of water use in the average weather year (10th driest year in 20) were derived by modeling different irrigated farming systems (Morris et al. 2004). The marginal value of water shows the change in profitability due to applying one more (or less) unit of water (£/m^3) at a given level of use, inclusive of the costs of the irrigation system itself. This indicates a 'shadow' or hidden price of water. The marginal value of water varies according to amounts applied. In the average year, farmers are typically applying about 70% of the irrigation water needed in the dry design year (maximum water need).

The marginal value of water varies amongst different farming systems according to crop yield and quality response to irrigation, the crop mix, the underlying profitability of crops and irrigation system costs. Irrigation water in the UK gives the greatest returns in market-oriented potato and vegetable farming systems where the emphasis is principally on quality assurance.

Figure 10 shows, for example, marginal value of water over the relevant range of applications at around £1.50/m^3 (2004 prices), which equates to about £1.70/m^3 based on 2012 prices. The estimates of water value shown in Fig. 10 are consistent with estimates derived in this updated study.

It is apparent, however, that returns to irrigation margins have reduced somewhat given a real rise in irrigation costs associated with energy and winter storage compared with relatively steady commodity prices. This cost-based squeeze has prompted the need for increased efficiency in irrigation design and operation.

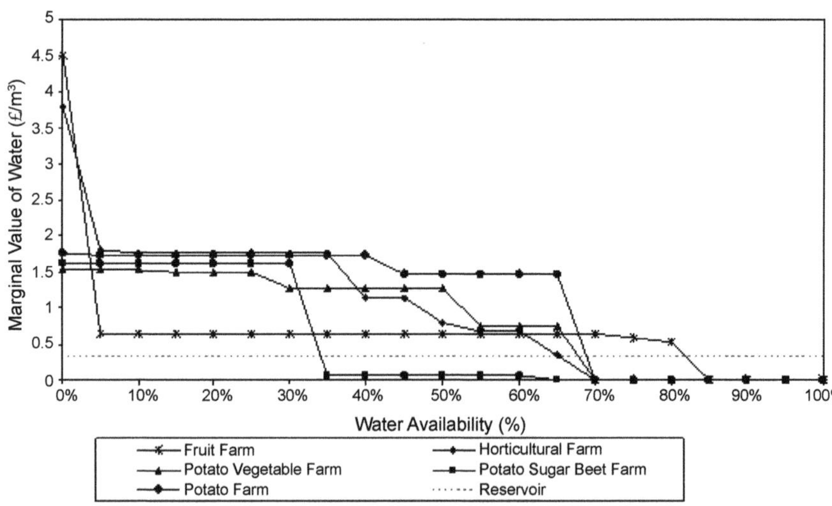

Figure 10. Marginal value of water by irrigation farming system in England. Water availability expressed as a % of that required in 5th driest year in 20 (Source: Morris et al. 2004).

Conclusion

Irrigated potatoes account for around half of all water used for agricultural in irrigation in England, and is justified by farmers in terms of the benefits accrued through extra crop yield and quality. Therefore, the economics of potato irrigation provide a useful benchmark for irrigation economics in England as a whole.

In a humid climate, irrigation needs and crop responses vary considerably between years, as do the benefits of irrigation. Farmers commonly express the view that markets for fresh and processed potatoes demand assurance of yield and quality that can only be reliably delivered when the crop is irrigated. Increasingly, especially on lighter soils, rainfed potato production is perceived to be too risky. Recent research on the impacts of climate change on land suitability confirm that under conditions of increased aridity and particularly droughtiness, many areas of current rainfed potato production would become commercially unfeasible unless irrigation is available, increasingly supported by on-farm reservoirs (Daccache et al. 2012). From a national economy perspective (HMT 2011) this questions the 'additionality' and/or 'substitution' aspects of potato irrigation and the appropriate counterfactual against which to assess irrigation investment. Here the counterfactual might involve a switch to a different rainfed crop (such as wheat) and the movement of potatoes to another location where they can either be irrigated without constraint using direct abstraction, or be rainfed. Alternatively, the displaced potatoes could be substituted by imports valued at cost.

Viewed in these terms, the economic appraisal of irrigation investments is complex and may involve a range of agricultural commodity groupings and regions, all set in the context of rising water values in other sectors. The capital costs of irrigation range between £2500/ha and £7000/ha with the highest cost occurring where lined storage reservoirs are needed. Representative costs of water applied to the crop range from between £0.40/m^3 to £0.70/m^3, depending on site conditions and the need for storage. However, climate change projections and the underlying increase in water demand from within agriculture and competing sectors will be a major driving force for promoting on-farm reservoir investments, for both existing and new developments (Knox et al. 2010b).

This study confirms that the benefits of irrigation for potatoes can be substantial, ranging in the Shropshire case reported here of an average annual benefit of between about £2,000/ha and £3,500/ha relative to a rainfed potato crop, with considerable variation between years. This gives a gross value of about £0.70/m^3 to £2.00/m^3 at current rates of water use on potatoes (of about 1500 to 1700 m^3 per ha). Irrigation costs range between £0.40/m^3 and £0.70/m^3, depending on scale and storage requirements, giving an average value added of between £0.50/m^3 to £1.50/m^3 or more, allowing for likely yield variation in weather years and market conditions. Where there is a clear need for irrigation, the return on irrigation as an 'incremental' investment appears favorable, between 10% and 25% IRR with winter storage for the assumptions made here.

The key factor and main source of uncertainty affecting the viability of irrigation at the farm-scale is the size of extra crop benefits that vary year to year due to climate and market conditions. Irrigation costs vary considerably between sites according to the need for reservoirs and the extent of supply infrastructure, notably distributional mainline pipes and access. There is probably less uncertainty about costs for a given irrigation development once site conditions are known. Future irrigation development will require investments in on-farm storage due to pressure on water resources. The scope and justification for irrigation is closely attuned to the underlying profitability of the crops selected for irrigation, which by their nature tend to be high cost and high risk. As farmers quickly point out, irrigation adds value by attenuating risk. It appears that irrigation costs have risen in real terms in recent years relative to many crop commodity prices, mainly due to rising energy prices and the need for storage. This will prompt industry wide improvements in irrigation management and water use efficiency.

Acknowledgement

The authors acknowledge the many growers, irrigation suppliers, reservoir design engineers, and other key informants involved in this study, including those in the potato supply chain and the potato levy board (PCL) for their support, provision of data and active involvement in the research. Funding support from Defra is kindly acknowledged.

References

ABC. 2012. The Agricultural Budgeting & Costing Book, 74th ed. Agro Business Consultants, Melton Mowbray, England.

Daccache, A., Weatherhead, E.K., Stalham, M.A. and Knox, J.W. 2011. Impacts of climate change on irrigated potato production in a humid climate. Agricultural and Forest Meteorology 151: 1641–1653.

Daccache, A., Keay, C., Jones, R.J.A., Weatherhead, E.K., Stalham, M.A. and Knox, J.W. 2012. Climate change and land suitability for potato production in England and Wales: impacts and adaptation. Journal of Agricultural Science 150(2): 161–177.

Defra. 2009. Agriculture in the United Kingdom 2009. http://www.defra.gov.uk/.

Environment Agency. 2008. Water resources in England and Wales—current state and future pressures. EA Bristol, 21 p.

Godwin, D.C. and Singh, U. 1998. Nitrogen balance and crop response to nitrogen in upland and lowland cropping systems. pp. 55–77. *In*: Tsuji, G.Y., Hoogenboom, G. and Thornton, P.K. (eds.). Understanding Options of Agricultural Production. Kluwer Academic Publishers and International Consortium for Agricultural Systems Applications, Dordrecht, The Netherlands.

Griffin, T.S., Johnson, B.S. and Ritchie, J.T. 1993. A simulation model for potato growth and development: SUBSTOR-Potato. IBSNAT Research report series 02-05/93 (500).

Hess, T.M., Knox, J.W., Kay, M.G. and Weatherhead, E.K. 2010. Managing the water footprint of irrigated food production in England and Wales. pp. 78–92. *In*: Hester, R.E. and Harrison,

R.M. (eds.). Issues in Environmental Science and Technology 31: Sustainable Water. Royal Society of Chemistry, London.
HMT. 2011. The Green Book: Appraisal and Evaluation in Central Government. Her Majesty's Treasury. The Stationery Office, London.
Jones, J.W., Hoogenboom, G., Portera, C.H., Boote, K.J., Batchelor, W.D., Hunt, L.A., Wilkens, P.W., Singh, U., Gijsman, A.J. and Ritchie, J.T. 2003. The DSSAT cropping system model. European Journal of Agronomy 18: 235–265.
Knox, J.W., Weatherhead, E.K. and Bradley, R.I. 1996. Mapping the spatial distribution of volumetric irrigation water requirements for main crop potatoes in England and Wales. Agricultural Water Management 31: 1–15.
Knox, J.W., Morris, J., Weatherhead, E.K. and Turner, A.P. 2000. Mapping the financial benefits of spray irrigation and potential financial impact of restrictions on abstraction: a case study in Anglian Region. Journal of Environmental Management 58(1): 45–59.
Knox, J.W., Weatherhead, E.K., Rodriguez-Diaz, J.A. and Kay, M.G. 2009. Developing a strategy to improve irrigation efficiency in a temperate climate: a case study in England. Outlook on Agriculture 38(4): 303–309.
Knox, J.W., Rodriguez-Diaz, J.A., Weatherhead, E.K. and Kay, M.G. 2010a. Development of a water strategy for horticulture in England and Wales. Journal of Horticultural Science and Biotechnology 85(2): 89–93.
Knox, J.W., Morris, J. and Hess, T.M. 2010b. Identifying future risks to UK agricultural crop production—putting climate change in context. Outlook on Agriculture 39(4): 249–256.
Knox, J.W., Kay, M.G. and Weatherhead, E.K. 2012. Water regulation, crop production and agricultural water management—understanding farmer perspectives on irrigation efficiency. Agricultural Water Management 108: 3–8.
Morris, J., Weatherhead, E.K., Mills, J., Dunderdale, J.A.L.D., Hess, T.M., Gowing, D.J.G., Sanders, C. and Knox, J. 1997. Spray Irrigation Cost benefit Study. Report to Environment Agency. Cranfield University, Silsoe, Bedford.
Morris, J., Vasileiou, K., Weatherhead, E.K., Knox, J.W. and Leiva-Baron, F. 2003. The sustainability of irrigation in England and the impact of water pricing and regulation policy options. Advances in Water Supply and Management. Proceedings of the International Conference on Advances in Water Supply Management, September 15–17, 2003, London, pp. 623–632.
Morris, J., Weatherhead, E.K., Knox, J.W., Vasileiou, K., de Vries, T.T., Freeman, D., Leiva, F.R. and Twite, C. 2004. The Sustainability of European Irrigated Agriculture under Water Framework Directive and Agenda 2000: The Case of England and Wales. Cranfield University, Silsoe, England.
Morris, J., Weatherhead, E.K., Knox, J., Daccache, A. and Kay, M.G. 2013. An updated assessment of the economics of on-farm irrigation reservoirs. Final Report (May 2013). Defra funded project FFG1112: Efficient supply of water for agriculture. Cranfield University, Bedford.
Nix, J. 2011. Farm Management Pocketbook, 42th (2012) ed. The Pocketbook, Melton Mowbray, UK.
Palutikof, J.P. and Lister, D.H. 1999. Proportion of potato crop that is irrigated. pp. 44–45. *In*: Cannell, M.G.R., Palutikof, J.P. and Sparks, T.H. (eds.). Indicators of Climate Change in the U.K. Department of Environment, Transport and Regions, London.
Potato Council. 2010. GB potatoes: market intelligence 2010–11. Agriculture and Horticulture Development Board (AHDB) 2010, Stoneleigh Park, Warwickshire, UK, 32 p.
Ritchie, J.T. 1981. Water dynamics in the soil-plant-atmosphere. Plants and Soil 58: 81–96.
Rodriguez-Diaz, J.A., Weatherhead, E.K., Knox, J.W. and Camacho, E. 2007. Climate change impacts on irrigation water requirements in the Guidalquivir river basin in Spain. Regional Environmental Change 7(3): 149–159.

CHAPTER

5

Irrigation Water Resource: Economic Evaluation and Scenario Analysis in a Rice-cultivated Area[#]

Gian Battista Bischetti,[a] *Enrico A. Chiaradia,*[b] *Claudio Gandolfi,*[c] *Federica Monaco** and *Guido Sali*[d]

Water Management in Agriculture: General and Critical Aspects

Water represents a fundamental element for all sectors of economic, social and environmental interest. For the agricultural sector in particular, it plays a role as a key productive input for the conduction of related activities, both and especially in arid and semi-arid regions, but more and more often

Dipartimento di Scienze Agrarie e Ambientali – Produzione, Territorio, Agroenergia, Facoltà di Scienze Agrarie e Alimentari, Università degli Studi di Milano, Via G. Celoria 2, 20133 Milano.
[a] Email: bischetti@unimi.it
[b] Email: enrico.chiaradia@unimi.it
[c] Email: claudio.gandolfi@unimi.it
[d] Email: guido.sali@unimi.it
* Corresponding author: federica.monaco@unimi.it

[#] Research funded by Regione Lombardia, fondo per la promozione di Accordi Istituzionali, project BIOGESTECA 15083/RCC.

in temperate ones. In the former, water allows to obtain a sufficient crop production, in the latter to maintain production at high levels and reduce the risk of loss of the product (Tarimo et al. 1998; Iglesias et al. 2005; IPPC 2012).

Water resource as an element independent from climate trends, will become more and more important in relation to climate change (Fischler et al. 2007).

Focusing on strictly irrigated agriculture, which represents a widespread activity in such areas where water availability for the primary sector has not traditionally been limiting, but only a factor of exaltation and stabilization of yields, we must recognize and emphasize that things are changing.

Also in these areas, however, the decrease of water availability in agricultural sector is nowadays a concrete possibility. Global phenomena of climate change, increasing population and rapid urbanization are progressively causing a quantitative reduction of the resource, emphasizing the conflict of use among different sectors (agriculture, civil and domestic uses, industry and hydroelectric) as a result of an increasing demand on the part of each of them. In this context, an exceeding demand to what is actually available—especially in relation to particular periods coinciding with those associated with crop requirements—determines an insufficient satisfaction of needs of the various sectors, exacerbating the effects of decreased usability (UNEP 1999). The resulting consequences inevitably have repercussions on productive and economic performances of farms, changing in a long-term period, their competitiveness and burdening on their possibility in continuing the activity.

This situation is to be added to the need of reducing the waste of the resource and promoting an efficient and equal use of the resource itself, aimed to its protection, as long as it is an economic asset with a limited availability (ICWE 1992). A more efficient allocation is possible only by attributing a fair price to irrigation water. The quantification of water supplying and use costs, is therefore becoming more and more important, and the estimation of water irrigation costs undoubtedly plays a role in supporting regulatory decisions about water and has a strong power of direction, allowing decision makers to make aware choices to face water shortages.

In agriculture, different modes of delivery of water service can represent an incentive for a sustainable use of the resource. Water supply at a farm level is in fact subjected to the payment of fees charged by the supplier and the types of tariff applied differ from each other according to the efficiency in promoting a more rational use of the resource, avoiding wastes. Among the most commonly adopted methods of payment, a fixed fee set per irrigated or irrigable hectare (€/he), which may vary by crop, tends not to encourage such practices, but is relatively easier to adopt and may in some cases represent the most recommended solution (Giannoccaro et al. 2007); volumetric fees (€/m^3), whether calculated on the basis of the area booked at the beginning of the season or applied to consumption actually measured in farms by counters (Dono and Severini 2006), determine a more aware exploitation, but more often have unit costs well below the actual cost of the resource.

The assignment of a political price to water service supply has revealed itself as an inefficient management system, not stimulating a proper use (Rogers et al. 2002). On the other hand, efficient pricing may determine undesirable effects on farmers' productive decisions, affecting their incomes and their economic capacity, or environmental implications not immediately anticipated. In this sense, then, it would be advisable to analyse costs and benefits from time to time, considering economic, social and environmental impacts of management choices adopted.

Along with economic farm-level aspects, is to be considered that in districts where irrigation has a long tradition, such as the one analysed, water supply systems are accompanied by environmental aspects and aspects of multiple use of the resource that must be considered (Cadario and Bischetti 2006). In these contexts, in fact, techniques of water distribution and use in primary sector are generally obsolete and technically inefficient, since they are characterised by huge losses due to filtration. The complex stability created over the centuries, however, managed to create an optimised system, reusing in the valley territories the losses of the mountain ones, both through springs and groundwater wells, and through the conveyance of leakages. The network of irrigation ditches and channels inherited from the past, moreover, allowed the creation of valuable paranatural aquatic environments (sometimes of such a high quality to be defined as Site of Community Importance).

Irrigation methods and their consequences

The traditional method (1960s–2000s) implied that flooding was more or less contemporaneous in the entire district during April and supplier consortium had to satisfy demands. In the last ten years, a lengthening of the flooding-period has occurred, in which farmers submerge the sections of paddy fields even in May. This is partly due to the commercialisation of short-cycle varieties that can be planted later, and on the other hand to the spreading of soil-sowing and postponed flooding. This trend seems to be increasing in extension also because it allows the work to be spread, to spread the work avoiding peak periods in farmers' works and some problems related to algae proliferation during rice emergence-stage.

Semi-traditional cultivation

Semi-traditional cultivation implies less pressure on requirements during April and a progressive increase of the demand, until the complete submersion occurs, maybe in June; water requirement is then quite constant until July, when corn water requirements cause a higher demand (flood maintenance for rice along with corn). It is something similar to conditions in the 1960s' what was used to be in 1960s, when rice was first planted in suitable sections (but on a smaller area, and therefore with less flow demand) and then transplanted by rice-weeders in the field, leading to an increasing flow demand.

From the physical point of view, with the semi-traditional cultivation there is a delay in loading the aquifer (thus reducing losses that, in the early season, are high). This delay means that the water, which in April–May comes from snowmelt, fails its purpose of loading the water table and a portion of it, that before was only used to maintain the submersion of the sections, now goes into groundwater. At the moment, there are no studies indicating this behaviour, and it would be interesting to analyse it for some seasons.

Operatively, this method should not have negative consequences; on the contrary, it helps to satisfy the needs with less difficulty and pressure, even in the case of breakdowns and interruptions of channels and ditches.

Dry cultivation

The decisive shift to dry cultivation (also called periodic irrigations or intermittent submersions) involves several changes. It is not a furrow irrigation because the terrain modelling carried out over the past decades has led to the creation of sections as large as possible and perfectly horizontal (levelled with laser and special machines). It is therefore likely to proceed with short submersions, which may last a few days, alternate to a period of about 8–10 days of dry.

From the physical point of view it is likely to notice a strong lowering of the aquifer level; this situation will lead to the dry of springs, or at least to the maintenance of a water flow similar to the one observed during winter period.

Regarding water supply, there will be a reduction in field demand, but it must be considered that for the provision of about 265 m^3/s on the area of the district, only about 200 m^3/s are from external sources, and the rest comes from internal recirculation (springs, wastes, leakages).

It is therefore necessary to estimate the decreasing of demand. In addition, there would be a greater peak of demand than the current one, because in July (in addition to corn, as it is today) it will be necessary to provide water also to a rice that is not submerged. This conduct of the paddy field will result in a soil that is not soaked, and then, if a furrow irrigation would be adopted, it will have a lower irrigation efficiency.

Alternatively, a sprinkler irrigation could be carried out, but this would lead, besides the necessary investment and a higher energy charge, to a risk of lodging of the cereal.

From the management point of view, there could be problems with flooding because the supply is continuous and once started, the system does not require extraordinary interventions. On the contrary, as for corn, periodic irrigation requires the presence of the keeper also during the night or on holidays. It is therefore likely that there is greater need for labour, even though farmers may not accept these cadences. Currently a single keeper manages about 100 ha, and it can be expected that the dry cultivation would take 1.5–2 keepers for the same area. Alternatively, there would be unused water during nights and on holidays, increasing the overall need.

Cost of the resource

The allocation of a fair price for irrigation water is an aspect emphasized by *Water Framework Directive* (from now on WFD) 60/2000/EC (European Parliament and Council 2000). In order to pursue this logic, the legislation calls on Member States to introduce the concept of full cost and to adopt economic instruments to improve allocation efficiency of the resource itself.

Regarding economic evaluation of water management and use, the adoption of appropriate economic instruments and the "polluter-pays" principle allow to determine and apply in practice, the concept of the so-called *"full cost"* (Fig. 1), which, taking into account financial costs, opportunity and environmental costs, causes to end users of water, the

<table>
<tr><td>Environmental costs not related to water</td><td rowspan="2">Enivronmental costs (external)</td><td rowspan="6">Economic costs</td></tr>
<tr><td>Environmental costs related to water</td></tr>
<tr><td>Opportunity costs (scarcity)</td><td>Cost of the resource (external)</td></tr>
<tr><td>Other direct costs</td><td rowspan="3">Financial costs (including environmental and opportunity costs already internalized)</td></tr>
<tr><td>Administrative costs</td></tr>
<tr><td>Capital, operating and maintenance</td></tr>
</table>

Figure 1. Structure of the *full cost* (modified from WATECO 2003).

payment of a price high enough to recover all the items of expense resulting from the use and which reduces wastes and non-virtuous behaviours caused by an underestimation of the resource.

In agriculture, however, the quantification of environmental costs (EEB 2001) of irrigation water is not immediate and this means a difficult determination of the exact amount of full cost; in any case, in agriculture applied fees are much lower than those hypothesized by regulators. Also for this reason, it is reasonable to think that, in a medium-term period, an increase of the amount of irrigation costs shall occur. Thereafter, paradoxically, the farmer, as the end user of the resource, would be in the condition of having less water at a higher cost.

Irrigation water supply from consortia or other entities occurs according to different fare types, direct and indirect (OECD 1999), covering distribution costs according to different criteria. Responding to the intention to fulfil its guide lines, the WFD suggests using preferentially a volumetric rate, and requires its implementation as an economic instrument able to reduce water consumption and cover the costs of water service, although it needs high costs for the precise control of consumption and more often, it fails to fulfil the *"polluter pays"* principle. Based on the water quantity actually given, it would represent a method of pricing more transparent and efficient (Tsur et al. 2003), in contrast to the indirect tariffs, unable to empower end users and encourage them to the rational use of the resource (Johansson 2000). On a larger scale, however, can arise different opinions about the increase in the cost of the resource. This choice can in fact be easily interpreted as a principle of marketing not acceptable ethically (Solanes and Gonzalez-Villareal 1999), while the use of incentives is reported as absolutely not encouraging of good behaviours (Rogers et al. 2002; Rogers et al. 1998; De Carli et al. 2007; Dinar and Mody 2004).

On the other hand, as demonstrated in previous studies about this theme (Dono et al. 2006; Giannoccaro et al. 2007; Bartolini et al. 2007), the effects both of a different tariff level and of the increase of irrigation water cost on farmers' choices, lead to a substantial reduction of consortium water consumption, at the expense of withdrawals from wells and private water sources, as well as the need for management and/or productive changes, ranging from reducing the irrigated area, to crop diversification towards less water-demanding crops, to an increase in the efficiency of distribution and a different method of water application, which can finally result in a significant decrease in farm income.

Irrigation water value is, moreover, connected and associated with that of agricultural production to which it contributes to, and consequently a higher cost is inevitably reflected on water use efficiency, meant as productivity (Kassam and Smith 2001; Molden 1997), both in physical terms (units of product per unit of water consumed) and economic terms (value of the product per unit of water consumed) (Seckler et al. 1998). An increase of this index, therefore, can contribute to represent the best way to achieve an efficient use of water, also through new management and/or productive strategies.

Aims and Analysis Methodology

The purpose of the study has been the identification and implementation of a mathematical programming model, in order to get to an economic evaluation of water resource according to the features of the district analysed, a Lombardy rice-cultivate district of ancient irrigation characterised by peculiar uses of the resource itself and particularly suited for this analysis.

The collection of information started from the development of specific questionnaires submitted to the farms identified and sampled within the area of interest.

The selection of the farms operating in the area of San Giorgio di Lomellina (PV), in Northern Italy (Fig. 2 and Fig. 3), began from their extraction from the regional database SIARL (*Sistema Informativo Agricolo della Regione Lombardia*), their subsequent classification on the basis of the

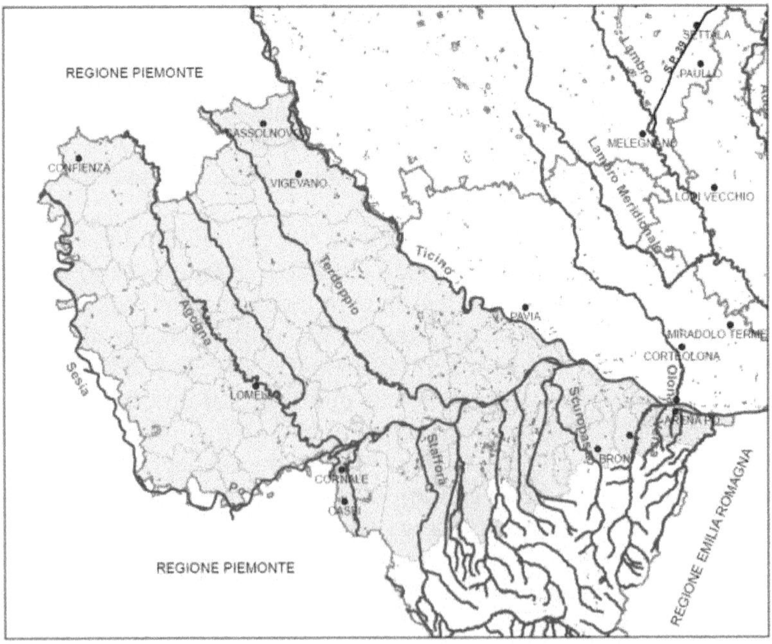

Figure 2. Location of the supplier consortium (www.territorio.regione.lombardia.it/shared/ccurl/97/362/allegato2,0.pdf).

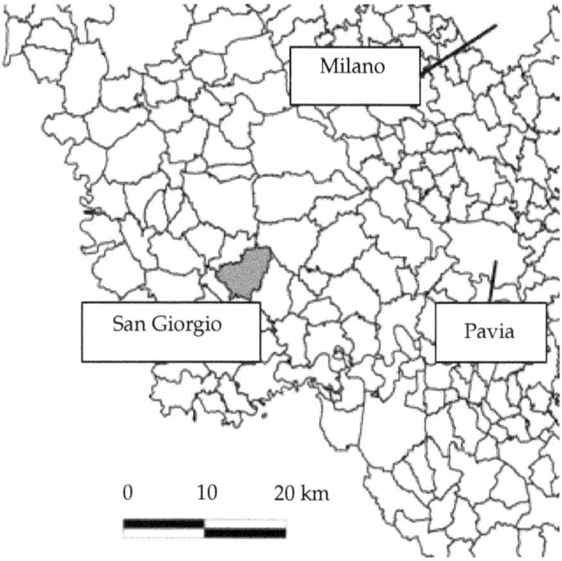

Figure 3. Location of the case study area.

Utilized Agricultural Area (from now on UAA) dedicated to rice only in the municipality (Table 1) and sampling inside of each size class. For each farm thus identified, questionnaires were filled through direct survey of farmers, for a total of 19 surveys carried out and a total rice-cultivated area of 808.5 hectares.

The compilation of questionnaires has allowed to obtain information about crop year 2010–2011. In particular, it has been designed to provide a general overview on matters related to:

- Origin and sources of water (from consortia or private sources, from rivers, wells and leakages);
- Methodology for the supply of the resource (continuous or not), with any specification of shift and time;

Table 1. Classification of sampling farms by size.

UAA of rice class	N. of farms	N. direct surveys
<1 he	8	0
1–9.99 he	17	3
10–29.99 he	15	6
>30 he	16	10
Total	*56*	*19*

- Type of tariff and fee;
- Cultivated and irrigated area, both total and by each type of water management (traditional flooding with sowing seeds both in water and soil, periodic irrigations), specifying crop yields;
- Calendar of irrigation and volumes of water provided for each type of crop;
- Execution modalities of each irrigation intervention, taking into account the method of irrigation, the use of technical means and labour during the interventions, their duration, any chain of working machines used and the work force.

The study area: general characteristics and framework

The sampled farms are mainly family-run, with minimal or occasional use in external labour; their total area is not a single body, but it consists of many fields, separated and not contiguous.

Overall, agricultural area is mainly rice-planted, with a marginal portion dedicated to arable crops (e.g., corn, soybean) and poplar. The plots are irrigated with water derived from Cavour Canal, Arbogna River and leakages, even though there are supplies from private sources. Distribution of water is mostly continuous and, for a lesser part, it refers to pre-established rotating shifts, that vary from 8 to 15 days and make water available for a time based on irrigated area. In any case it is flowing irrigation, often followed by flooding and carried out using retrofitted pumps on tractor.

In the case study area the agricultural activity is carried out in different water management and agronomic strategies, to which correspond the same product but different cultivation typology, as shown in Table 3.

Table 2. Overview on the study area.

District	**San Giorgio di Lomellina**
N. farms surveyed	19
Total irrigated area (he)	808.5
Average irrigated area (he)	42.55
Supplier	Associazione Irrigazione Est Sesia
Origin of water	consortia (river or leakages)
Supply methodology	Continuous, discontinuous
Fee Type	Monomial (€/ha)
Main crops	Rice, corn, poplar
Irrigation methodology	Flooding (continuous and intermittent)

Table 3. Different managerial typologies on rice-fields: general characteristics.

<table>
<tr><th>Crop type code</th><th>Water dispensation</th><th colspan="2">Water management</th><th>Agronomic management</th><th>Farms (n.)</th><th>% UAA</th></tr>
<tr><td>CFW</td><td rowspan="2">Continuous</td><td colspan="2" rowspan="2">Continuous flooding. Water flows continuously for the whole duration of the crop cycle.
Submersions are interrupted by 3 or 4 dries in correspondence of certain phases of the cycle or treatments with herbicides or fertilizers.</td><td>Water-seeding after the submersion of the field</td><td>10</td><td>50.63</td></tr>
<tr><td>CFS</td><td>Soil-seeding; the ground remains dry until the rice has reached the stage of 4th–5th leaf, then it is restored the normal regime of submersion</td><td>5</td><td>8.95</td></tr>
<tr><td>SCFW</td><td rowspan="4">Rotating shifts</td><td rowspan="4">Intermittent flooding. Water is available continuously only during predefined shifts.</td><td rowspan="2">Continuous flooding during the shift</td><td>Water-seeding after the submersion of the field</td><td>1</td><td>8.02</td></tr>
<tr><td>SCFS</td><td>Soil-seeding before the first irrigation</td><td>2</td><td>10.02</td></tr>
<tr><td>SIW</td><td rowspan="2">Flowing irrigation, trying to maintain water on the ground until the next shift</td><td>Water-seeding after the first irrigation</td><td>1</td><td>0.47</td></tr>
<tr><td>SIS</td><td>Soil-seeding before the first irrigation</td><td>6</td><td>21.91</td></tr>
</table>

Almost 60% of the rice-growing area is conducted with the traditional method (continuous flooding), with both water- (CFW, around 52%) and soil-seeding (CFS, almost 9%).

In some farms, irrigation on rotating shifts binds them to irrigate periodically, according to water availability during shift and time. In these cases, the farmers neglects a shift only in case of rain and if the general trend of the season allows it. In general, irrigation here is carried out mainly for flowing, but it differs from the usual interventions on other arable crops (e.g., corn). Two farms show irrigation with a greater similarity with the conventional technique, with more detailed description of irrigation period and duration, both water- and soil-seeding (SCFW and SCFS), while in other farms periodic irrigations are observed, with which however, farmers try to keep the water on the field until the next shift, assimilating the operation, as far as possible, to a sort of submersion (intermittent flooding) (SIS and SIW).

According to what has been declared, crop production doesn't seem to be linked to water and agronomic management, since there are no substantial differences among yields (Table 4).

With regard to water consumption, from the estimation of the quantity of seasonal irrigation water for each type of crop, it is confirmed that the traditional method requires more water; differentiation of sowing techniques shows a lower overall water consumption in the case of soil-seeding (Table 5). At the same delivery typology, the determining factor increasing water consumption is the resource management typology during the growing season.

If observed together with production data, these values demonstrate that water quantity doesn't affect the amount of rice produced, but it's possible to achieve the same result preferring water saving techniques.

Table 4. Average yield for each crop type.

Crop type	**Average yield (t/ha)**
CFW	6.47 ± 0.31
CFS	6.33 ± 0.48
SCFW	6.70 ± 0.00°
SCFS	6.60 ± 0.00
SIW	6.50 ± 0.00°
SIS	6.79 ± 0.27
Average rice-paddy	*6.54 ± 0.17*

°only one data available

Table 5. Seasonal water consumption for each crop type.

Crop type	Water consumption (m^3/he)
CFW	37,183
CFS	29,900
SCFW	22,219°
SCFS	11,677
SIW	6,048°
SIS	16,949

°only one data source

Modelling for Irrigation Water Management

The issue of irrigation water management requires appropriate tools capable of providing adequate supports in the decisional processes.

Through models of mathematical programming, this important aspect is taken into account, providing information not directly observable and allowing simulations of different scenarios in agricultural policies, resource management or changes and developments in market. These approaches can indeed guide the decision maker to identify the most suitable interventions to achieve economic and environmental targets of water policies.

In the economic analysis of irrigation water, several studies based on formalization and implementation of econometric and programming models have been carried out, both at a farm and a local level. Considering the studies conducted and the regulatory about water resources, it emerges that analysis tools at a regional level better answer to the requirements of WFD, which states that catchment area is the unit for the analysis and the integrated management of water resources.

On several occasions, the econometric approach, which needs less informative inputs, has allowed the estimation of function of operating costs of water distribution in irrigation districts and consortia (Dono 2003; Giraldo 2010; Dono and Giraldo 2010; Dono et al. 2011), but often economic analysis of irrigated agriculture focuses on the evaluation of the effects caused by alternative conditions, internal and external to the system. The methodology based on linear programming models (mono-objective, multicriteria, stochastic discrete) is widely used for this purpose as well as for responding to the needs and suggestions introduced by the WFD; also, it allows resolving optimization problems. Each simulation generates a new solution showing the effects of the changes themselves on crops, technological choices, use of productive inputs, and economic performances of farms (Dono 2003; Giraldo 2010; Dono and Giraldo 2010; Dono et al. 2011; Giannoccaro et al. 2008; Dono et al. 2008; Bazzani et al. 2005; Bazzani and

Zucaro 2008; Bazzani and Scardigno 2008). These models, which require the collection and processing of a large amount of information on the economic and productive farm system, allow to understand the features of an agricultural system by identifying relationships between the use of inputs and productivity, but the results of simulations are strongly affected by the constraints imposed on the model.

More recently, in the same context the Positive Mathematical Programming (PMP) (Howitt 1995; Paris and Howitt 1998) is spreading. This approach requires a limited amount of data used to perfectly calibrate the model for the reference period, according to three phases: specification of a linear programming model that uses all the information available, reconstruction of a total variable cost (Arfini and Paris 1995), formulation of a non-linear programming model to be used to perform simulations. The application of this methodology to the analysis concerning water resources is currently underdeveloped. In particular, it recalls the work of Blanco et al. (2004) in which is considered the impact of pricing policies on two irrigation districts in Spain by specifying a cost function for each of them, and what has been developed by Cortignani and Severini (2008a,b, 2009) with reference to territorial analysis, also following the introduction of tariffs differentiated according to the season.

The potential of this type of model lies in the fact that they can be used to face issues related to the variation in the cost of the water and its availability, but they do not allow to take into consideration new and different production activities compared to the reference situation, thereby limiting the possibility of analyzing future scenarios.

The Implemented Model

General structure and model input

Starting from data and information collected by direct surveys in selected farms, and using results of *ad hoc* experimentations conducted in an experimental farm, a non-linear programming descriptive model has been developed, allowing both to summarize observed data and making them easier to understand, and solve optimization problems through the use of efficient methods which identify optimal solution for a subsequently defined objective-function that depends on variables identified from time to time.

Model output

The model is solved through software GAMS (General Algebric Modeling System; Brooke et al. 1988). The software has facilitated writing simply in programming language, the model elaborated and its elements (equations, variables, parameters, objective functions). Once set, it then has solved rapidly the model and the equations stated.

Nevertheless, the performances of the software depend, essentially, on the programming instructions set through the formalization of the model; they respond, therefore, to the purpose for which the model itself has been elaborated and to the type of data to be obtained.

Scenario analysis finally carried out has returned data and information that cannot be considered as absolute and certain values, but have to be interpreted according to the conditions of the wider context. The software becomes, then, only a means to process data and simulate alternative situations; it is in fact possible that some results it returned are not feasible from a practical or economic (or any other) point of view.

For our purposes, non-linear programming modelling has given quite satisfactory results; however, it is possible to set other resolution options, that means solving the model by applying different algorithms which could generate other results or improve the overall performances.

The model has allowed the generation and the display of a large amount of data output, returning information about structural features of farms, their productive inputs (land, water, labour, technical means and machinery) and productive and economic performances (revenues, supply and water management costs, gross margins, water cost, water productivity) of each of them and for every type of culture, allowing comparison between farms and cultural types, homogeneous or not.

Distributed water volumes

Starting from the calendar of irrigation described by each farm, seasonal water distribution has been estimated, in relation to the crop water requirement calculated by the consortium supplier (*i*) and to the duration of watering season (from April to August); for flooding irrigation the flow rate is multiplied by the duration of the various submersions, while in the presence of periodic irrigations it is instead considered continuous for the only time of the shift stated in the interview.

Total costs related to irrigation

Total costs (CT, in €) related to irrigation practice are divided into two components (Castellani et al. 2008): cost of water supply (C_f) and cost of water management (C_g).

$$CT = C_f + C_g = C_f + C_{mr} + C_{cL} + M + Q_d + S_v + C_u$$

Where C_{mr} maintenance and repair costs (for machinery), C_{CL} consumables costs, M labor cost, Q_d cost of deterioration of machinery, S_v various expenses (related to buildings for shelter and surveillance of machinery, taxes, insurance), C_U further costs (interests on working capital, use cost of the buildings for shelter of machinery used in water management, other taxes), set equal to 5% of total costs.

Water cost

In order to evaluate economic performances and features of farms, the price (PU) of irrigation water (€/m³) during data analysis has been quantified, as the ratio between costs—intended both as supply costs (C_f) and total costs (CT)—and water available to farm:

$$\text{PU}\ (\text{€/m}^3) = \frac{\text{costs (€)}}{\text{irrigation water}\ (\text{m}^3)}$$

Total revenues

Total farm revenues (RT, in €) include revenues from the sale of paddy-rice at market price (V) and CAP subsidies (C):

$$RT - V + C$$

The former derives from the attribution of a medium market price for crop year 2010/2011, while for CAP subsidies is set a fixed contribution per hectare (r), with allowance of 8% for the portion in excess of € 5,000:

$$V = p * yield$$

$$If\ C > 5000\ \text{€}\ then\ C_n = C - (0.08 * (C - 5000))$$

Farm gross margins

Gross margins (ma, in €) represent the difference between total revenues and total costs related to the irrigation practice:

$$ma = RT - CT$$

Water productivity

In the final analysis, an estimation of Crop Water Productivity is reported. It is defined as the ratio between total yield of each crop (in tons) and its water consumption (m³) during season due to evapotranspiration (Kassam and Smith 2001; Molden 1997); for the case study, the yield is compared to the total amount of water used during watering season not considering line losses, namely the amount potentially distributed in a year according to the available resource. This kind of productivity can be named Irrigation Water Productivity (IWP):

$$IWP\ (g/kg) = \frac{total\ yield\ (t)}{distributed\ water\ (m^3)}\ x\ 1000$$

Additionally, the Economic crop Water Productivity (EWP) is also considered, based on market value of the crop itself (Igbadun et al. 2006; Palanisami and Suresh Kumar 2006; Teixeira et al. 2008; Vazifedoust et al. 2008):

$$EWP\ (€/kg)\ =\ \frac{crop\ economic\ value\ (€)}{distributed\ water\ (m^3)}$$

Scenario analysis, variables and objective functions

In order to evaluate the effects of new managerial and/or productive strategies on cultivated surfaces (possible reduction of the irrigated surface, crop diversification, increase of the distribution efficiency and different method of water provision), decisional variable is represented by the farm rice-growing area ($xcrop_{f,c}$), subject to irrigation according to the different ways of water supply and management.

The first scenario (scenario #1) optimises surfaces starting from initial condition in the district, as observed in the surveys. Alternative situation (scenario #2) provides the evaluation of economic and productive performances as a result of the application of a volumetric fee, to replace the current one, *ceteris paribus*.

Referred to both scenarios, economic and productive features resulting from the adoption of optimum allocation of rice-cultivated areas suggested (scenarios #3 and #4) have been recalculated.

The decision variable chosen is included in the defined objective function (Z), maximising farm gross margins (ma), intended as the difference between revenues and costs and subjected to farm-level and consortium-level constraints, related to land, water and water crop requirements estimated for the area.

Each equation in the model refers to each farm (*f* index) and different water management strategies (*c* index) identifying different crop types (see Table 2). In particular, the presence of a *k-1*–degree equation related to water managing costs, gives the entire model the characteristic of non-linearity, justifying the use of the specific algorithm.

In particular, Z takes the following form:

$$Z = \Sigma_{f,c}($$

1. $r * xcrop_{f,c} - 0.08 * (r * xcrop_{f,c} - 5000)$
2. $+(p * y_{f,c} * xcrop_{f,c})$

$-100/95*$

3. $((xcrop_{f,c} * (w + wc * i * 3.6 * dur_{f,c}))$
4. $+(m_f * ((n * xcrop_{f,c} * hhat_{f,c}) ** (k-1)) * xcrop_{f,c} * hhat_{f,c})$
5. $+\left(xcrop_{f,c} * (pwr_f / 69) * (hhat_{f,c} * (9.4 * fp + 0.04 * op) + 7.65 * int_{f,c})\right)$
6. $+(2 * l * xcrop_{f,c} * inttot_f)$
7. $+\left(Vo_f * ((1-td)/n)\right)$
8. $+(o * Vo_f))$

The first two rows represent farm revenues from

1. CAP subsidies for the crop, where *r* average premium of 850 €/ha of rice;
2. sale of paddy, where *p* selling price (€/ton)
   $y_{f,c}$ crop yield (tons/he);

The following refer to different elements of cost related to irrigation practice:

3. water supply cost (€ per yr), where
   w current fee adopted (€/he)
   wc volumetric tariff fee (€/m^3);
4. cost for maintenance and repair of technical means used for irrigation (€ per yr) (Lazzari and Mazzetto 2005)
   $m_f = FR * Vo(f) / D_f^k$;

where FR repair and maintenance factor (%),
V_o value of a new machine (€),
n economic life of the machine (yrs),
D_f physical life of the machine (hrs),
k (–) exponent coefficient for repair and maintenance,
$hhat_{f,c}$ working capacity of the pump used for irrigation (hrs/he);

5. costs for consumables (fuel and oil) (€ per yr) (Lazzari and Mazzetto 2005), where
   pwr_f power of the tractor machine used for irrigation (kW),
   $hhat_{f,c}$ working capacity of the pump used for irrigation (hrs/he),
   fp fuel price,

op oil price,
$int_{f,c}$ number of irrigation interventions during season for each crop type;

6. labour cost (€ per yr), where
   l hourly labour cost,
   $inttot_f$ total number of irrigation interventions during season;
7. share of deterioration of the machinery used for irrigation (€ per yr) (Lazzari and Mazzetto 2005), where
   V_o value of a new machine (€),
   Td depreciation rate,
   n economic life of the machine (yrs);
8. other expenses (€ per yr) (Lazzari and Mazzetto 2005):
   V_o value of a new machine (€),
   o coefficient of various expenses.

The first simulation (scenario #2) has required a preliminary step to determine the volumetric fee level (in €/m^3) to be then applied to the objective function above.

This value was defined starting from a new additional objective function that includes the same parameters explained above and intends to find the limit of the fee for which there is a change of water management. For this reason, the margins calculated on the basis of known values of areas have been equalled to those in which they appear as variable; in addition to the variable *xcrop (f,c)* has also been introduced the variable *tv*, replacing the previous parameter *wc*.

$$\begin{aligned}
&\sum_{f,c}\left(-xcrop_{f,c} * (w + tv * i * 3.6 * dur_{f,c})\right)\\
&\quad = \sum_{f,c}(r * xcrop_{f,c} - 0.08 * (r * xcrop_{f,c} - 5000) + (p * y_{f,c} * xcrop_{f,c})\\
&\quad - 100/95\\
&\quad * \left((o * Vo_f) + (Vo_f * ((1 - td)/n)) + (2 * l * xcrop_{f,c} * inttot_f\right)\\
&\quad + \left(xcrop_{f,c} * (pwr_f/69) * (hhat_{f,c} * (9.4 * fp + 0.04 * op) + 7.65 * int_{f,c})\right)\\
&\quad + \left(m_f * \left((n * xcrop_{f,c} * hhat_{f,c}) ** (k-1)\right) * xcrop_{f,c} * hhat_{f,c})\right) - ((r * a_{f,c}\\
&\quad - 0.08 * (r * a_{f,c} - 5000) + (p * y_{f,c} * a_{f,c}) - 100/95 * ((a_{f,c} * (w - wc * i\\
&\quad * 3.6 * dur_{f,c})) + (o * Vo_f) + (Vo_f * ((1 - td)/n)) + (2 * l * a_{f,c}\\
&\quad * inttot_f) + (a_{f,c} * (pwr_f/69) * (hhat_{f,c} * (9.4 * fp + 0.04 * op) + 7.65\\
&\quad * int_{f,c})) + (m_f * ((n * a_{f,c} * hhat_{f,c}) ** (k - 1)) * a_{f,c} * hhat_{f,c}))))))
\end{aligned}$$

Constraints

- *Land constraints*

The first constraint ensures that no more land than the total available in each farm (*land*$_f$, in he) is cultivated. It is expressed by the formula

$$\sum_c xcrop_{f,c} \leq land_f$$

Other two constraints ensure that cultivated areas ($a_{f,c}$, in he), still maintain the same water dispensation, whether continuous (first row) or not (second row):

$$xcrop_{f,CFW} + xcrop_{f,CFS} \leq a_{f,CFW} + a_{f,CFS}$$

$$xcrop_{f,SCFW} + xcrop_{f,SCFS} + xcrop_{f,SIW} + xcrop_{f,SIS} \leq a_{f,SCFW} + a_{f,SCFS} + a_{f,SIW} + a_{f,SIS}$$

With regard to the scenario analysis, for the secondary objective function, new and more suitable constraints have been defined, where is shown the equality between known and unknown areas:

$$\sum_c xcrop_{f,c} \leq land_f$$

$$xcrop_{f,CFW} = a_{f,CFW}$$

$$xcrop_{f,CFS} = a_{f,CFS}$$

$$xcrop_{f,SCFW} = a_{f,SCFW}$$

$$xcrop_{f,SCFS} = a_{f,SCFS}$$

$$xcrop_{f,SIW} = a_{f,SIW}$$

$$xcrop_{f,SIS} = a_{f,SIS}$$

- *Water constraints*

Water balance ensures that water flow resulting from the model is not higher than the one currently provided by the consortium:

$$\sum_c i * xcrop_{f,c} \leq \sum_c i * a_{f,c}$$

Results and Comments

Cultivated areas optimisation

In current conditions, according to direct surveys, optimal areas allocation able to maximise overall margin, is mostly far from what is really applied.

The margin maximization prefers water saving techniques (Table 6), not to bring down water supply cost (also because in this case there is a fixed fee per hectare and the total area remains the same), but rather to reduce water management costs. In particular, model elaborations tend to encourage soil-seeding over water-seeding, even though for some farms, allocation of areas would not change. Also for periodic irrigation, margin maximization leads to a change in irrigation technique, with an increase in areas with intermittent as compared to continuous flooding; exceptions are some farms not using technical means for irrigation and where continuous flooding may instead be practiced.

The secondary objective function results in a volumetric fee of 0.12 €/m^3, which is the average of the cost of the resource of each individual farms weighted on their own water consumption. This tariff level, once matched with the cancellation of the current one, represents again the input parameter *wc* to proceed with the optimisation according to the main objective function. It leads to the identification of cost levels that favour different irrigation techniques; the imposed value determines the cancellation of gross margin for farms with continuous supply, suggesting that the total price to pay would be too high to front the high water consumption required. In fact, maximum price levels these farms can go through, are in the range between 0.107 €/m^3 and 0.117 €/m^3.

The surfaces optimising the overall margin therefore make reference only to those with discontinuous supply, whose tariff level threshold is much higher, more than 3.125 €/m^3.

For about a third of farms, it can be worthwhile to adopt dry or semi-dry cultivation; this result is also confirmed by previous surveys carried out in the same area and referred to as soil-seeding (which includes also what is then submerged): during the season 2004–2005, 5.4% of the denounced

Table 6. Surfaces allocation, in the initial condition and after optimizations.

Crop type	% UAA		
	Initial condition	Scenario #1	Scenario #2
CFW	50	42	-
CFS	9	17	-
SCFW	8	10	-
SCFS	10	-	-
SIW	1	-	100
SIS	22	31	-

rice-growing areas in S. Giorgio di Lomellina were soil-seeded, while from 2008 this value has risen to almost 30% (with peak values of 37%).[1]

Costs analysis

Since current fee type is a fixed value per hectare, water supply cost at farm-level is directly proportional to irrigated areas and independent from the amount of water actually available, contrary to what happens with a volumetric fee.

In continuous flooding, water supply costs are slightly higher for soil-seeding, and higher for a unit size than other water management modalities. Paradoxically adopting productive and managerial water saving techniques seems to determine for farms (and the district) an increase in supply costs, which in itself could mean, in the future, a change in farming systems. Actually, the adoption of periodic irrigation is independent of the farmers' will; moreover, these considerations derive from the type of fee set by the consortium and a change in this direction would also modify these aspects.

The application of a volumetric fee doesn't modify the related water supply cost, both before and after the optimal allocation of surfaces (Table 7) and in both situations it is higher than the cost resulting from a fee per hectare.

Total costs related to irrigation (Table 8) differ from each other according to operative procedures adopted by each farm during irrigation practice. Differences observed, also for each crop typology, are due to the fact that for farms not using technical means, the costs don't cover expenses related to tractors and operating machinery.

In these cases the allocation of the expenses is achieved with the remaining costs. The higher outputs are due to increased labour for periodic

Table 7. Irrigation water supply cost (€/m^3), for each crop type in different cases analyzed.

Crop type	Scenario 1	Scenario 2	Scenario 3	Scenario 4
CFW	0.007	0.12	0.007	-
CFS	0.008	0.12	0.008	-
SCFW	0.013	0.12	0.013	-
SCFS	0.025	0.12	-	-
SIW	0.230	0.12	-	0.12
SIS	0.073	0.12	0.038	-

[1] These results derive from a study carried out at the experimental farm of Centro Ricerche sul Riso (*Rice Research Center*) in Castello d'Agogna (PV), which pertains to Ente Nazionale Risi (www.enterisi.it).

irrigations and this component prevails on supply cost (Fig. 4; 57% of total expenditures compared to 38%). Where technical means are used (Fig. 5), supply cost increases to 43% of total expenditures related to irrigation and labour costs decrease proportionally.

Table 8. Total costs (€/ha) related to irrigation in different scenarios.

Crop type	Scenario 1	Scenario 2	Scenario 3	Scenario 4
CFW	983.17	430.34	774.46	-
CFS	708.96	704.55	891.95	-
SCFW	398.62	120.00	120.00	-
SCFS	459.16	180.54	-	-
SIW	638.62	360.00	-	400.66
SIS	578.62	300.00	374.11	-

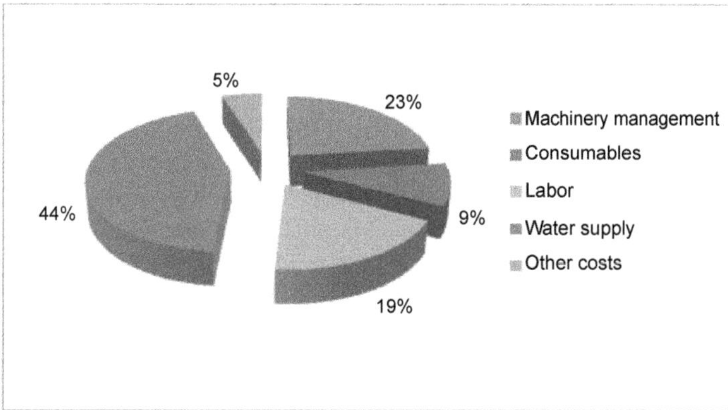

Figure 4. Allocation (%) of costs in farms not using technical means.

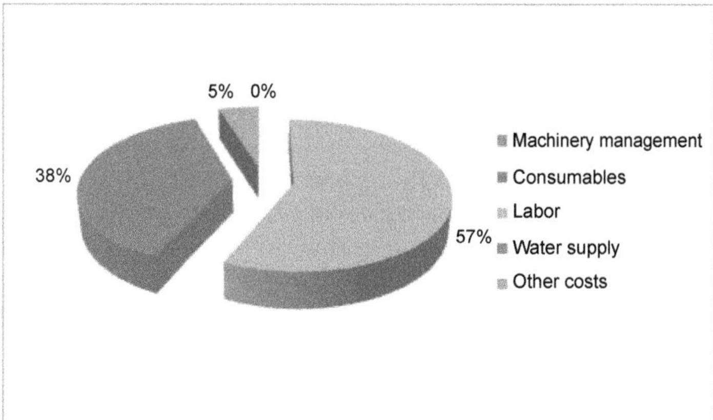

Figure 5. Allocation (%) of costs in farms using technical means.

Considering irrigation water costs compared to each item of cost, a significant increase in the price of the resource (€/m³) is observed, with a general higher amount for periodic irrigation (Table 9).

Table 9. Irrigation water cost (€/m³) deriving from total costs, for each crop type in different cases analyzed.

Crop type	Scenario 1	Scenario 2	Scenario 3	Scenario 4
CFW	0.027	0.019	0.021	-
CFS	0.020	0.012	0.026	-
SCFW	0.009	0.003	0.003	-
SCFS	0.013	0.005	-	-
SIW	0.264	0.149	-	0.166
SIS	0.079	0.041	0.026	-

Revenues analysis and water productivity

Since homogenous yields in the district are observed, repercussions on total revenues (€/ha) in the case study area (Table 10) are due to the amount of cultivated area. Differences among values are in fact observed only in the case of adopting optimal allocation of areas, which determines a slightly different yield and thus different costs and revenues (and consequently, margins).

Table 10. Revenues (€/he) for each crop type.

Crop type	Scenario 1	Scenario 2	Scenario 3	Scenario 4
CFW	3004.57	3004.57	2957.89	-
CFS	2917.30	2917.30	2541.10	-
SCFW	3055.48	3055.48	2266.81	-
SCFS	3026.09	3026.09	-	-
SIW	3095.43	3095.43	-	-
SIS	3108.57	3108.57	2819.00	2301.90

Crop water productivity (WP) depends essentially on seasonal water amount and thus on water management of the rice-paddy. WP, in terms of unit of product, is in fact proportional to the water volume used for irrigation, with a reasonable significance ($r = 0.74$) (Fig. 6).

Moreover (Table 11), being independent from the resource price, it doesn't show differences among scenarios using different types of fee

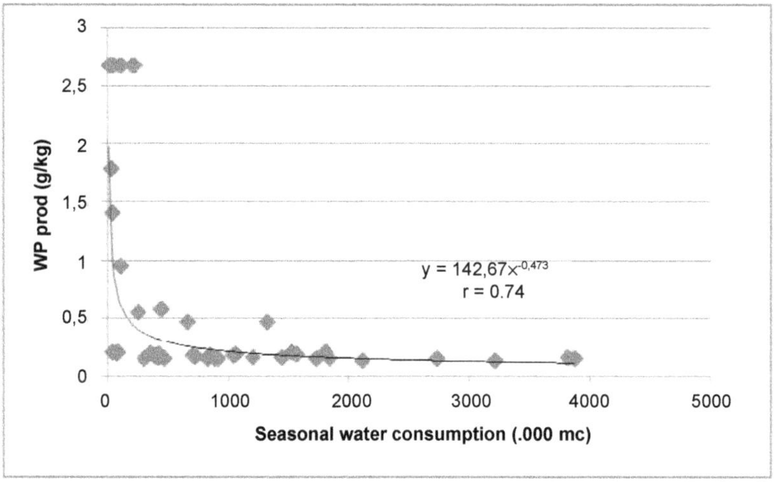

Figure 6. Relation between seasonal water consumption (m^3) and WP (g/kg).

Table 11. Water Productivity (g/kg) for each crop type in different scenarios.

Crop type	Scenario 1	Scenario 2	Scenario 3	Scenario 4
CFW	0.17	0.17	0.17	-
CFS	0.18	0.18	0.51	-
SCFW	0.15	0.15	0.15	-
SCFS	0.19	0.19	-	-
SIW	2.69	2.69	-	2.69
SIS	0.91	0.91	0.51	-

(scenario #1 vs. #2); slight deviations are instead described when it is calculated on areas optimising margin (scenarios #3 and #4), since their changes cause a variation both in yields and water consumptions. Different land managing leads to an increase of the WP value: in every case it is evident that higher values correspond to periodic irrigation.

In the same way, the economic water productivity also (€/m^3) depends on water amount during irrigation season. Even if differently expressed, it shows the same trend as the previous one, with which it is fully correlated (r = 1, Fig. 7).

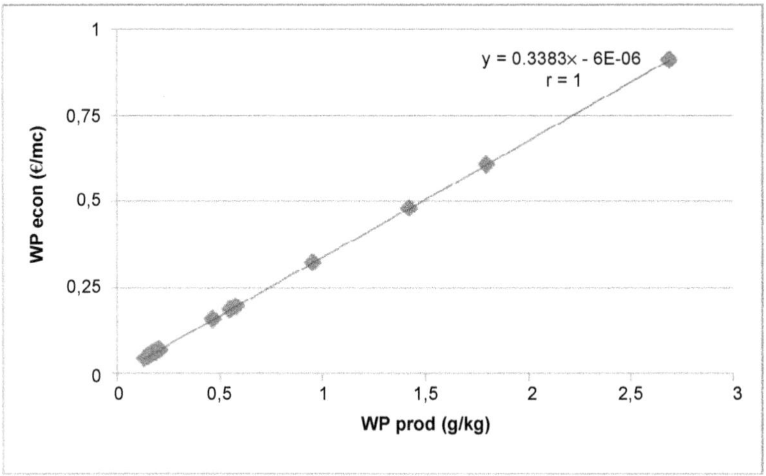

Figure 7. Relation between different WPs.

Conclusions and Operational Guidelines

In rice-paddy fields, the adoption of non-traditional techniques, which is becoming more and more common, allows the achievement of positive targets both in terms of water saving and economic conditions for farmers, without substantially modifying their overall margins. The effectiveness of the choices oriented in this way, is pointed out by a better water use efficiency, that could also be expressed as water productivity, corresponding to water saving techniques.

From an economic point of view, the adoption of these methods even though they don't affect significantly margins of farms, in particular for dry cultivation, could necessitate an increase in workforce and higher manpower costs. The increase in water supply cost could also determine a better allocation of the resource, but it is, therefore, not entirely bound by farmers' decisions and its change could impact on ecological and environmental aspects not immediately identifiable.

Alternative methods of water management in paddy fields, which are favoured both in simulations and real agricultural practices, act in this way. In fact, the adoption of semi-traditional irrigation leads to a lower water demand at the beginning of the season that increases gradually and shows a peak when it has to satisfy the water requirements of other crops, with the advantage, however, of ensuring operatively, the provision of the resource in exceptional cases (e.g., breakdowns of channels).

The ability of the consortia suppliers to reduce the amount of water to farms, or increase its cost (consequently causing a decrease in demand), besides the change of their managerial aspects and their farming systems, could also lead, in medium and long-term period, to a less efficient allocation of the resource. In fact, decreased water availability may affect the hydrological cycle on a local scale, interfering and changing the water returns to farms, surface water bodies and groundwater. Changing irrigation methods may in fact result in a delay in the loading of the water table, as well as a lowering in the water table itself due, in particular, to dry cultivation.

Similarly, a higher technical and infrastructural efficiency able to reduce losses occurring during distribution, can act in the same direction, with implications that affect recharge and supply of water sources and eliminating the potential benefits of reallocation. However, in many cases a large part of the water flow available to farms comes from internal recirculation and this quantity may be a contrasting element to the reduction of the water demand.

In conclusion, these considerations about environmental aspects must be properly weighed in order to make a complete economic evaluation of water resource. In particular, they can play a role as a starting point to the quantification of the *full cost* of the resource itself.

References

Arfini, F. and Paris, Q. 1995. A positive mathematical programming model for regional analysis of agricultural policies. EAAE Proc., 40th Seminar of the European Association of Agricultural Economists, Ancona, Italy, pp. 17–35.

Bartolini, F., Bazzani, G.M., Gallerani, V., Raggi, M. and Viaggi, D. 2007. The impact of water and agricultural policy scenarios on irrigated farming systems in Italy: An analysis based on farm multi-attribute linear programming models. Agricultural Systems 93: 90–114.

Bazzani, G.M. and Scardigno, A. 2008. Un modello di simulazione territoriale per l'analisi economica dell'uso dell'acqua e della riforma della PAC: una proposta metodologica e prime applicazioni all'agricoltura pugliese. Paper presentato al XLV convegno di studi SIDEA, Portici, Italy.

Bazzani, G.M. and Zucaro, R. 2008. Scarsità idrica e Direttiva Acque. Politiche e metodologie di analisi: un caso di studio in Italia centrale. pp. 67–83. *In*: Casini, L., Gallerani, V. and Viaggi, D. (eds.). Acqua, agricoltura e ambiente nei nuovi scenari di politica comunitaria Franco Angeli Editori, Milano, Italy.

Bazzani, G.M., Di Pasquale, S., Gallerani, V., Morganti, S., Raggi, M. and Viaggi, D. 2005. The sustainability of irrigated agricultural systems under the Water Framework Directive: first results. Environ. Modeling & Software 20: 165–175.

Blanco, M., Iglesias, E. and Sumpsi, J.M. 2004. Environmental and socioeconomic effect of water pricing policies: key issues in the implementation of the water framework directive. 13th Annual EAERE Conference, Budapest, Hungary.

Brooke, A., Kendrick, D. and Meeraus, A. 1988. GAMS: A User's Guide. The Scientific Press, Redwood City, California, USA.

Cadario, D. and Bischetti, G.B. 2006. Caratteri e funzioni del reticolo idrografico della pianura lombarda. pp. 58–61. *In*: Valutazione Ambientale. ISSN 1826-2201. – 10, luglio 2006.

Castellani, L., Mancuso, T. and Massaglia, S. 2008. Aspetti normativi ed economici in materia di gestione e tutela delle risorse idriche ad uso irriguo in Piemonte. pp. 31–48. *In*: Casini,

L., Gallerani, V. and Viaggi, D. (eds.). Acqua, agricoltura e ambiente nei nuovi scenari di politica comunitaria, Franco Angeli Editori, Milano, Italy.

Cortignani, R. and Severini, S. 2008a. Introducing deficit irrigation crop techniques derived by crop growth into a positive mathematical programming model. Comunicazione, XII Congress EAAE, Ghent, Belgium, 26–29 August 2008.

Cortignani, R. and Severini, S. 2008b. Direttiva quadro n°60/2000 ed aumento del costo dell'acqua in agricoltura: una valutazione d'impatto mediante modelli di PMP. Paper presented at the XLV Convegno di Studio SIDEA, Portici, Italy.

Cortignani, R. and Severini, S. 2009. L'impatto delle riforme OCM zucchero ed ortofrutta sull'uso dell'irrigazione: un'analisi condotta mediante un modello di PMP senza prima fase. Comunicazione, XLVI Convegno di Studi SIDEA Piacenza, Italy, 16–19 settembre 2009.

De Carli, A., Masarutto, A. and Taccagni, V. 2007. La valutazione economica delle politiche idriche: dall'efficienza alla sostenibilità. http://users.unimi.it/ambiente/800/1024/materiali20007.htm.

Dinar, A. and Mody, J. 2004. Irrigation water management policies: allocation and pricing principle and implementation experiences. Natural Research Forum, Oxford, England, Blackwell Publishing Ltd. 28: 112–122.

Dono, G. 2003. Costi della distribuzione idrica per l'irrigazione nell'Italia meridionale e problemi della formazione dei prezzi dell'acqua per l'agricoltura. Rivista di Economia Agraria, n. 1.

Dono, G. and Severini, S. 2006. Il recupero del costo pieno nella direttiva quadro delle acque: problemi per l'agricoltura italiana. Agri Regioni Europa, Anno 2, n. 7. Dicembre 2006. www.agriregionieuropa.univpm.it.

Dono, G. and Giraldo, L. 2010. Un'analisi dei costi per la distribuzione dell'acqua in agricoltura. Rivista di Economia Agraria, vol. 64, n. 3–4.

Dono, G., Marongiu, S. and Severini, S. 2006. Gli effetti della Riforma della PAC sull'uso delle risorse idriche di falda e di superficie. Un modello di analisi territoriale. pp. 85–103. *In*: Casini, L., Gallerani, V. and Viaggi, D. (eds.). Acqua, agricoltura e ambiente nei nuovi scenari di politica comunitaria, Franco Angeli Editori, Milano, Italy.

Dono, G., Marongiu, S., Severini, S., Sistu, G. and Strazzera, E. 2008. Studio sulla gestione sostenibile delle risorse idriche: analisi dei modelli di consumo per usi irrigui e civili. ENEA editore. Collana Desertificazione—Progetto RIADE, 257 p.

Dono, G., Giraldo, L. and Severini, S. 2011. I costi della distribuzione irrigua: un tentativo di riconciliare i concetti di copertura dei costi. Economia e Diritto Agroalimentare, Italia, 16 Marzo 2011. www.furpress.net/index.php/edaarticle/view/9373.

European Parliament and the Council of EU. 2000. DIRECTIVE 2000/60/EC establishing a framework for community action in the field of water policy. OJEC L 327, 22/12/2000.

European Environmental Bureau (EEB). 2001. Proposals to strengthen the environmental action programme. http://www.eeb.org/publication/6th%20EAP%20final.pdf.

Fischler, G., Tubiello, F.N., van Velthuizen, H. and Wiberg, D.A. 2007. Climate change impacts on irrigation water requirements: effects of mitigation, 1990–2080. Technological Forecasting and Social Change 74: 1083–1107.

Gallerani, V. and Viaggi, D. 2006. Il valore dell'acqua per il territorio e l'ambiente rurale. Italian Journal of Agronomy 3: 569–576.

Giannoccaro, G., Prosperi, M. and Zanni, G. 2007. Analisi dell'efficienza tecnica ed ecologica dei metodi di tariffazione diretti ed indiretti dell'acqua irrigua. Paper presentato al XLIV convegno SIDEA, Taormina.

Giannoccaro, G., Prosperi, M. and Zanni, G. 2008. DEA application to evaluate the technical and ecological efficiency of water pricing policies. *In*: Bartova, L., M'Barek, R. and Ratinger, T. (eds.). Modelling Agricultural and Rural Development Policies. Proceedings of the 107th Seminar of the EAAE, Sevilla, Spain, 29 January–1 February, 2008. European Communities.

Giraldo, L. 2010. I costi operativi per la distribuzione dell'acqua a scopo irriguo. Un'analisi empirica con una forma funzionale flessibile. DEAR, Università della Tuscia.

Howitt, R.E. 1995. Positive mathematical programming. American Journal of Agricultural Economics 77: 329–342.
ICWE. 1992. The Dublin Statement and Report of the Conference. International Conference on Water and the Environment: Development Issues for the 21st century. 26–31 January. Dublin, Ireland.
Igbadun, H., Mahoo, H., Tarimo, A. and Salim, B. 2006. Performance of two temp.-based reference evapotranspiration models in the Mkoji subcatchment in Tanzania. Agric. Eng. Int.: the GIGRE J. Manuscript LW 05 008. Vol. VIII.
Iglesias, A., Estrela, T. and Gallart, F. 2005. Impacts on Hydric Resources. A Preliminary General Assessment of the Impacts in Spain Due to the Effects of Climate Change. Ministerio de Medio Ambiente, Spain.
IPPC. 2012. Renewable Energy Sources and Climate Change Mitigation—Special Report of the Intergovernmental Panel on Climate Change. Cambridge University Press, New York, USA.
Johansson, R.C. 2000. Pricing Irrigation Water. The World Bank, Washington, USA.
Kassam, A. and Smith, M. 2001. FAO methodologies in crop water use and crop water productivity. Expert meeting on water productivity, FAO, Rome 3–5 December 2001.
Lazzari, M. and Mazzetto, F. 2005. Prontuario di meccanica agraria e meccanizzazione. Reda Edizioni, Torino, Italy.
Molden, D. 1997. Accounting for water use and productivity. International Irrigation Management Institute—System Wide Initiative for Water Management (SWIM) paper 1.
OECD. 1999. The Price of Water Trends in OECD Countries. OECD, Paris, France.
Palanisami, K. and Suresh Kumar, D. 2006. Challenges in Impact Assessment of Watershed Development: Methodological Issues and Experiences. *In*: Palanisami, K. and Suresh Kumar, D. (eds.). New Associated Publishing Company Ltd., New Delhi, India.
Paris, Q. and Howitt, R.E. 1998. An Analysis of Ill-posed Production Problems Using Maximum Entropy. American Journal of Agricultural Economics 80(1): 124–138.
Rogers, P., Bathia, R. and Huber, A. 1998. Water as social and economic good: how to put the principle in practice. Tac background papers n. 2, Stockholm, Sweden, GWP-Swedish international Development Cooperation Agencie.
Rogers, P., de Silva, R. and Bathia, R. 2002. Water in an economic good: how to use prices to promote equity, efficiency and sustainability. Water policy 4: 1–17.
Seckler, D., Amarasinghe, U., Molden, D., de Silva, R. and Barker, R. 1998. World Water Demand and Supply, 1990 to 2025: Scenarios and Issues. International Water Management Institute Research Report 19, Sri Lanka.
Solanes, M. and Gonzalez-Villareal, F. 1999. The Dublin principles for water as reflected in a comparative assessment of institutional and legal arrangement for integrated water resource management. Tac background papers n. 3, Stockholm, Sweden, GWP-Swedish international Development Cooperation Agencie.
Tarimo, A.K.P.R., Mdoe, N.S. and Lutatina, J.M. 1998. Irrigation water prices for farmer-managed irrigation systems in Tanzania: a case study of lower Moshi irrigation scheme. Agricultural Water Management 38: 33–44.
Teixeira, A.H. de C., Bastiaanssen, W.G.M., Moura, M.S.B., Soares, J.M., Ahmad, M.D. and Bos, M.G. 2008. Energy and water balance measurements for water productivity analysis in irrigated mango trees, Northeast Brazil. Agricultural and Forest Meteorology 148: 1524–1537.
Tsur, Y., Roe, T., Doukkali, R. and Dinar, A. 2003. Pricing Irrigation Water: Principles and Cases from Developing Countries. Resources for the Future, Washington DC, USA.
United Nations Environment Program (UNEP). 1999. Global Environmental Outlook 2000. Earthscan Publications Ltd., London, England.
Vazifedoust, M., van Dam, J.C., Feddes, R.A. and Feizi, M. 2008. Increasing water productivity of irrigated crops under limited water supply at field scale. Agricultural Water Management 95: 89–102.
WATECO. 2003. Common Implementation Strategy for the Water Framework Directive (2000/60/CE)—Guidance Document n. 1, Bruxelles.

CHAPTER

6

Economic Analysis of Reducing Diffuse Nutrient Discharge into Water Bodies*

Philip Journeaux

Introduction

This chapter is based on two studies of the cost of reducing diffuse discharges of contaminants from farms into water bodies:

A) The cost effectiveness of reducing diffuse nitrogen discharges from dairy farming into the **upper Waikato River in New Zealand** along with the environmental benefits that accrued.

B) An economic analysis regarding reduction of diffuse discharges of nitrogen, phosphate, microbes, and sediment into the **Aparima River Catchment** in the Southland Region of New Zealand. This second study is included solely as a means of discussing the ability of farmers to pay for reducing discharges.

The approach in each study was to determine the cost of reducing farm nutrient output required so as to achieve the required discharge limit. The typical farm system was identified and scenarios developed as to how

Consulting Agricultural Economist, AgFirst Waikato, PO Box 9078, Hamilton 3240, New Zealand; Email: phil.journeaux@agfirst.co.nz

farmers might use mitigation strategies to reduce contaminant flows. The costs and benefits of these strategies were determined, using averages of long term costs and prices inflated to present values in New Zealand dollars. The main index used to compare the economic benefits and costs of the various strategies was Economic Farm Surplus (EFS). EFS is defined as *gross revenue plus change in livestock values less farm working expenses less depreciation less wages of management*.

Background

Agriculture is the primary driver of the New Zealand economy, accounting for 47 percent of mercantile exports (Statistics NZ 2012). Pastoral agriculture is carried out on approximately half the land area in New Zealand, and comprises mostly dairying, sheep, beef and deer farming. Due to the temperate climate, farm animals are grazed outdoors on pasture all year.

The general New Zealand philosophy on resource management is one of effects-based management. As a result, farmers are expected to manage the effects of any contaminant discharges from their lands into water bodies. The government provides no subsides to the sector and farmers are therefore also expected to meet the cost of mitigating any discharges.

The national expectation that farmers will both manage and pay for mitigation of contaminant discharges is raising some complex management issues because a significant source of contaminant flows into water bodies are diffuse discharges. These discharges include nitrogen, which mostly leaches through the soil profile, and over-land run off of phosphate, microbes and sediment. The issue of contaminant discharge has been compounded over recent years by a general trend to more intensive farming systems, particularly the conversion of easier topography land into dairy farming which has been driven by the greater profitability of dairying as compared to other types of agriculture.

Environmental Benefits & Economic Analysis of Management of Diffuse Nitrogen Discharges into the Waikato River

Problem definition

Located in the upper North Island, the Waikato region is one of the largest regions in New Zealand, based on the catchment of the Waikato River (Fig. 1). It contains the largest concentration of dairy farming in the country, with 35 percent of national dairy herds located there (LIC 2012). Gross regional product (GRP) for the region in 2010 was $14.2 billion, or 8.5 percent

Figure 1. Location of Waikato Region.

of national GDP. Within the Waikato, dairy farming contributes 9.6 percent of GRP, with all other agriculture contributing 4.6 percent (WRC 2012_1).

The upper Waikato River catchment is located in the southern area of the Waikato region (Fig. 2). It covers an area of 657,000 hectares, comprising 724 dairy farms, 516 sheep and beef farms, and significant forestry areas. The 724 dairy farms cover an area of 130,505 hectares; the average farm is 175 effective hectares, milking 468 cows (MPI 2012_1).

The Waikato Regional Council is the regional authority, with the primary purpose of overseeing and promoting environmental stewardship under the auspices of the Resource Management Act 1991 (RMA 1991). One of its prime concerns is the discharge of contaminants from farming into water bodies, and it has recently imposed restrictions on nitrogen leaching in the Lake Taupo catchment in order to protect the lake (WRC 2012_2). Lake Taupo is the primary source of water for the Waikato River and is located at the head of the upper catchment. The Regional Council has also identified the upper Waikato River catchment as a "sensitive" catchment

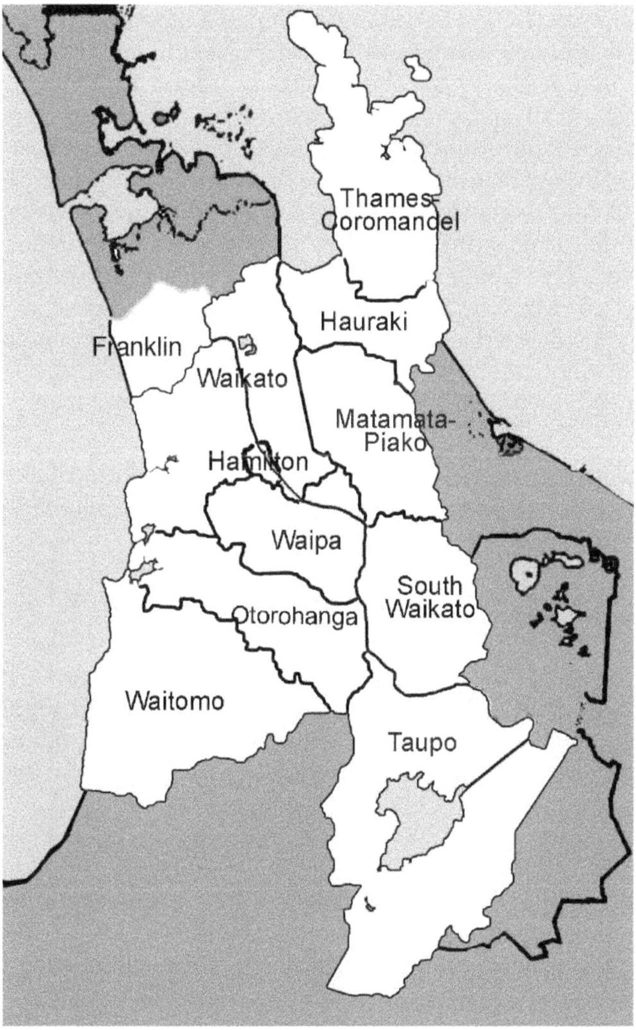

Figure 2. Location of Upper Waikato catchment.

due to nitrogen inflows resulting in algal blooms, and a number of studies have been carried out on the catchment (WRC 2012_3).

Farming is the main source of nitrogen and dairy farming the main contributor. The main cause of nitrogen leaching from dairy farms is cows urinating in the field, where the very high ammonia concentration within the urine patch leads to high nitrate leaching, especially over the winter where high rainfall is coupled with low pasture growth.

Vant (2006, 2007) found the average leaching rate of nitrogen from dairy farms in the upper Waikato River catchment was 36 kg nitrogen per hectare per year (KgN/Ha/yr), increasing at 1.5 percent per year due to the intensification of farming in the catchment. In contrast, the average leaching rate from sheep and beef farms is around 12–15 KgN/Ha/year mainly because sheep have much lower nitrogen leaching rates due to a much lesser concentration of urine patches relative to cattle. Vant projected that by 2030 the average discharge from dairy farms would be 51 KgN/Ha/year. The Council is considering "capping" the nitrogen discharge into the river from dairy farms at 30 kgN/Ha/year, in order to bring the river back to 2006 levels. This would require a 42 percent reduction in nitrate leaching rates from dairy farms by 2030, compared to 2006. For purposes of simplicity the study considered dairying only, and ignored other farming systems.

Management scenarios summary

Two management scenarios were considered which could reduce nitrate leaching rates from dairy farms by 42 percent: A) a combination of Best Management Practices and B) Conversion to Forestry.

Best Management Practices

A combination of "Best Management Practices" (BMPs) could achieve the reductions required. This scenario was analyzed as it represents the most likely approach to reducing nitrogen outflows, particularly given that the suggested management practices are already known and used by farmers.

The Best Management Practices considered in the study are:

(i) Storage of dairy effluent to allow for better application to the land when soil moisture levels are not saturated;
(ii) No application of nitrogen fertiliser over the winter months (May/June/July);
(iii) Stock exclusion—fencing off streams and development of riparian margins;
(iv) Construction of a wintering pad/off-paddock feeding facility;
(v) Use of nitrification inhibitors;
(vi) Constructed and facilitated wetlands;
(vii) Reduced stocking rates.

A further practice of grazing cows off the farm over winter (in another catchment) was not considered. While this is very effective in reducing nitrate leaching, it merely shifts the issue from one catchment to another and hence only mitigation strategies which resulted in permanent reductions were considered.

Conversion to forestry

Production forestry is a low nitrogen output land use. Therefore, in this scenario, a straightforward swap of dairy land for production forestry was considered. Land was progressively taken out of dairying and converted into forestry so as to eventually achieve the necessary reduction in nitrate discharge. This scenario was analyzed as some environmental commentators had suggested it as a more radical solution to the issue.

Management scenario description

Best management practices

There are a range of known technologies or approaches that can be used by farmers to reduce contaminant input into water bodies. These are variously described as best management practices (BMPs—utilized here), good agricultural practices (GAPs), and good environmental management practices (GEPs). Many of these BMPs have mitigation impacts across a range of contaminants while others are more specific, especially for nitrogen. Within New Zealand these practices have been organized into two tiers by the main New Zealand pastoral research organization (AgResearch) relative to their efficacy and cost (Monaghan et al. 2009). These are illustrated in Table 1. The term "Tiers" relate to the cost effectiveness: Tier one strategies are cheaper and/or more cost effective, whereas Tier two strategies are more expensive and/or less cost effective. Within the study, not all the mitigation strategies outlined below were modeled.

The costs and benefits of the BMPs listed below were analyzed for the upper Waikato River case study. Within the study catchment a proportion of farms had already instigated a number of the BMPs. This was considered a sunk cost, and therefore that proportion of farms was excluded from the analysis with respect to the BMP in question.

A description of the BMPs and their general costs and benefits is as follows. The efficacy of BMPs is summarized in Table 2.

(i) *Effluent storage*

Description: Installation of a storage pond with the capacity to store 90 days of effluent over the winter/early spring period, for subsequent irrigation onto the farm later into the spring/early summer, so as to prevent run-off of effluent when soil moisture levels are high.

Costs: Initial cost for consent to build the pond and irrigate the effluent and the capital cost of the pond. Ongoing maintenance cost.

Benefit: The value of the nutrients saved that would normally run off.

Table 1. Best management practices arranged by tier (Source: Monaghan et al. 2009).

Tier	Best Management Practice	Targeted contaminant
Tier One	Improved farm dairy effluent management • Storage (i) • Low application rate	Phosphate, Microbes, Ammonium-N
	Stock exclusion (iii) from: • Streams • Wetlands • Swales and wet gullies (especially on winter crops)	Phosphate, Microbes, Ammonium-N, Sediment
	Facilitated wetlands	Microbes, Nitrogen, Sediment, Phosphate
Tier Two	Nitrification inhibitors (v)	Nitrate-N
	Wintering cows in herd shelters (iv)	Nitrate-N, Ammonium-N, Microbes, Phosphate
	Restricted autumn grazing	Nitrate-N
	Substituting N-fertilised pasture with low N feeds	Ammonium-N
	Tracks and lanes sited away from streams & lane (runoff diverted to land)	Phosphate, Microbes, Ammonium-N, Sediment
	Constructed wetlands	Nitrate-N, Ammonium-N, Microbes, Sediment
	Grass buffer strips (iii)	Nitrate-N, Ammonium-N, Microbes, Phosphate, Sediment
	Limiting nitrogen fertilizer use	Nitrate-N

Table 2. Efficacy of BMPs (Source: Waugh 2012).

Mitigation Strategy	Nitrogen Reduction (%)	Phosphate Reduction (%)
(i) Deferred effluent		10–30
(ii) No winter N	0–15	
(iii) Stream Stock Exclusion	2–5	3–30
(iii) Riparian margin	4–14	0–62
(iv) Wintering Facility	18–40	3–15
(v) Nitrification Inhibitor	0–35	
(vii) Constructed Wetlands	24–50	26–77

(ii) *Nil nitrogen fertilizer application over winter*

Description: The main period of nitrogen leaching occurs over the high risk/rainfall months of May, June, and July. The management strategy here can be to apply nitrogen fertilizers in the autumn and spring, but missing out the winter period. A reduction in N fertilizer application over these critical months has two effects:

i) It will avoid direct leaching of the applied fertilizer N, which can be up to 30 percent of the N applied, depending on the N rate, rainfall, and any specific conditions within that year (Ledgard et al. 1988); and
ii) An indirect effect of less N fertilizer applied overall, resulting in less increase in pasture nitrogen percent, which reduces urine N excreted, but also less pasture growth (Ledgard et al. 1999).

Cost: Reduction in winter nitrogen fertiliser application will result in loss of pasture growth.

Benefit: The benefit from reduction in winter fertilizer is reduced direct cost.

(iii) *Stock exclusion/riparian margins*

Description: Streams are fenced off so as to exclude stock. Within New Zealand, most existing dairy farms have already fenced the relevant streams, as a supply requirement to Fonterra who is the main dairy company. The width of riparian strips, to be effective, depends on soil type and slope (J. Quinn, pers com, as cited in Journeaux et al. 2011)). The greater the slope, the greater the velocity of run-off and thus the greater the width of riparian strip required. For the purposes of the Waikato River studies, it was assumed that new dairy farmers fenced off streams 5 m back from the edge of the stream and planted up the margin in a variety of (mostly) native plants. To control nutrient run-off, grass strips are often more effective. With a 5 m margin, the amount of light entering the strip would be high, and a reasonable level of grass growth could be expected. There are added complexities in practice, e.g., a potential need to allow temporary grazing during certain times of the year to manage weeds and avoid excessive grass growth which will reduce the effectiveness of the strip. For practical reasons, these complexities were excluded from the analysis. The 5 m width noted above would be considered the minimum required to achieve some degree of environmental effectiveness, without taking up significant areas of productive land. To be most effective in reducing nutrient run-off, the optimum width of riparian strips need to be in the order of 10 m–20 m wide (J. Quinn, pers com), but at these widths there could be significant opportunity costs due to the loss of grazing land.

Cost: Given that it is a supplier requirement, and the majority had already complied with this, the cost of fencing the remaining streams on existing dairy farms was considered a sunk cost and ignored. There is an ongoing maintenance cost for these fences. Costs involved on new dairy farms and sheep and beef farms include the capital cost of the fencing, the opportunity cost of capital involved, and an ongoing maintenance cost. For riparian strips there is a capital cost in planting, maintenance costs and an opportunity cost of lost grazing for riparian strips.

Benefit: There is no direct economic benefit.

(iv) *Wintering facility*

Description: A wintering facility is a physical structure used as an on-off grazing system over the winter months. This involves grazing stock (usually dairy cows) for several hours during the day on pasture, with the remainder of the time spent on the wintering facility, or with the stock solely on the facility for some weeks over the winter period. The facility can range from a sawdust-bark stand-off pad through to a free-stall facility or herd home.

Cost: There is an up-front capital cost of the structure, ongoing maintenance and operating cost for the facility, i.e., the cost of any bought-in feed.

Benefit: The saved costs of not grazing cows off over winter on contract, reduction in pugging damage on run-offs, reduced travel costs to run-offs, increased milking period, better cow condition (and hence enhanced milk production), and a reduction in the number of dry/empty cows.

(v) *Nitrification inhibitor*[1]

Description: Nitrification inhibitors are a relatively recent technology in New Zealand farming and involve the use of dicyanimides applied to paddocks just prior to winter (usually in May) and in late winter (August). The dicyanimides act on soil microbes such that the ammonia excreted in the urine by cows is converted more slowly to nitrates. The effect of a nitrification inhibitor applied in late winter could well be nullified by the use of a wintering facility. Even allowing for on-off grazing using the wintering facility, very little urine would be deposited on the pasture over the winter period and hence the efficacy of the second application of an inhibitor could be significantly less. The recommendation in this situation would still be to apply the two applications, but both in the autumn. Research has shown that an application in March and another in late April/early May were quite effective at reducing nitrate leaching over that period (Monaghan et al. 2009).

Cost: The main cost is the up-front one of applying the inhibitor.

Benefit: Pasture response from the increase in available nitrogen within the soil profile.

---

[1] In late 2012 the use of Nitrification Inhibitors was temporarily banned in New Zealand until food safety limits are established.

(vi) *Constructed wetlands*

Description: This involves the physical construction of a wetland to intercept drainage off a catchment (McKergrow et al. 2007). In some regions this could well be an opportune strategy to intercept tile drainage outflows which would normally bypass all other mitigation strategies.

Cost: There is a capital cost in the wetland (e.g., earthworks, fencing, plants, drainage into the wetland), an ongoing maintenance cost, and the opportunity cost of any productive land which was taken out of grazing and incorporated into the wetland.

Benefit: While there are biodiversity and environmental benefits from these constructed wetlands, there is no direct economic benefit.

(vii) *Facilitated wetlands*

Description: These are natural wetland areas that intercept run-off, especially from ephemeral streams, that are fenced off and planted. The performance of the wetland is enhanced as the protected plants then soak up discharges.

Cost: There are up-front fencing and planting costs as well as ongoing maintenance costs.

Benefit: There are no direct economic benefits.

(viii) *Optimizing the farm system (reduced stocking rates)*

Description: Recent modeling (Ridler et al. 2010) has indicated that many farms could reduce their stocking rate by 10 percent, reduce their level of bought-in supplementary feed by around 50 percent, but maintain their current level of production, provided they can lift their grazing management to ensure that pasture quality is maintained. This latter assumption is quite critical; maintaining a higher level of grazing pressure at a lower stocking rate requires good grazing management skills. Within the study, the assumption of intensive technology transfer would assist in helping farmers acquire this skill.

The reduction in nitrogen leaching is not necessarily significant, as the cows are essentially substituting pasture for the bought-in feed in order to maintain production, and hence the overall urine/nitrogen output from the cows is similar. A farm was modeled under this system using the *Overseer™* Nutrient Budget software program,[2] with the stocking rate reduced by 10 percent and the amount of bought-in feed reduced by 45 percent. This resulted in a 7 percent reduction in nitrogen leaching.

---

[2] A computer program which summarizes all nutrient inputs and outputs from a farming system: www.overseer.org.nz.

Cost: A reduction in income from having a lesser number of livestock to sell.

Benefit: Savings in bought-in feed. As noted, the critical assumption here is that grazing management lifts to ensure no drop in pasture quality due to the lower stocking rate. A secondary saving is made in reduced variable costs due to the lower numbers of cows.

Technology transfer

A critical part of enabling farmers to adopt new innovations or systems is technology transfer programmes. The upper Waikato River study assumed that the rate of adoption for the Best Management Practices (BMPs) analyzed, and the associated costs and benefits, followed the pattern illustrated by Fig. 3 below. This adoption curve is based on significant research (summarized in Journeaux 2009) about the value of farm extension in assisting farmers to adopt innovations or new systems, particularly environmental practices.

As the curve illustrates, approximately 50 percent of farmers have adopted (i.e., taken up and are implementing the innovation) over a 5 year period, with the remainder somewhat more slowly over the remaining 15 years. This represents quite a fast (but not impossible) rate of adoption, and an intensive extension effort would be required to achieve this.

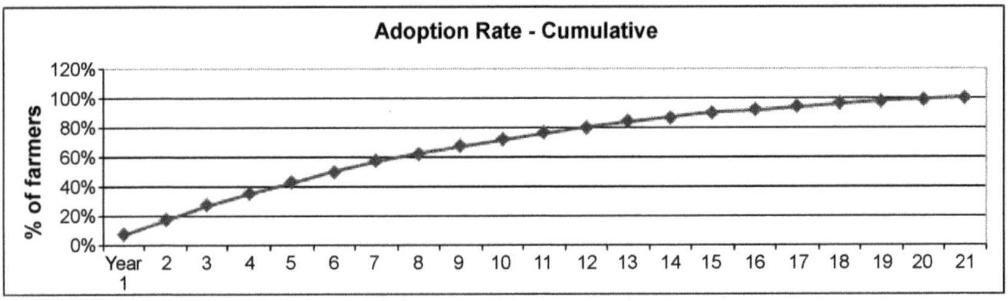

Figure 3. Assumed Adoption Rate (Source: Journeaux 2009).

Conversion to forestry

This involved converting dairying land (a high nitrate discharge land use) into commercial forestry (a low nitrate discharge land use).

Management scenario costings

Discount rates

The discount rates used in the study were as follows. Both were expressed in constant terms throughout the analysis and all figures are expressed in 2009 New Zealand dollars.

(i) The New Zealand Treasury guideline rate of 8 percent real, which is based on the government opportunity cost of capital and which represents the "risk based rate". This corresponds to a 11.3 percent nominal rate, and incorporates an inflation rate of 3 percent; and

(ii) A "risk-free alternative" or social time preference rate, which was taken as the ten year average of the ten-year government bonds, which gave a real rate of 3 percent. This corresponds to a nominal rate of 6.2 percent, and again incorporates an inflation rate of 3 percent.

BMPs

Table 3 summarizes the costs and benefits of using BMPs excluding stocking rates while Table 4 shows the same figures with stocking rates included. A breakdown of costs and benefits is shown in the Table 5.

The analysis indicates that the required 42 percent reduction in nitrogen output by farm can be achieved via the implementation of a range of BMPs. However all the strategies, with the exception of nitrification inhibitors and reduced stocking rate, carry a net cost to the farmer. As can be seen from Tables 3 and 4, an economic benefit can be achieved if BMPs are implemented alongside a strategy to optimize overall farm performance. The reduction in stocking rate is restricted to 10 percent as, based on the recent modeling work carried out, this results in the greatest net gain. If a greater reduction in stocking rates had to be made, then the costs rapidly outweigh the benefits.

The crucial issue in using BMPs successfully is whether farmers have sufficient skills and knowledge to improve grazing management so as to

Table 3. Use of BMPs excluding stocking rate reduction* (Source: Journeaux et al. 2011).

	$ NZ Million		€ Million	
	8.0%	3.0%	8.0%	3.0%
PV of Costs	291.3	409.5	131.4	184.7
PV of Economic Benefits	153.0	265.7	69.0	119.8
Nett Benefit	–138.3	–143.8	–62.4	–64.9

*Discounted period 20 years.

Table 4. Use of BMPs including stocking rate reduction* (Source: Journeaux et al. 2011).

	$ Million		€ Million	
	8.0%	3.0%	8.0%	3.0%
PV of Costs	312.8	447.2	141.1	201.7
PV of Economic Benefits	378.3	661.3	170.6	298.2
Nett Benefit	65.5	214.1	29.5	96.5

*Discounted period was 20 years.

Table 5. Cost and benefits disaggregated [NZ $million (€ million)] (Source: Journeaux et al. 2011).

	Economic Cost		Economic Benefit		Net Economic Benefit	
	$PV_{8.0\%}$	$PV_{3.0\%}$	$PV_{8.0\%}$	$PV_{3.0\%}$	$PV_{8.0\%}$	$PV_{3.0\%}$
Effluent storage	19.1 (8.7)	25.2 (11.4)	5.5 (2.5)	9.6 (4.3)	–13.6 (–6.2)	–15.6(–7.1)
Fencing streams/riparian margins	66.3 (29.9)	92.3 (41.6)	0	0	–66.3 (–29.9)	–92.3 (–41.6)
Nitrification inhibitors	14.3 (6.5)	19.4 (8.8)	18.9 (8.5)	25.6 (11.5)	4.6 (2.0)	6.2 (2.7)
Wintering facilities	144.9 (65.4)	196.1 (88.5)	90.4 (40.8)	158.8 (71.6)	–54.5 (–24.6)	–37.5 (–16.9)
No winter nitrogen	29.5 (13.3)	51.8 (23.4)	18.4 (8.2)	32.3 (14.5)	–11.1 (–5.1)	–19.5 (–8.9)
Reduced stocking rate	21.5 (9.7)	37.7 (17.0)	225.2 (101.6)	395.6 (178.4)	203.7 (91.9)	357.9 (161.4)
Technology transfer	17.2 (7.7)	24.6 (11.1)	19.9 (8.9)	39.5 (17.8)	2.7 (1.2)	14.9 (6.7)

ensure no loss in milk production with reduced stocking rate. The study assumed an intensive extension system was in place which would improve management capability but there is still no guarantee that the full benefit would occur.

The overall conclusion is that farmers must improve the efficiency and profitability of their farming systems, of which stocking rate is one aspect, in order to pay for the costs of the environmental improvements. It could be argued that there is some scope for this; in 2009/10 the top 10 percent of farmers had an Economic Farm Surplus per hectare (EFS/Ha) of $4,200, compared to the average EFS/Ha of $2,400, and the bottom 10 percent had an EFS/Ha of $800/Ha (MAF 2010). This indicates there is room for average/below average farms to improve their profitability.

Many farmers are implementing BMPs despite them resulting in an apparent net cost; for example riparian margins and wintering facilities. This is because farm management decision drivers are not only economic; many farmers seek to be good stewards of the land and the water while others implement strategies for lifestyle, aesthetic and/or animal welfare reasons.

Conversion to forestry

In the conversion to forestry scenario, dairy land within the catchment was assumed to be progressively converted over to forestry, a low nitrogen output land use, until the combined total forestry and dairying areas leached the target average rate of 30 kg/N/ha/yr.

The key assumptions within this scenario are:

- The forestry system is a conventional New Zealand commercial forestry regime growing pine trees (*pinus radiata*) on a 28 year rotation period.
- The total land converted, in equal annual increments over 28 years, is 66,400 hectares, equivalent to 51 percent of the area in dairying.
- Nitrogen leaching under pine trees is 3 KgN/Ha/yr.
- The resultant land use mix of dairying and forestry achieved the desired average nitrogen leaching from the catchment of 30 KgN/Ha/yr.
- The discount period is 56 years; this allows for the planting of forest over 28 years and then one full rotation thereafter.

There would be an additional issue of legacy nitrogen; once the area was planted in trees, nitrogen levels in the soil would take some years to drop back to a status quo level compatible with trees. There is nothing that can be done to mitigate this, and it was ignored in the analysis.

The cost of this scenario (Table 6) is the cumulative net loss of income from dairying over this period, and into the future. This was costed assuming an Economic Farm Surplus, in effect the free cash flow, of $1164 per hectare (being the 10 year average of the EFS from the MAF Waikato/Bay of Plenty Dairy Farm Monitoring model (MAF 2009a).

Table 6. Reduction in dairying income (Source: Journeaux et al. 2011).

$NPV_{8\%}$	$403.5 million (€182 million)
$NPV_{3\%}$	$1.69 billion (€761.7 million)

The gain is (shown in Table 7):

- the returns from forestry (timber plus pulp) over this period and into the future,
- the value of carbon credits at $25 per tonne of CO_2.[3] The CO_2 sequestration rates used were based on the MAF look-up tables for the Waikato region (MAF 2009b), with an assumed decay rate for the slash at harvest (waste wood left behind).

The inclusion of a carbon price has a major effect on the profitability of the forestry regime.

Overall therefore, there is an economic loss as a result of the land use change, unless carbon prices increase above $50/tonne.

Table 7. Returns from forestry (Source: Journeaux et al. 2011).

	Carbon price = $0/T	Carbon price = $25/T	Carbon price = $50/T
$NPV_{8\%}$	–$8.8 million (–€4.0 million)	$120.3 million (€54.2 million)	$249 million (€112.3 million)
$NPV_{3\%}$	$798 million (€359.9 million)	$1.28 billion (€577 million)	$1.759 billion (€793 million)

Note: These figures include the value of timber and pulp and value of carbon discounted over a 56 year period (i.e., 2 rotations).

Multiplier effect

The impacts of conversion to forestry would, in fact, be much wider than just the economic benefits summarized above. Therefore multipliers based on national input/output tables were used to calculate the wider implications of such a change. The multiplier effect is where spending in one area of the economy stimulates spending in other areas; if there is an increase in final

[3] This was the price at the time of the study. Current 2013 carbon prices are $2 per tonne.

demand for a particular product, there will be an increase in the output of that product as producers react to meet the increased demand: this is the "direct effect". As these producers increase their output, there will also be an increase in demand on their suppliers and so on down the supply chain: this is the "indirect effect" (i.e., Type I multipliers). As a result of the direct and indirect effects, the level of household income throughout the economy will increase due to increased employment. A proportion of this increased income will be re-spent on final goods and services: this is the "induced effect" (i.e., Type II multipliers) (Butcher 1985).

Value-Add multipliers provide estimates of value added to products resulting from the sale of a good or service to another sector. This Value Add includes the cost of employee compensation, indirect business taxes, and proprietary and other property income.

There are both backward and forward multipliers—backward multipliers relate to services provided into the farm/forest, while forward multipliers relate to services and processing beyond the farm/forest gate. Both need to be added together to gain a total multiplier impact, as illustrated in Table 8. So, for example, an extra $1 produced by dairy farming creates a further $5.05 in economic activity nationally while an extra $1 produced by forestry creates a further $6.67 in economic activity nationally.

The total multipliers were used across both the forestry and dairy cash flows, to indicate the wider impact across the region, as shown in Table 9.

Table 8. National multipliers (Source: Journeaux et al. 2011).

Dairy farming	**Backward**	**Forward**	**Total**
Type II Value Add	2.70	2.35	5.05
Type II Employment	2.72	2.58	5.30
Forestry and logging			
Type II Value Add	2.95	3.72	6.67
Type II Employment	4.23	5.93	10.16

Table 9. Value Add: Multiplier results (Source: Journeaux et al. 2011).

	$ NZ Million	**€ Million**
Dairy (loss)		
$NPV_{8.0\%}$	3 214	1 449
$NPV_{3.0\%}$	13 451	6 066
Forestry (gain)		
$NPV_{8.0\%}$	1 193	538
$NPV_{3.0\%}$	9 350	4 217
Net Benefit		
$NPV_{8.0\%}$	–2 021	–911
$NPV_{3.0\%}$	–4 101	–1 849

These figures show that the negative economic impact of the land use change would be accentuated across the wider economy, resulting in considerable net economic loss. In addition to this, a net 920 jobs would also be lost.

Environmental benefits

The earlier sections provide an estimate of the potential cost to the catchment from reducing nitrogen to reach the environmental target. The next step is to weigh that up alongside how much the community values the environmental improvement. If an individual values an environmental improvement more than every day consumables, such as a box of beer, then they ought to be willing to sacrifice those consumables for an improved environmental outcome. There are a variety of ways in theory to assess societies' preference for environmental goods and services. In practice however, assessing the environmental benefits from improved water quality in the Waikato River proved to be somewhat problematic, due to the lack of suitable data and the long lag timeframes involved in nitrogen moving through aquifers and into the river. Nevertheless, the following is a discussion of these issues and some figures were calculated, although these should be regarded as illustrative at best.

Calculating the total economic value of water

The total economic value of an environmental resource incorporates all of the environmental, financial and social benefits associated with the use of that resource. These values reflect the well-being of society in relation to that resource. This is illustrated in Fig. 4 below:

Direct use values of a resource are those for which a value can easily be inferred from the market in which they are produced and/or traded, e.g., production of agricultural goods such as dairy or wood products as discussed in the last section. Indirect use values are where the resource provides a function which is less able to be directly valued, e.g., ecosystem services or recreational use, while passive values are the more intangible values such as aesthetic values and bequest/existence values. Indirect/passive values include the value people gain from the basic life support functions associated with ecosystem health or biodiversity, to the enjoyment of a scenic vista or a wilderness experience, to appreciating the option to fish or bird watch in the future, or the right to bequest those options to grandchildren. It also includes the value people place on simply knowing that (for example) giant pandas or whales exist (King and Mazzotta 2000).

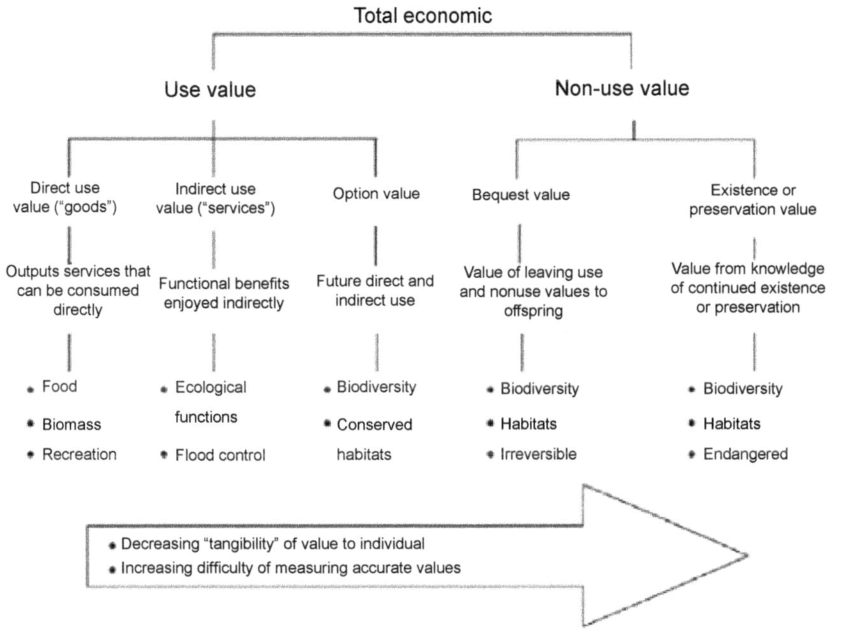

Figure 4. Constituents of Total Economic Value.
Source: EVRI (www.evri.ca).

As we move along the spectrum, valuation of these uses becomes progressively more difficult. Techniques have been developed to understand and measure individuals' preferences for the indirect and passive use values of environmental resources, where a tradable market does not exist (i.e., non-market goods and services). The two categories of techniques are revealed preference techniques and stated preference techniques.

Revealed preference techniques such as the travel-cost method (TCM) collect data on number of trips taken and the financial outlay a user incurs to obtain an experience related to the environmental resource. For example, the amount a fisherman may spend to travel to, and stay near, a favored fishing spot. The TCM tends to be restricted to site specific studies such as the use of recreation parks and it fails to capture the full value society may attach to the resource. In other words, there is likely to be a consumer surplus over and above the cost incurred to experience the resource.

Stated preference techniques such as choice modeling (CM), on the other hand, are able to elicit the full value society attaches to these indirect use and passive values. Both methods involve respondent surveys that elicit their willingness-to-pay (WTP) or willingness-to-accept compensation (WTA) for certain environmental changes. Values derived using stated preference WTP techniques in studies relevant to the Upper Waikato situation were identified and assessed as to whether they would be appropriate to use in a benefit transfer context. Benefit transfer is the process of applying the results of existing studies to new situations (values thus obtained are called transfer values).

Benefit transfer is a complex process and is still an evolving discipline. There are a number of issues involved, including; physical site characteristics (e.g., similarity between the study site and the new site), population characteristics (e.g., socio-demographic characteristics), framing issues within the choice survey (e.g., scale and scope differences, welfare measures), and value transfer methods (e.g., direct transfer, or functional transfer), as discussed in Navrud (2008), Bell et al. (2011) and Carson (2000).

In addition, there are issues around aggregation of the values obtained across the relevant population, as discussed by Morrison (2000), and Bateman et al. (2006). These include:

- The response rate to the survey—the greater the response rate the lower any aggregation bias, and vice-versa;
- The similarity of respondents and non-respondents. If non-respondents are randomly distributed, the simple extrapolation of estimates across the population will be valid. However, this is difficult to determine with non-use values; and
- The correlation between preferences and socio-demographic characteristics. This relates to both their socio-economic status—individual wealth often influences the "willingness to pay", and distance from the issue. While often, the amount people are willing to pay decays with distance from a particular issue, this is not always the case. One study (Morrison 2000) cites the case of people throughout Australia being more willing to pay to preserve the Kakadu Conservation Zone from mining than people in the Northern Territory. Bell et al. (2009) noted that while they had found differences in WTP in their study decayed with distance, this was not statistically significant. Often distance decay can be high for active values, and low for passive values.

Adjusting for some of these factors, Morrison (2000) noted that mean WTP could be 0–50 percent less than stated in the original survey. In the absence of data to accurately calculate these aggregation biases, an arbitrary figure of 25 percent (as a mid-point) was used within the analysis to reduce the WTP figures.

A key component that has to be considered within the upper Waikato catchment is the lag effect of nitrogen already in groundwater, i.e., the time period it takes for the nitrogen in the groundwater to move through the aquifer and into water bodies. The time period of these lags are very variable, and means that any reduction in contaminant flows via groundwater into the Waikato River due to reductions in nitrate leaching from farms can take some time. In this respect, while this study looks at reducing nitrogen into the river, these reductions will not significantly impact on the river itself until the "legacy nitrogen" has moved through the ground water aquifers. This issue is important, as while the economic costs and benefits are relatively immediate, significant environmental benefits may not be apparent for some years, which needs to be taken into account when discounting these future benefits.

Within the upper Waikato catchment, the time periods involved in ground water lags have not been researched significantly. Simulation modeling indicates the time for the effects of an instantaneous, regional scale land use change to reach a new equilibrium to be in the order of 350–400 years, though 90 percent of the change is predicted to occur after approximately 160 years[4] (Weir and Moore 2012). These lag periods also vary spatially, depending on how close farms are to water bodies. For example, a farm alongside the Waikato River may have a lag of only 1–5 years, whereas a farm at the top of the catchment may have a lag period of 50–200 years.

The spatial effect is also important in that any reduction in N leaching on farms close to the river will have a more immediate effect in N reductions within the river, and it can be assumed that the dairy farms within the catchment are generally located on the easier contour land closer to waterways. Given this, the assumption made for the study was a mean lag period for nitrogen leached from dairy farms of 15 years.

Any reduction in nitrogen leaching from farms will also have an impact, albeit minor, on the nitrogen flows within the groundwater. A slow decrease in nitrogen levels within the groundwater flows into the river will occur, as opposed to nothing happening for 15 years before an effect is apparent (K. Rutherford, pers com). For the purposes of this study, a minor, curvilinear reduction in groundwater N flows was assumed through to year 14, with a significant drop assumed in year 15. This affected the benefits as calculated for the recreation/in-river ecology values and the ecosystems services value as discussed below.

[4] This report was not available at the time the study was undertaken.

Within the study, four areas of non-market benefit were considered using three studies

(i) Recreation (ability to swim)
(ii) Passive use values (in-river ecology)
(iii) Passive use values (riparian biodiversity)
(iv) Ecosystem service (waste treatment)

(i) & (ii) *Recreation and passive use values*

This is based on a Choice Modeling exercise carried out by Marsh (2010) who investigated the community's willingness to pay for improvements in the water quality of the Karapiro and Arapuni hydro lakes (on the upper Waikato River). Respondents were asked a series of questions around their WTP with respect to suitability for swimming, water clarity, the ecological health of the lakes, and potential job losses in dairying.

A summary of the results is shown in Table 10.

Table 10. WTP for environmental factors for Karapiro and Arapuni Lakes (Source: Marsh 2010).

Compensating surplus: welfare gain for change from status quo to improved outcome (NZ$ per household per year over 10 years)				
Attribute	**Status Quo**	**Policy 1**	**Policy 2**	**Policy 3**
Swim (Chance of algal bloom)	50%	20%	10%	2%
Clarity (metres)	1 m	1.5 m	2 m	4 m
Ecology (% excellent)	40	50	60	80
Median welfare gain (assuming no job losses)		$26/yr	$51/yr	$86/yr

As can be seen from Table 10, the lowest level of improvement gave a WTP of $26 per household for 10 years which is the figure used in the study. A Present Value of the WTP figure was calculated at the two discount rates, adjusted for aggregation bias, applied across the number of households in the Waikato, the lag effect allowed for, and then applied in the same pattern as the rate of adoption discussed earlier (in the section on Technology Transfer), which would equate with the rate of improvement. The result is shown in Table 11.

(iii) *Biodiversity values*

The planting of riparian margins within the study area would result in a significant area (1,951 ha) being planted in native plants, resulting in a major network of native plant "corridors" throughout the catchment, which is very likely to attract native birds. In this respect, therefore, the plantings would result in a biodiversity benefit, which is a measure of the health of an ecosystem. Again the approach is to use choice modeling in order to

Table 11. Environmental Benefits (Source: Journeaux et al. 2011).

	$PV_{8.0\%}$	$PV_{3.0\%}$
Recreation/in-river ecology values	\$1.1 million (€0.5 million)	\$7.8 million (€3.5 million)
Biodiversity Benefits	\$16.6 million (€7.5 million)	\$56.9 million (€25.7 million)
Ecosystem Services Benefits	\$1.0 million (€0.45 million)	\$7.3 million (€3.3 million)
Total Environmental Benefits	\$18.7 million (€8.4 million)	\$72.0 million (€32.5 million)

gain an appreciation of the preferences and values of the community with respect to the biodiversity issue in question.

In addition to the probability of attracting native birds, the planting of riparian strips would also have an impact on in-stream biodiversity, where the shading and cooling of the stream by the trees, and the addition of leaves, etc. would result in an increase in invertebrates and fish species. These would include brown trout, bullies (*Gobiomorphus* spp.), eels, *Glaxias* spp., mayflies (*Ephemeroptera*), and caddisflies (*Trichoptera*) (Glova and Sagar 1994).

The choice modeling approach needs to deal with the unique issue in question, but unfortunately no such study existed for the upper Waikato catchment. To illustrate the point though, proxy values have been used based on a study with respect to the planting of shrubs and trees within the riparian margins. No such studies were readily available around in-stream biodiversity, so this aspect is excluded in this study.

Yao and Kaval (2008) considered the willingness to pay for councils to encourage plantings of native plants in order to enhance native biodiversity. The sample for this study was drawn from various regions, including in the Waikato. The study showed a median WTP of \$42 per ratepayer. Normally, in such studies the WTP is for a set period, but in this case the respondents weren't asked to consider a time period, so the inference is that the WTP on the extra rates is in perpetuity. The \$42 was converted to a present value, and extrapolated over the 20 year period relative to the adoption rate, again adjusted for aggregation bias.

The riparian margins assumed in this study would represent a relatively narrow corridor (being 5 m wide on either side of the stream), and as such they would suffer from strong edge effect gradients, and in all probability they would be of use for only a minority of generalist bird species. In this respect, therefore, the biodiversity effect would be greatly diminished relative to if the 1,951 ha was in one contiguous block, and consequently the value calculated has been reduced by 50 percent. Using the figures and assumptions outlined above, the value of increased biodiversity was calculated, as shown in Table 11.

(iv) *Ecosystems services*

Ecosystem services refer to the many goods and services emanating from the functioning of local ecosystems. They include all market values, anthropocentric and ecocentric non-market use-values, and anthropocentric and ecocentric non-market non-use values that function in nature and are necessary to sustain ecosystems (Kaval 2006). The community benefits from many different ecological functions, from water purification services within water bodies, to wild pollination (Coleman 2009). Ecosystems are natural assets and provide services that, if not vital to human existence, at least contribute to our welfare (van den Belt et al. 2009).

Within this study, the adoption of best management practices within the upper Waikato catchment would result in an improvement in ecosystems services, in that the reduction of nutrients into the water system would improve its capacity to assimilate nutrients. The value of this has been extrapolated from a study done in the Manawatu (van den Belt et al. 2009), looking at the value of ecosystem services across that region. The values used by van den Belt have, in turn, been extrapolated from overseas published literature, so caution is needed in interpreting the results.

The Manawatu report notes that the total ecosystems value for dairy farming, incorporating both direct and indirect values, is $1,796 per hectare. The direct benefits have already been incorporated within this study via the economic analysis, and therefore the main values of interest are the indirect values. The indirect ecosystem value for dairying is $404 per hectare. This is a 2006 value, which when updated to 2010 by the CPI (11.5 percent over the period) equals $450/ha. Of this "waste treatment"—the assimilation of nutrients by the environment—provides 7.6 percent. Given that the reduction in nutrient outflows as a result of the introduction of the best management practices is 42 percent, the value of this was calculated as:

$450 x 7.6% x 42% = $14.33/hectare

This was then applied to the effective area of dairying in the catchment, following the rate of adoption curve, and allowing for the lag effect in groundwater. Results are shown in Table 11.

A summary of the combined environmental benefits (recreation + biodiversity + ecosystem services) shows total benefits of $18.7 or $72 million, depending on the discount rate adopted. The level of inherent assumptions means these figures are very generally illustrative only, but the overall environmental benefit as calculated is relatively minor compared to the economic costs and benefits.

Economic Analysis of Management of Diffuse Contaminant Discharges in the Aparima Catchment

Problem definition

The Aparima catchment is in the southern-most region of NZ, covering an area of 153,700 hectares, with the predominant land use being pastoral agriculture (MPI 2012_2). The major pressure on waterways comes from the increasing level of conversion of sheep and beef land into dairying, with a resultant increase in the discharge of contaminants.

The issue is similar to that of the upper Waikato catchment, where the Regional Council is considering limiting nutrient discharges from farms.

Analysis method

Analysis of Aparima (Journeaux and Wilson 2014) was carried out in a very similar way to the upper Waikato River study except that it considers a wider range of contaminants than the Waikato; a particular focus on nitrogen and phosphate and, to a lesser extent, microbes and sediment, but did not consider the environmental benefits. The study also considered the ability of farmers to pay for the mitigation strategies, and it is this aspect which is discussed below.

Within the studies the main indice used to compare the economic benefits and costs of the various strategies was Economic Farm Surplus (EFS), as this allows for easy comparison and amalgamation across differing farming systems. However, at a farm level it is important to consider other costs, such as debt servicing and taxes, which must also be paid.

The impact on farm profitability was measured using "Farm Surplus for Reinvestment" (FSR). FSR is defined as *the cash surplus from the farm business after all costs are accounted for, including interest, tax, and personal drawings, and then available for expenditure on farm development, capital purchases, and debt reduction*. The average FSR for a model Southland dairy and sheep & beef farm (MPI 2012_3) over 10 years deflated through to 2009 values is \$761/hectare or \$230/cow for the dairy farm, and \$133/hectare for the sheep and beef farm. This represents the annual amount of money available for the expenditure noted above, including environmental BMPs.

Impact of BMP adoption on FSR

A summary of the results for Southland dairy farms is shown in Table 12 and sheep and beef farms in Table 13.

Table 12. Capital Cost and Net Benefit of BMPs for a Southland Dairy farm per cow (Source: Journeaux and Wilson 2014).

	Capital	Interest	Operating cost	Benefit	Net
Effluent	279.60	22.37	4.66	4.07	–$22.96
Fencing	26.40	2.11	1.91		–$4.03
Riparian	222.12	17.77	4.62		–$22.39
DCD			86.03	12.87	–$73.16
Wintering facility	1,864.00	149.12	192.74	290.49	–$51.36
Winter N			27.34	24.00	–$3.34
Facilitated Wetlands	12.59	1.01	0.62		–$1.62
Constructed Wetlands	842.79	67.42	3.89		–$71.31
Nutrient management plans	5.70	0.46	1.14		–$1.60
	$3,253.20				–$251.77
	(€1,467)				(–€114)

Table 13. Capital Costs and Net Benefit per hectare of BMPs for a Southland Sheep & Beef Farm (Source: Journeaux and Wilson 2014).

	Capital	Interest	Operating	Net
Fencing	21.02	1.68	1.05	–2.73
Riparian	38.62	3.09	0.38	–3.47
Facilitated Wetland	63.07	5.05	2.63	–7.67
Nutrient Management Plan	8.76	0.70	1.75	–2.45
	$131.47			–$16.33
	(€59)			(–€7)

The net annual cost of the BMPs at $252 per cow is greater than the average FSR of $230/cow, meaning that at least half the dairy farms in Southland (i.e., average and below) could not afford to implement all BMP strategies. The capital cost of all the BMPs at $3,253/cow means that, if the farmer had to borrow this to implement the mitigation strategies, it would take the average total debt per cow from an already high $8,300 (MPI 2012_4) to $11,790. This again means that the average farm cannot afford to introduce all BMPs.

There are two aspects flowing from Table 12; if farms need to introduce all BMPs they would need to do so over an extended time period, and the financial performance of the average farm needs to increase, reinforcing the concept of improving the profitability of the average farm in order to pay for the environmental BMP's required.

For sheep and beef farms (also studied in the Aparima analysis) the situation is different as shown in Table 13. In this case the cost of the BMPs is well within the capability of farmers to pay.

Sheep and beef farms are different in that the main discharges are phosphate, microbes, and sediment, and the BMP mitigation strategies for these tend to be more cost effective, while the more expensive BMPs are associated with nitrogen mitigation, which is not such an issue on sheep and beef farms.

Conclusion

The two studies showed that there are a number of competing issues with respect to mitigation of diffuse contaminants from farms:

- Over time, land use is likely to shift to more intensive uses due to economic pressure, and there is significant economic gain for the community from the land use change;
- Land use change into more intensive systems will result in an increased contaminant flow into water bodies, and while BMP mitigation strategies will reduce the contaminant loading from individual farms, they will not eliminate it;
- This has implications for the catchment as a whole, as the increase in total contaminant loading from intensification is likely to be more than the reduction from implementing Best Management Practices (BMPs);
- Regional Authorities, driven by community concern at deteriorating water quality and central government legislation, are actively considering placing limits on nutrient discharges from farming activities;
- Nutrient discharge limits, and BMP mitigation strategies to achieve these, cost money. However, there is significant variation in both cost and effectiveness for different BMPs. Some strategies are highly cost-effective, while others much less so;
- Mitigation strategies to reduce over land run-off (e.g., phosphate, sediment) are generally more cost effective, whereas strategies to reduce nitrogen leaching tend to be much more costly;
- A relatively intensive technology transfer programme is required to assist farmers in achieving adoption of the BMPs in a reasonable time span. Such a programme is likely to have a double positive spin-off in improved farm incomes;
- While there is an obligation on farmers to mitigate contaminant flows, farm profitability limitations means that there is a limit to the number of BMPs that many farmers can implement and remain viable, especially

over a short time period. The best performing farmers in a financial sense will find it easier to adopt BMPs and remain viable;

- Implementation time frames are critical. Faster implementation of limits will result in higher overall costs and more economic and social disruption. But the longer any decisions are delayed, the greater the likelihood that any cleanup will be bigger and longer;
- Essentially, there is a need for political or community decisions to determine any trade-offs across these issues;
- It is important in these analyses to incorporate both economic and environmental costs/benefits. Often the issue is the availability of robust monetised environmental data, and a methodology to handle long time-lags until environmental benefits are achieved.

Acknowledgements

The author would like to acknowledge Charlotte Cudby, New Zealand Ministry of Primary Industries, and Jane Shearer, Resolutionz Consulting, for their reviews of this chapter.

References

Bateman, I., Brett, H.D., Georgiou, S. and Lake, I. 2006. The aggregation of environmental benefit values: welfare measures, distance decay, and total WTP. http://researcharchive.lincoln.ac.nz/dspace/bitstream/10182/390/3/cd_dp_114.pdf.txt.

Bell, B., Yap, M. and Cudby, C. 2009. Valuing indigenous biodiversity in the freshwater environment. Paper to the 2009 NZARES Conference. http://www.nzares.org.nz/.

Bell, B., Yap, M. and Cudby, C. 2011. Biodiversity Valuation Manual—A Technical Manual for BNZ. Internal report to Ministry of Primary Industry.

Butcher, G.V. 1985. Regional Income Output and Employment Multipliers: Their Uses and Estimates of Them. Cost Benefit Handbook, Vol. 4. A commissioned study for the Economics Division, Ministry of Agriculture and Fisheries, Wellington.

Carson, R. 2000. Contingent valuation: auser's guide. Department of Economics, University of California, Environmental Science and Technology 34: 1413–1418. http://pubs.acs.org/doi/pdf/10.1021/es990728j.

Coleman, D. 2009. Whangarei District Ecosystems Services Background Report. www.wdc.govt.nz/.../ecosystem-service-final-report-background-sustainable-futures-30-50-3.pdf.

Glova, G.J. and Sagar, P.M. 1994. Comparison of fish and macro invertebrate standing stocks in relation to riparian willows (*Salix* spp.) in three New Zealand streams. New Zealand Journal of Marine and Freshwater Research 28(3): 255–266.

Journeaux, P.R. 2009. Developing an institutional model for the extension and adoption of environmental best management practices by pastoral farmers in New Zealand. Unpublished masterate dissertation.

Journeaux, P.R. and Wilson, K. 2014. Economic analysis of the impact of limiting the discharge of agricultural contaminants on farming; A Case Study of the Aparima (Southland). In Print.

Journeaux, P.R., Schischka, T. and Phillips, Y. 2011. Economic analysis of reducing nitrogen inputs into the upper Waikato River catchment. MAF Technical Paper 2011/98, ISSN2230-2794, www.mpi.govt.nz.
Kaval, P. 2006. Valuing ecosystem services: a new paradigm shift. Working Paper in Economics 6/10. Department of Economics, University of Waikato, Hamilton, New Zealand.
King, D.M. and Mazzotta, M.J. 2000. Ecosystem Valuation. Methods, Section 6, Contingent Valuation Method. http://www.ecosystemvaluation.org/contingent_valuation.htm.
Ledgard, S.F., Steele, K.W. and Feyter, C. 1988. Influence of time of application on the fate of 15N-labelled urea applied to dairy pasture. New Zealand Journal of Agricultural Research 31: 87–91.
Ledgard, S.F., Penno, J.W. and Sprosen, M.S. 1999. Nitrogen inputs and losses from clover/grass pastures grazed by dairy cows, as affected by nitrogen fertilizer application. Journal of Agricultural Science, Cambridge 132: 215–225.
LIC. 2012. 2011/12 New Zealand Dairy Statistics. http://www.lic.co.nz/lic_Publications.cfm.
MAF. 2009a. Waikato/Bay of Plenty Dairy Model Farm Monitoring Report. July 2009. http://www.maf.govt.nz/mafnet/rural-nz/statistics-and-forecasts/farm-monitoring/2009/pastoral/.
MAF. 2009b. Look-up Tables for the Emissions Trading Scheme. March 2009. Schedule 6, Pg 29. http://www.maf.govt.nz/sustainable-forestry/tools/.
MAF. 2010. National Dairy Farm Monitoring Report. http://www.maf.govt.nz/forestry/statistics-forecasting/statistical-publications.aspx.
Marsh, D. 2010. Water resource management in New Zealand: jobs or algal blooms? Paper to the NZ Association of Economists Conference, July 2010.
McKergrow, L.A., Tanner, C.C., Monaghan, R.M. and Anderson, G. 2007. Stocktake of diffuse pollution attenuation tools for New Zealand pastoral farming systems. NIWA Client Report HAM2007-161. http://www.niwa.co.nz/our-science/freshwater/tools/pollution-attenuation-tools.
Monaghan, R.M., Smith, L.C. and Ledgard, S.F. 2009. The effectiveness of a granular formation of dicyandiamide (DCD) in limiting nitrate leaching from a grazed dairy pasture. NZ Journal of Agricultural Research 52: 145–159.
Morrison, M. 2000. Aggregation biases in stated preference studies. Australian Economic Papers 76: 215–230. http://www3.interscience.wiley.com/journal/119032203/abstract?CRETRY=1&SRETRY.
MPI (Ministry for Primary Industry). 2012_1. AgriBase database. www.mpi.govt.nz.
MPI. 2012_2. FarmsOnLine database. http://www.biosecurity.govt.nz/biosec/camp-acts/farmsonline.
MPI. 2012_3. Southland Diary and Sheep and Beef Farm Monitoring Reports 2002–2012. www.mpi.govt.nz.
MPI. 2012_4. Southland Dairy Farm Monitoring Report. www.mpi.govt.nz.
Navrud, S. 2008. The Transfer and Aggregation Challenges in Biodiversity Valuation. The Economics of Global Loss of Biological Diversity 5–6 March 2008, Brussels Session B3. http://www.ecologic-events.eu/eco-loss-biodiv/documents/01-B3_Navrud.pdf.
Quinn, J. Principal Scientist, New Zealand National Institute of Water and Atmospheric Research (NIWA). Personal Communication.
Ridler, B.J., Anderson, W.J. and Fraser, P. 2010. Milk, money, muck and metrics; inefficient resource allocation by New Zealand dairy farmers. Paper presented at NZARES conference 2010. www.nzares.org.nz
RMA. 1991. Resource Management Act. http://www.legislation.govt.nz/act/public/1991/0069/latest/DLM230265.html.
Rutherford, K. Principal Scientist, New Zealand National Institute of Water and Atmospheric Research (NIWA). Personal Communication.
Statistics NZ. 2012. Mercantile trade data. www.stats.govt/nz.

van den Belt, M., Chrystall, C. and Patterson, M. 2009. Rapid Ecosystem Service Assessment for the Manawatu-Wanganui Region. Report prepared for Horizons Regional Council.

Vant, B. 2006. Water quality effects of land use change in the catchments of the Waikato River Hydrolakes. December 2006, revised May & July 2007. Internal Environment Waikato paper.

Vant, B. 2007. Water quality effects of land use change in the catchments of the Waikato River Hydrolakes. November 2007. Internal Environment Waikato paper.

Waugh, N. 2012. Factors influencing nutrient losses. Paper to MPI (in print).

Weir, J. and Moore, C. 2012. Groundwater Modelling of the Upper Waikato Catchment: Stage 2. Waikato Regional Council Technical Report 2012/13. www.waikatoregion.govt.nz/PageFiles/21715/TR201213.pdf.

WRC. 2012_1. Waikato Regional Economic Profile. Waikato Regional Council technical report 2012/37. http://www.waikatoregion.govt.nz/Services/Publications/Technical-Reports/TR-201237/.

WRC. 2012_2. Nitrogen Management in the Lake Taupo Catchment. http://www.waikatoregion.govt.nz/Community/Your-community/For- Farmers/Taupo/Nitrogen-management-in-the-Lake-Taupo-catchment/.

WRC. 2012_3. Reports on the Upper Waikato River Catchment. http://www.waikatoregion.govt.nz/Forms/Searches/Environmental-information-GCS1/?q=upper+waikato+river.

Yao, R.T. and Kaval, P. 2008. Valuing biodiversity enhancement in New Zealand. PhD paper, University of Waikato, Hamilton, New zealand.

CHAPTER

7

Water Trading with Multiple Water Sources: A Case Study in the Reno Basin, Italy

Zavalloni Matteo,[1,a,*] *Raggi Meri*[2] and *Viaggi Davide*[1,b]

Introduction

Background and objective

Challenged by increasing demand and by the modification in the structure of supply due to climate change, water resources worldwide have become an issue of primary importance in both the political and institutional realm (e.g., EC 2012).

Water markets (WMs) are one of the tools proposed to substitute or complement command and control policies, where and when water is a scarce resource, especially for the agricultural sector (Easter et al. 1999). Allowing for a flexible allocation mechanism, WMs enable water to

[1] Department of Agricultural Sciences, University of Bologna, viale Fanin 44, 40127 Bologna (It).
[a] Email: matteo.zavalloni3@unibo.it
[b] Email: davide.viaggi@unibo.it
[2] Department of Statistical Sciences, University of Bologna, via delle Belle Arti 44, 40126 Bologna (It).
Email: meri.raggi@unibo.it
* Corresponding author

follow the most profitable use, thus leading to an economically efficient management characterised by the equalization of the profitability of the resource among users (Chong and Sunding 2006; Easter et al. 1999). However, only a few countries have formally institutionalised the possibility of water trading, for example Australia (Tisdell 2011), Chile (Hearne and Easter 1997), USA (Howitt 1998), and Spain (Garrido 2000).

The source of the resources at stake can highly affect the flows of transactions within a given market. Connection to surface water entails different issues than the connection to groundwater with respect to the typology of the economic structure faced by farmers, the managerial setup, and the necessary institutional arrangements (Negri 1989; Wang 2011).

Moreover, a number of studies highlight the need for a conjunctive approach where surface water and groundwater are interconnected, and water transfers are taken into account. For instance, Knapp et al. (2003) find that involuntary cutbacks on surface water increase the pressure on groundwater, hence reducing the water tables. On the other hand, assuming surface water is more variable than groundwater, conjunctively managing the water resources enables addressing the "stabilization value" of the groundwater in reducing the total water variability (Gemma and Tsur 2007). In addition, Brennan (2008) analyses the inefficiency that is generated in case the spatial water transfers occur within a system that lacks clearly defined property rights on seasonally stored water. The same typology of inefficiencies can arise when there are interconnections between different water sources. In such cases groundwater resources can play a role as storage.

Since the institutional, managerial and technical environment in which WMs are introduced plays a major role in the subsequent functioning of the water trading mechanisms, via path dependence patterns, the type of water source present in a given area is an important element to take into account for the WM initial design (Harris 2011; Libecap 2011).

In Italy, the law does not envision any type of water trade among private users. However, given concerns related to climate change, relevant stakeholders are showing a tepid yet increasing interest for water trading mechanisms. An analysis of the potential effects of the institutionalization of WMs appears to have significant policy importance. The potential for WM in the Italian context, with a particular focus on different typologies of transaction costs, was studied by Pujol et al. (2006) in a case study in a southern region. However, the authors did not take into account the impact of the heterogeneity of water sources on the water trading flows.

In this paper, we focus on the implication of the heterogeneity of water sources on the functioning of potential formal WMs. Hence, we investigate how the attribution of water rights in an area where water sources are heterogeneous affects water trading flows. To investigate the issue, we develop a mathematical programming model which is applied to the Reno Basin area, located in the north of Italy, in the Emilia-Romagna region, where farms are either connected to groundwater, or surface water. In the model, the different water sources are then characterised by a different structure of the extraction cost. Moreover, we simulate the potential effect of involuntary cutbacks and local water scarcity that heterogeneously reduces water availability.

Case study description

The case study area is located in the Emilia-Romagna Region, between the province of Bologna and the province of Ravenna, where a local irrigation board, the "Consorzio Romagna Occidentale", manages irrigation water and hydrological issues. Two thousand five hundred and seventy (2570) farms are present in the case study area, for a total Utilized Agricultural Area (UAA) of 16940 ha. Despite the national legislation, relevant stakeholders indicate that forms of water exchanges based on barter are in place. The case study area has been chosen because of the coexistence of two types of water sources. A first group of farms are connected to a system of pressurised water coming through the "Canale Emiliano-Romagnolo" (Cer group, hereinafter), a canal that diverts water from the Po River. The remaining farmers rely mainly on groundwater resources (Ncer group, hereinafter).

While technically not a river, the structure of the problem faced by the Cer group farmers can be considered conceptually similar. The quantity of water itself is a minor issue given the volumes present in the Po River, but the timing of extraction in case of severe water scarcity might be managed on rotation basis that is under the administration of the local irrigation board. The managerial principle in such a case is usually a priority rule, under which permanent crops (mostly fruit trees) have the precedence over seasonal crops. Water costs are on a volume basis.

The Ncer group faces different issues, with respect to the previous group. Unlike the issues analysed in the "conjunctive management" literature, given the volumes of water in the Po River, water availability is characterised by greater variability than in the Cer group. To a certain extent, water extraction might be subject to inter-temporal trade-offs since the extraction rate might be higher than the refill rate. Given the difficulty of controlling the water volumes extracted, the costs are differentiated from the Cer group as they are set according to the size of irrigated land, and are hence considerably lower than those of the Cer group. On the other hand, farmers face pumping costs that are absent in the Cer group.

A cluster analysis was performed to identify noteworthy farm typologies (internally characterised by similar managerial and decisional behaviour) that were subsequently used in the mathematical programming model

(Gallerani 2009). The Ward method was employed to assess the significance of the discriminatory variables, namely land size and share of land allocated per crop (Ward Jr 1963). The clusters have been further validated through several interviews with local farmers and relevant stakeholders. Five farm typologies have been identified. The farm typologies do not take into consideration the sources of water, but they are further classified according to the connection to the source of water. Table 1 summarizes the main information for the 5 farm typologies.

Table 1. Cluster analysis results.

	cl1	cl2	cl3	cl4	cl5
Farm structure	*Hh* part time*	*Hh* intensive*	*Hh* professional*	*Hh* professional, with labourers*	*Market oriented with labourers*
arable crops (%)	63	19	40	48	70
vineyard (%)	32	24	20	13	6
Orchard (%)	0	53	36	34	16
Others (%)	1	0	1	2	3
average UAA (ha)	3.4	3.76	12.54	32.59	70.14
Cer group (nr.)	159	254	89	13	5
Ncer group (nr.)	432	1135	412	62	9

Source: (Gallerani 2009).
*Hh = household.

Model and Scenario Description

We develop a mathematical programming model to simulate the potential impact of the implementation of WMs in the area. The general structure of the model follows. The model maximizes the sum of the gross margins of the farms in the area (cf. Pujol et al. 2006):

$$Max \sum_j \sum_k \lambda_{jk} \left(\sum_i GM_{ijk} x_{ijk} \right) \tag{1}$$

s.t.

$$\sum_i x_{ijk} c_{iz} \leq v_{zj} \tag{2}$$

$$\sum_i x_{ijk} w_i \leq a_{jk} \tag{3}$$

with:

GM_{ijk} = gross margin for crop i on farm typology j connected to the k ϵ*(Cer, Ncer)* water source; assume for simplicity each group of farms can either be connected to surface water (Cer) or to groundwater (Ncer), with no overlaps;

x_{ijk} = land allocated to crop i by farm j; connected to water source k;

c_{iz} = coefficient for the technical constraints z;

v_{zj} = availability of resource for the constraint z;

w_i = water use for crop i;

a_{jk} = amount of water available for farm j connected to the water source k;

λ_{jk} = number of farms for the j typology connected to the water source k;

We develop 2 institutional scenarios: 1) "Benchmark", with equations (1) to (3), where the water trade is not allowed, and 2) "Water Trade" (WT, hereinafter) where the water availability is flexible given the possibility of water transfers among farmers. While equation (2) applies here too, equation (1) becomes:

$$Max\sum_j\sum_k \lambda_{jk}\left(\sum_i GM_{ijk}x_{ijk} + p^w w_{jk}^s - p^w\left(w_{jk}^{P,Cer} + w_{jk}^{P,Ncer}\right)\right) \quad (4)$$

Equation (4) represents the theoretical mathematical notation. In the operational model formulated in GAMS, equation (4) is coded as $Max\sum_j\sum_k \lambda_{jk}\left(\sum_i GM_{ijk}x_{ijk}\right)$, since p^w is endogenously chosen by the model, and shown by the marginal value of equation (6). However, p^w is introduced to compute the gross margins for the single farms.

Moreover, equation (3) becomes:

$$\sum_i x_{ijk}w_i \leq a_{jk} + w_{jk}^{P,Cer} + w_{jk}^{P,Ncer} - w_{jk}^S \quad (5)$$

$$\sum_j\sum_k\left(\lambda_{jk}w_{jk}^{P,Cer} + \lambda_j w_{jk}^{P,Ncer}\right) = \sum_j\sum_k \lambda_{jk}w_{jk}^S \quad (6)$$

$$x_{ij}, w_j^P, w_j^S \geq 0 \quad (7)$$

with

$w_{jk}^{P,Cer}, w_{jk}^{P,Ncer}, w_{jk}^S$ = water purchased (whose origin can be both Cer and Ncer) and sold by farm type j, connected to the source k;

p^w = price of water determined by the market.

The institutional scenarios are tested with a sensitivity analysis that changes the amount of water available (a_{jk}). In the "Full Capacity" the water availability per hectare is equal among the groups of farmers

$$a_{j,k \in Cer} = a_{j,k \in Ncer} \tag{8}$$

In the "Cer Water" scenario we have:

$$a_{j,k \in Cer} \geq 0 \text{ and } a_{j,k \in Ncer} = 0 \tag{9}$$

In the "Ncer Water" scenario we have

$$a_{j,k \in Cer} = 0 \text{ and } a_{j,k \in Ncer} \geq 0 \tag{10}$$

The different water availability levels simulate the impact of different seasonal climatic conditions and the possibility of involuntary cutbacks in surface water coming from the Po River. Table 2 summarizes the relevant equations for each scenario.

Given the hydrological and physical structure of the area, water transfers within the Ncer group, and in-between the two groups, occurs only by means of tanker truck. On the other hand, water transfers within the Cer group do not entail any physical movement of the resource, but only longer (shorter) periods of extraction for those who purchase (sell) water. These elements are translated in the simulation model by assuming relatively low transaction costs for water exchanges within the Cer group, and relatively high transaction costs (transportation costs) for water exchanges within the Ncer group and for *inter* group exchanges.

Moreover, the hydrological structure of the area is characterised by two different systems that are practically disconnected from each other. For instance, the water from the Cer group could be subject to voluntary cutbacks to divert water toward strategic sectors like tourism or industrial uses. For this reason we implemented a sensitivity analysis that takes into account the potential differentiation in the availability of the water between the two groups.

Table 2. Relevant equations for each scenario.

	Water availability		
Institutional setup	*Full Capacity*	*Cer Water*	*Ncer Water*
Benchmark	(1) to (3), (7)	(1) to (3), (8)	(1) to (3), (9)
Water Trade	(1), (2), (4) to (8)	(1), (2), (4) to (7), (9)	(1), (2), (4) to (7), (10)

Results

Water transfers

As we can observe in Fig. 1, the flows of water transfer for the "Full capacity" scenario first increase with the water availability and then decrease after a certain threshold, until there are no exchanges at a level of 1600 m^3/ha and beyond. Different patterns emerge in the "Cer Water" and "Ncer Water" scenarios. The "Ncer Water" flows are similar in the shape of the curve to the "Full capacity", but they stabilize at higher levels, since the Ncer group supplies the resource to the whole area. In the "Cer Water" scenario the water transfers increase with the amount of water available, reaching the highest volumes of water transfer of all scenarios stabilizing at even higher levels (not shown in the Figure). The two patterns between the "Cer Water" and "Ncer Water" are due to the differences in the relative size of the groups (the Cer group is smaller than the Ncer group, so demand for water is much higher in the "Cer Water" scenario than in the "Ncer Water" scenario).

The different patterns are clearer if we observe the directions of the water flows (Fig. 2). In the "Full Capacity" scenario, the presence of relatively high transportation costs subdivides the market in two submarkets where water transfers occur within each group. Most of the transfers occur within the Cer group where farmers face relatively low transaction costs. Not surprisingly, in the other two water availability scenarios, much of the transfers occur in between the groups.

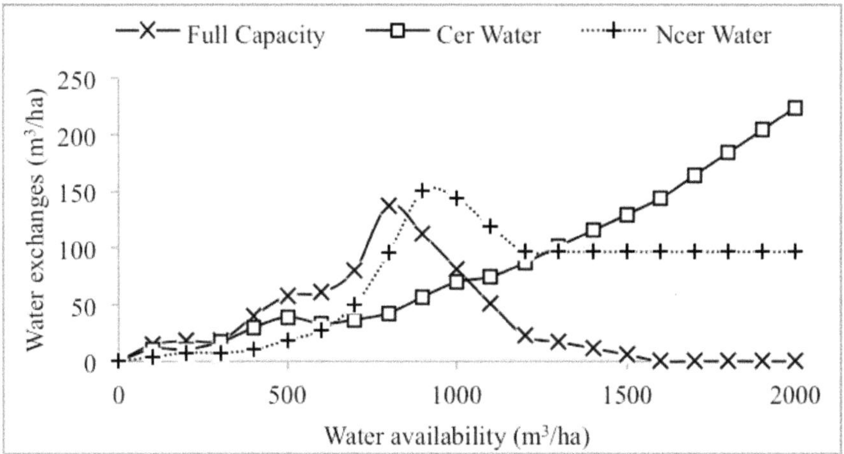

Figure 1. Water exchanges—volumes.

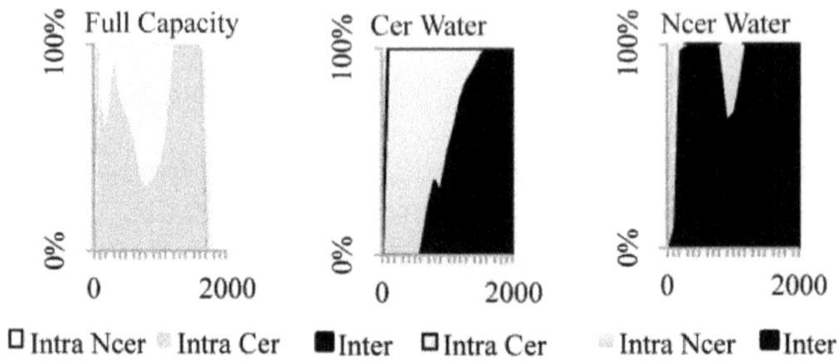

Figure 2. Water exchanges—direction of transfers.

Water consumption

Figure 3 shows the water extracted from the different sources, in the various scenarios. In general the WT scenarios show an increase in the amount of water that is utilized in total, with the highest percentage increases occurring in the surface water coming from the Po river (Cer). The involuntary cutbacks increase the pressure on the groundwater, increasing the extraction by a maximum of 30% with respect to the "Benchmark".

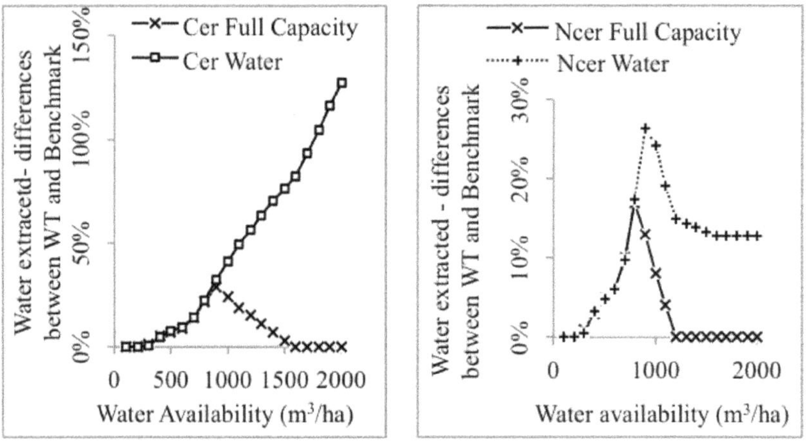

Figure 3. Water extracted—percentage difference between WT and Benchmark.

Irrigated land

The "Full Capacity" shows the highest share of irrigated land, followed by the "Ncer Water", and the "Cer Water", according to the number of farms in each group that have access to the water resource (Fig. 4). The possibility of water trading on average increases the share of irrigated land, up to 10% in the "Full Capacity", 7% in the "Ncer Water", and 6% in the "Cer

Water" scenarios (Fig. 5). For water availability levels of up to 400 m^3/ha, the institutionalization of WM leads to a decrease in the share of irrigated land with respect to the "Benchmark" scenarios due to a change in the crop choice toward crops that are more water demanding and more profitable (mostly kiwi).

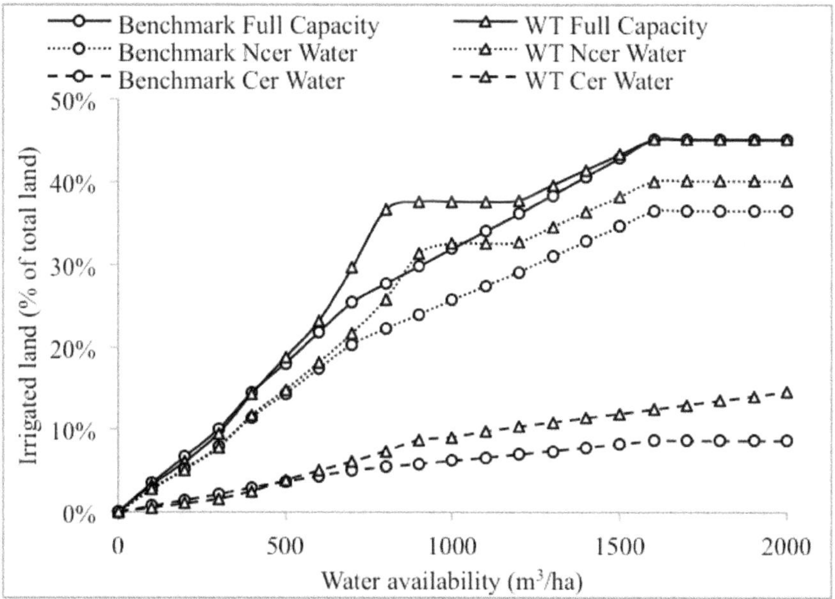

Figure 4. Share of irrigated land.

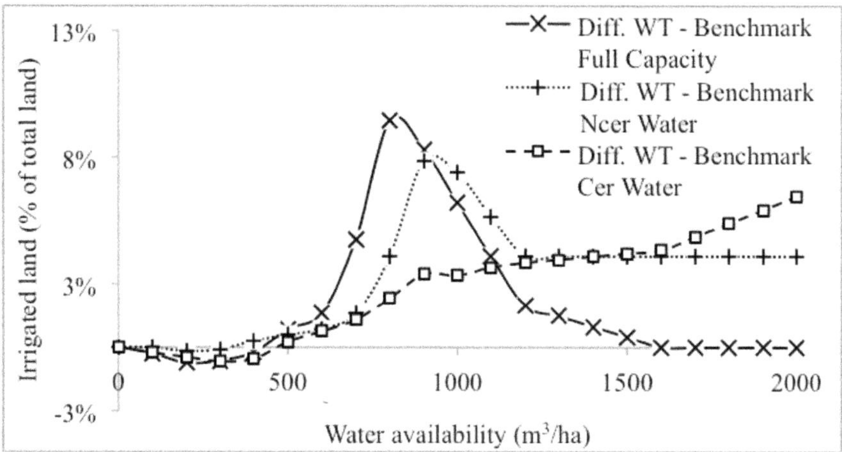

Figure 5. Share of irrigated land—differences among scenarios.

Gross margins

The gross margins increase with the amount of water available and then stabilize at a fixed level at around 1200 m³/ha for all scenarios (Fig. 6). In all cases, the WT scenarios increase the gross margins of the area (Fig. 7). In percentage terms, the patterns differ among the water availability scenarios. In the "Full Capacity", the gains from the water trade first increase with the water availability, then decrease after a level of 600 m³/ha. In the other two scenarios, the gains from the trade increase with larger volumes of water availability, and then stabilize. The highest percentage increase from the institutionalization of WM in the area occurs in the "Cer Water" scenario, where the Cer group supplies water to the whole area.

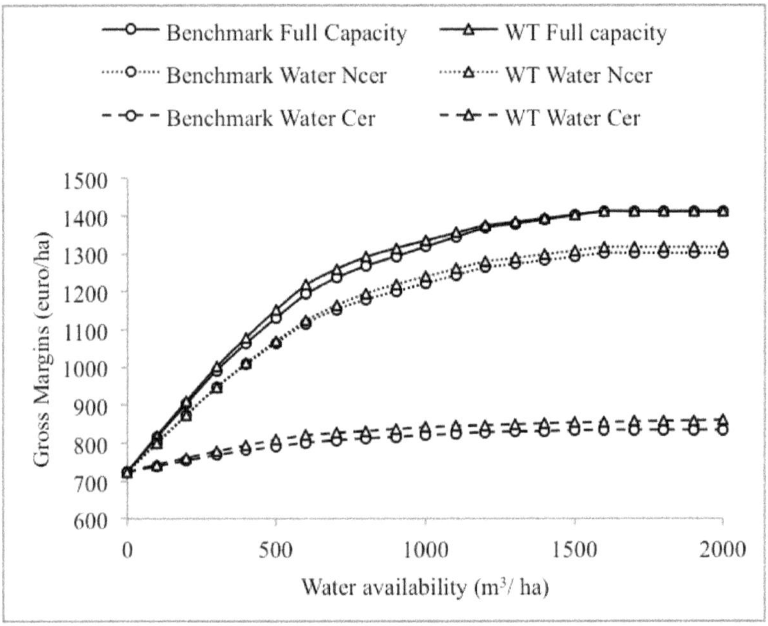

Figure 6. Gross margins.

Discussion

In this paper we analyse the potential effects of the institutionalization of water trading in the Reno Basin, in Emilia-Romagna, Italy. The water in the area is either extracted from groundwater resources, or from the Po River, channelled and then distributed by way of pipes. The characteristics of the area make it suitable to model the impact of different water sources on the potential implementation of WMs.

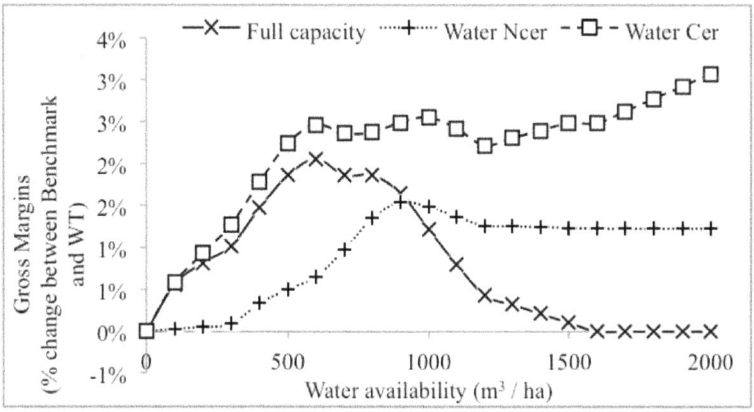

Figure 7. Gross margins—differences among scenarios.

The results show that the volumes of water traded in the area are lower than those found by Pujol et al. (2006) in the South of Italy, differences that might be due to the higher farm heterogeneity in the case study area of Pujol et al. (2006). In the case of involuntary cutbacks on the Po River surface water, the presence of water trading mechanisms might partially sustain the irrigated agriculture for the farmers connected to the Po by allowing groundwater to be used extensively in the whole area subject to cutbacks. A direct consequence of the transfers is increased pressure on the groundwater resources. Clearly, the opposite results occur when water availability is severely limited for the groundwater resources.

The WM could potentially improve the gross margins of the area: the gains are relatively low, in the order of 2%–4%, but they become relatively more important in cases of high heterogeneity of water available between the water sources. The water transfers are directed towards the most profitable and water demanding crops. The results regarding the gains from the water trade are in the same order of the estimates computed by Pujol et al. (2006).

Adding hydrological details and a temporal dimension would certainly improve the results of the analysis. First, with respect to the management of the groundwater resource, a dynamic analysis coupled with a better understating of the hydrology of the area would enable the assessment of the potential temporal trade-offs and "pumping race" triggering water trading effects. The higher pressure on groundwater resources due to the cutbacks in the surface water are likely to have major effects in the future that are assessable only if the temporal dimension is introduced (Knapp et al. 2003). Secondly, with respect to extraction from the Po river surface water, the main issue is the daily timing of the extraction: in case of conflicts, trading with the groundwater resource might smooth over temporary shortages.

Adding more precise crop-water production functions would also increase our understanding of the potential for water trade in the area.

Conclusion

Despite its simplicity, the model seems to find relatively clear patterns of differences, caused by the presence of heterogeneous water sources, in the effects of the institutionalization of water trading mechanisms. The results indicate that the presence of different water sources might increase the scope for a potential water market in the area, and in case of high heterogeneity of water availability such an institution would soften the effects of seasonal drought. However, further studies are necessary to assess if the institutionalization of water markets might require ancillary institutional arrangements to address the potential inter-temporal allocation inefficiencies and the environmental problems generated by the depletion of the groundwater resources.

Acknowledgements

The research from which this paper is derived is part of the project "Water Cap and Trade", funded by ISPRA—Istituto Superiore per la Protezione e la Ricerca Ambientale for the Italian partnership, in the context of the IWRM-net project "Towards a European exchange Network for integrating research efforts on Integrated Water Resources Management".

References

Brennan, D. 2008. Missing markets for storage and the potential economic cost of expanding the spatial scope of water trade. Australian Journal of Agricultural and Resource Economics 52: 471–485.

Chong, H. and Sunding, D. 2006. Water markets and trading. Annual Review of Environment and Resources 31: 239–264.

European Commission [EC]. 2012. A Blueprint to Safeguard Europe's Water Resources, COM (2012) 673 final, Bruxelles.

Easter, K.W., Rosegrant, M.W. and Dinar, A. 1999. Formal and informal markets for water: institutions, performance, and constraints. The World Bank Research Observer 14: 99–116.

Gallerani, L.V. 2009. Acqua e agricoltura in Italia. Valutazioni di scenari e strumenti di supporto alle decisioni. Franco Angeli, Milano, Italy.

Garrido, A. 2000. A mathematical programming model applied to the study of water markets within the Spanish agricultural sector. Annals of Operations Research 94: 105–123.

Gemma, M. and Tsur, Y. 2007. The stabilization value of groundwater and conjunctive water management under uncertainty. Applied Economic Perspectives and Policy 29: 540–548.

Harris, E. 2011. The impact of institutional path dependence on water market efficiency in Victoria, Australia. Water Resources Management 25: 4069–4080.

Hearne, R.R. and Easter, K.W. 1997. The economic and financial gains from water markets in Chile. Agricultural Economics 15: 187–199.
Howitt, R.E. 1998. Spot prices, option prices, and water markets: an analysis of emerging markets in California. pp. 119–140. *In*: Easter, K.W., Rosegrant, M.W. and Dinar, A. (eds.). Markets for Water. Kluwer Academic Publishers, Norwell, USA.
Knapp, K.C., Weinberg, M., Howitt, R. and Posnikoff, J.F. 2003. Water transfers, agriculture, and groundwater management: a dynamic economic analysis. Journal of Environmental Management 67: 291–301.
Libecap, G.D. 2011. Institutional path dependence in climate adaptation: Coman's "some unsettled problems of irrigation". American Economic Review 101: 64–80.
Negri, D.H. 1989. The common property aquifer as a differential game. Water Resources Research 25: 9–15.
Pujol, J., Raggi, M. and Viaggi, D. 2006. The potential impact of markets for irrigation water in Italy and Spain: a comparison of two study areas. Australian Journal of Agricultural and Resource Economics 50: 361–380.
Tisdell, J.G. 2011. Water markets in Australia: an experimental analysis of alternative market mechanisms. Australian Journal of Agricultural and Resource Economics 55: 500–517.
Wang, Y. 2011. Trading water along a river. Mathematical Social Sciences 61: 124–130.
Ward, J.H., Jr. 1963. Hierarchical grouping to optimize an objective function. Journal of the American Statistical Association 58: 236–244.

CHAPTER

8

Incentive Pricing for Irrigation Water with Information Asymmetries

Francesco Galioto,[1,*] *Meri Raggi*[2] and *Davide Viaggi*[3]

Introduction

In accordance with the European Water Framework Directive (WFD), the management of water resources should be delegated to appropriately established local bodies, referred to as River Basin Districts (Bogaert 2012). Basin Districts are responsible for the full cost of water usage in order to ensure the good hydrological status of the territory under their jurisdiction. Such costs should then be transmitted to end users through appropriate (higher) tariffs. The existence of economic tools, that enable local water

[1] Research Fellow in Agricultural Economics, at the University of Bologna. Email: francesco.galioto@unibo.it
[2] Researcher in Statistics, at the University of Bologna. Email: meri.raggi@unibo.it
[3] Associate Professor in Agricultural Economics, at the University of Bologna. Email: davide.viaggi@unibo.it
* Corresponding author

authorities to allocate costs among users and to control water uses in accordance with the WFD principles change considerably depending on several factors. Lack of metering is the main constraint. This condition hinders the ability to monitor volumes used and to promote volumetric pricing as a way to allocate costs and to ensure efficient water use (Viaggi et al. 2010; Johansson et al. 2002; Smith and Tsur 1997; Bowen and Young 1986). This issue is a challenge especially for the agricultural sector since the most common delivery system for irrigation water worldwide is open canals (Bogaert 2012). Here, the presence of a heterogeneous population of farmers poses implicitly the two typical problems of asymmetric information: a) adverse selection, due to the non-observability of a farm's type and its technology; b) moral hazard, due to the non-observability of a farm's actions, such as the actual use of irrigation water (Viaggi et al. 2011).

In the current regulatory framework, the implementation of the Full Cost Recovery principle (FCR) for those water authorities that are unable to control uses may exacerbate the lack of consistency with two fundamental pricing principles (article 9 comma 1, WFD): the Incentive Pricing Principle (IPP) and the Polluter Pay Principle (PPP).

In light of these principles, the study compares the discriminatory pricing strategies of an Italian water authority with a per area basis tariff (no discrimination) and with an alternative discriminatory pricing strategy that faces both the issues of adverse selection and moral hazard. The aim of the study is to analyse the relevant impact on users' benefits of both the current case study discriminatory strategies and the alternative discriminatory strategy with respect to no discrimination. This allows to criticise water tariffs as an instrument aimed at incentivising efficient water uses under conditions of non-metering. Thus, the study challenges the practicability of a tariff design that is able to deal with biases in information between the regulator and its beneficiaries.

The chapter is organised into four sections. Section II offers the literature background from which the present study draws inspiration, briefly describing the most interesting insights on the subject. Section III describes the case study's institutional framework and its pricing policies and outlines the structure of the optional tariffs. Section IV sets forth the methodology. The section is divided into three paragraphs: *Optional Tariff, Differential Tariff, Monitoring*. All paragraphs examine both public and private perspectives. The first step, presented in the first paragraph, is an analysis of farmers' behaviour in light of the water authority's set of tariff options. The paragraph highlights the identification of a threshold ratio between the two tariffs that guide farmers' preferences. In a second step, the regulator's objective function is described. In this case, the tariffs are endogenised into the public problem facing the WFD goals by introducing the Full Cost

Recovery constraint and comparing the regulator's discriminatory strategy with no discrimination. The second paragraph highlights the possibility to improve the actual tariff design by identifying tariff schemes for each farm typology. The third paragraph questions the regulator's ability to check for uses by analysing its monitoring strategies. Finally, Section V offers a numerical example based on available information collected in the RIB's area. The organisation of this section is analogous to that of the previous ones. A conclusion is provided in Section VI. The main concluding argument is that information asymmetries udermine the service provider ability to discriminate users. However, incentive theory, according to the heterogeneity in water uses, could significantly contribute to improve most of the current pricing strategies enforced by local water authorities under conditions of non-metering.

Literature Background

Smith and Tsur (1997) are the first authors to deal with the issue of asymmetric information for irrigation water in the absence of metering. They argue that under such circumstances, the need to identify a strategy in order to induce farmers to an optimal use of the resource arises due to the fact that part of the delivery cost and part of the externalities associated to water uses are borne by parties that are not direct users. This raises the need to properly design pricing schemes aimed at inducing farmers to make efficient use of irrigation water.

The issue of moral hazard has been directly tackled by Smith and Tsur (1997) and Zilberman (1997). The first authors argued that a tariff linked to output could reach conditions of 'first best' assuming zero transaction costs. Usually, information on output is costly for the regulator. As a consequence, transaction costs are higher than zero. Thus, the practicability of this tariff design rests on the regulator's ability to check for output levels.

In a wider perspective, Zilberman (1997) develops an optimal water pricing, allocation, and conveyance system over space to capture different upstream and downstream incentives which positively influence the level of compliance.

Finally, the last instrument (not yet discussed in literature) that enables indirect checking of water uses is satellite monitoring, and is suggested by the European Commission in order to detect illegal draws (EC 2012).

The issue of adverse selection has been tackled by Gallerani et al. (2005) and Viaggi et al. (2010). In the first study, farm characteristics, such as farm size and the type of crops, allow to discriminate farmers in clusters of producers that significantly differ from each other. This justifies the authors' attempt to develop a model that enables the regulator to differentiate tariffs for irrigation water by inducing users to reveal their tariff attitudes according to the farm typology to which they belong. The second study further improves on the former one, extending a model designed for two farm typologies only to a wider number of farm typologies.

The present chapter seeks to further develop this issue by integrating one of the most sophisticated discrimination strategies adopted by Italian water authorities, the *Tariff Option*, with tariff differentiation. In addition, the issue of moral hazard is tackled by identifying the monitoring cost threshold above which it is no longer convenient to enforce the incentive tariff. Moreover, the current allocation of monitoring costs among users is compared with an ideal design. Thus, the study puts into question the ongoing case study pricing schemes in order to verify under which conditions the ideal design guarantees the feasibility of more efficient strategies.

The Case Study

In Italy, water is usually delivered by open canals, particularly in the north. Under such conditions, local water authorities mostly enforce a two-part tariff characterised by a fixed component that each user pays for reclamation services and a variable component targeted to users that exploit water for private purposes. Pricing strategies arising from the use of irrigation water are currently designed autonomously by water authorities and vary with respect to the typology of the distribution system. However, most of the pricing schemes imposed on local users for irrigation are characterised by a per hectare tariff regardless of whether or not farmland is irrigated. There are also infrequent cases in which tariffs are enforced according to the irrigated farmland and even distinguished by the irrigation system and the type of crops (INEA 2011; Bazzani et al. 2004).

Italy counts almost 500 local irrigation water authorities, 30% of which are represented by water user associations (INEA 2011). In this case, irrigation water should be perceived as a common pool resource and property rights should be established in order to dissipate rent and efficiency gains unless the water is entrusted to residual claimants. Theoretically, water user associations are seen as endogenous institutions that are able to generate higher benefits with respect to what is otherwise the case of private control. These organisations, acting in the interest of water users, tend to substantially reduce the costs of implementing water pricing, such as monitoring and enforcement costs (Johansson et al. 2002; Schlagler and Ostrom 1992). The most relevant worldwide water user associations suggested in the literature are: the *warabandi system* in Pakistan and India (Bandaragoda 1998); the *guanajuatos unidades* in Mexico (Dayton-Johnson 2000); and the *reclamation and irrigation boards* (RIBs) in Italy (Coletta 2010; Viaggi et al. 2010).

The Italian institutional arrangement is an association of persons who own properties (land and buildings) in its jurisdiction. Land owners, according to the subsidiarity principle, contribute directly to the pursuit of the general interest. The case study considered in this paper is the RIB of Western Romagna (Consorzio di Bonifica della Romagna Occidentale—CBRO). The consortium covers an area of 195,000 ha of which 70,000 ha are plains. Fruit and wine grapes are the main crops grown in the area. Therefore, irrigation is both frequent and abundant. The study area is particularly attractive for the investigation of different delivery systems: part of it is served by pressure pipes and part is delivered by open canals. For each kind of delivery system at work in the study area, the RIB enforces a tariff both for reclamation and irrigation. For reclamation, the tariff is proportional to the landholding of each user, while for irrigation the tariff varies with the type of delivery system. Under metering conditions (farmers served by pressure pipes) the tariff for irrigation is clearly distinct from the ones intended for reclamation and it is proportional to water uses. In the absence of metering (farmers delivered by surface water), however, there is no connection between tariffs and water uses and tariffs for irrigation are often not distinguished from reclamation tariffs.

In a sub-area served by surface water, farmers are allowed to choose between a fixed tariff per hectare of harvested area and a tariff proportional to the irrigated farmland. We define this scheme as the *Optional Tariff*. This tariff scheme is implemented for the sole purpose of ensuring an equitable allocation of costs among beneficiaries since the share of irrigated crops tends to decrease with increasing farm size. Under such circumstances, a tariff for irrigation purposes proportional to the total harvested area would be too unbalanced for large farms.

However, the *optional tariff* does not guarantee to achieve compliance with the statutory principle of sharing water use costs among users on the basis of the degree of benefit because of its limited diffusion. This is due to the fact that the incentive tariff requires the farmer to self-report the quota of irrigated farmland, obliging the RIB to run controls in order to check for compliance. Moreover, costs arising from direct monitoring are charged via the water tariffs themselves on a per hectare value that is not explicit and which is shared by each user. As a result, a portion of the tariff paid by each farmer is due to other users' tariff attitudes. The dilution of monitoring costs among all users fosters the adoption of the incentive tariff to the detriment of some farms. This condition could lead to a *non-pareto* efficient allocation of monitoring costs among users as cost sharing could result in higher benefits for some farmers and greater decreases in benefits for others. By charging monitoring costs only to those users who comply with the incentive tariff, the regulator avoids this risk but could also fail to foster the adoption of the incentive tariff.

Methodology

On behalf of the local community of producers, the RIB is charged with controlling water uses and consequently seeks to share the costs associated with irrigation water according to the degree of benefit of each user. In compliance with the WFD, the methodology aims to identify and incentivise tariff schemes that enable the regulator to control water uses under conditions of non-metering, minimising the impact of water tariffs on profits. This is carried out by analysing benefit scenarios with different tariff schemes: *Optional Tariff, Differential Tariff, Monitoring*. In the remainder of this section, the description of each of the listed aspects is provided as well as the relevant theoretic implications.

Optional Tariff

The *optional tariff* is a third order discrimination strategy (Carlton and Perloff 1994). By allowing farmers to choose which tariff scheme to comply with, the regulator is able to recognise if the quota of irrigated farmland is higher or lower than a specific threshold given by the ratio between the unit value of the two tariff options. The first option is a per area based tariff, proportional to the total harvested area, characterised by a fixed per hectare value, T^a.[1] We define this pricing scheme as a *'no incentive tariff'* since it does not affect water uses. The second option, unlike the first one, is based on a tariff proportional to the irrigated farmland, such that $T^b(x^{irr}{}_i) = T^b * x_i^{irr}$, where x_i^{irr} is the quota of irrigated farmland. We define this pricing scheme as the *'incentive tariff'* since it affects water use. Here, farmers are required to declare the size of the irrigated area.

The first scenario follows the paper of Galioto et al. (2012). Here, assuming a profit maximising behaviour by the farmer, for each tariff option the farmer would consider the choice of the maximum profit between the ones allowed by the two tariffs described above:

$$\prod_i^k (x_i^{irr*}) = \max[\pi_i(x_i^{irr_a}) - T^a,\ \pi_i(x_i^{irr_b}) - T^b(x_i^{irr_b})] \tag{1}$$

$\prod_i^k (x_i^{irr*})$ corresponds to the higher value of the net profits among the two options; $x_i^{irr_a}, x_i^{irr_b}$ are respectively the quota of irrigated area optimising profit under each tariff option (resulting from the relevant first derivative of the difference between profits and the optional tariffs); accordingly, x_i^{irr*} is the optimal share of irrigated farmland resulting from the higher value of the net profit achievable among the two options.

---

[1] In fact this tariff mechanism is much more complex but doesn't add useful information for our purposes. For further detail see: Galioto et al. (2012).

From equation 1 we identify a tariff option threshold (Tr_i) that is compared with the optimal share of irrigated farmland resulting from the choice of the incentive tariffs ($x_i^{irr_b}$) plus the ratio between the relative gross profit differences and the incentive tariff ($[\pi_i(x_i^{irr_b}) - \pi_i(x_i^{irr_a})]/T^b$), the value of which increases the higher the difference in profit from irrigation water use. The value of this last item is always positive as $x_i^{irr_b} \le x_i^{irr_a}$.[2] Analytically, the threshold is given by the following:

$$\begin{array}{ll} \textit{if} & \pi_i(x_i^{irr_a}) - T^a(x_i) \ge \pi_i(x_i^{irr_b}) - T^b * x_i^{irr_b} \\ \textit{then} & T^a(x_i)/T^b \le x_i^{irr_b} + [\pi_i(x_i^{irr_a}) - \pi_i(x_i^{irr_b})]/T^b \quad \textit{and} \quad x_i^{irr*} = x_i^{irr_a} \\ \textit{else} & T^a(x_i)/T^b \ge x_i^{irr_b} + [\pi_i(x_i^{irr_a}) - \pi_i(x_i^{irr_b})]/T^b \quad \textit{and} \quad x_i^{irr*} = x_i^{irr_b} \end{array} \tag{2}$$

More generally, the threshold can be expressed as follows:

$$Tr_i = x_i^{irr_b} + [\pi_i(x_i^{irr_a}) - \pi_i(x_i^{irr_b})]/T^b \tag{3}$$

In brief, equation 3 demonstrates that the threshold is conditioned by both the public pricing policies and the intrinsic characteristics of each farm typology. Both the relative values of the price schemes ($T_a(x_i)/T_b$) and the absolute values of the incentive tariff (T^b) play a key role in conditioning private choices with different results depending on how farm typologies differ.

Assuming that the regulator is acting in compliance with the FCR principle, the cost of water use (*v, in* €/m3) does not depend solely on the current operating costs but is rather tied to other costs, including environmental and opportunity costs. Under conditions of scarcity, water quantity has a dominant role in conditioning the total cost of use. Here, we assume that the unit cost of water use (*v*) is proportional to irrigation water

---

[2] Indeed, by deriving the differences between the gross profit function and the relative tariffs with respect to the irrigated area for each farm type, the optimal solution is reached at the equality between marginal profits and marginal tariffs:

$$\partial \Pi_i^k / \partial x_i^{irr} = \pi_i'(x_i^{irr}) - T_i^{k'}(x_i^{irr}) = 0$$
$$\textit{if} \quad k = a,\ \pi_i'(x_i^{irr}) = 0; \quad \textit{if} \quad k = b,\ \pi_i'(x_i^{irr}) = T^b$$

The value of *k(a,b)* reflects the tariff choice subject to the optimal share of irrigated area (x_i^{irr*}). The first tariff scheme (T^a), being disconnected from the irrigated farmland, does not affect water consumption ($\partial T^a / \partial x_i^{irr} = 0$). In contrast, assuming a positive correlation between water consumption and irrigated areas, $T^b(x_i^{irr})$ is able to influence water uses ($\partial T^b(x_i^{irr}) / \partial x_i^{irr} = T^b$). Consequently, given the technology, producers subject to the first tariff scheme will maximise profits at higher shares of irrigated areas than would farmers who choose the other tariff. Then, from the first order condition of the above problem we achieve the farmland share at which producers direct their own choices, which in turn is a function of the marginal tariffs ($x_i^{irr*} = f(T_i^{k'})$ $\ with\ f'(T_i^{k'}) < 0$).

use, but the amount of water use per hectare of irrigated farmland varies across farm typologies. Hence, the total water use cost ($CW_i(x_i^{irr*})$) is directly connected to water use ($W_i(x_i^{irr*})$) that differ as both farm typologies (i) and irrigated farmland (x_i^{irr*}) differ, and such that: $CW_i(x_i^{irr*}) = v * W_i(x_i^{irr*})$.

The relationship between the regulator and its beneficiaries is addressed in a different perspective than what has been presented in recent papers (Galioto et al. 2012; Viaggi et al. 2010). In these studies, it was assumed that the regulator acts in the interests of the community by inducing farmers toward a rational use of the resource with the objective of maximising the differences between profits and water use costs. Here, the regulator is assumed to maximise the differences between profits and tariffs. Such an assumption is due to the fact that acting on behalf of its beneficiaries rather than the community, the RIB should seek to identify tariff schemes with a lower impact on farmers' profits. In this respect, according to the level of water use costs, the regulator adjusts the tariff value of both options conditioning producer behaviour in order to maximise the overall benefits.

Formalising the problem, we have:

$$\begin{aligned}
&\max && Z(x_i^{irr*}) = \sum_{i=1}^{n}[\gamma_i * \Pi_i^k(x_i^{irr*})] \\
&s.t.: && \\
&FCR: && T_i^{k*} \geq CW_i(x_i^{irr*}) \\
&PC: && \Pi_i^k(x_i^{irr*}) \geq 0 \\
&where, && T_i^{k*} = [T^a,\ T^b(x_i^{irr_b})],\ x_i^{irr*} = f(T_i^k),\quad T^a, T^b \geq 0
\end{aligned} \tag{4}$$

$Z(x^{irr*})$ represents the net social benefit; γ_i is the share of estimated land allocated to each farm typology; $\Pi_i^k(x_i^{irr*})$ is the net profit function resulting from the maximisation of the private problem; T_i^{k*} is the tariff choice resulting from the higher value of the net profit among the two options given in the private problem. $x_i^{irr*} = f(T_i^{k'})$ derives from the private problem and is the reaction function of the irrigated farmland for each farm typology that varies as the RIB's pricing policies change. An additional condition is $x_i^{irr*} \geq 0$.

FC is the constraint of full cost recovery for water use. It requires tariffs that are able to offset the cost of water use for each type of farm.

PC is the participation constraint. It prevents the application of a fee for water use exceeding farmers' profits.

Currently, as previously stated, the RIB ascribes only a part of the cost recovery to the recipients. The FCR constraint is due to the assumption that the regulator acts in compliance with the WFD. Hence, costs are indirectly shifted to producers by means of the tariff.

The awareness of farm typologies should allow the RIB to identify the optimal mix of tariff levels that minimise the impact of water use costs on beneficiaries, more or less justifying the optional strategy up to the application of the single *incentive* or *non incentive tariff*.

Differential Tariff

The ability to distinguish different kinds of producers should allow the regulator to differentiate tariffs according to users' characteristics, both for the *no incentive tariff* as for the *incentive tariff*. Under such circumstances the *optional tariff* will result as follows: $T_i^{k*} = [T_i^a,\ T_i^b(x_i^{irr_b})]$. Under perfect information, in order to minimise the impact on farmer profits, the regulator would induce each type of user to adhere to the incentive tariff. Consequently, the optional scheme would lose its *raison d'être*.

In fact, the RIB is unable to recognise to which farm typology each producer belongs. A producer could exploit this information asymmetry by choosing a tariff scheme intended for other types of producers, if less onerous than the tariff scheme intended for him, hence undermining the regulator's attempt to discriminate between users. In order to avoid adverse selection and thus to reach an efficient discrimination of users by way of the tariff, the regulator must enforce higher tariffs to less intensive water users than what would be expected under conditions of perfect information. This difference is an information rent that those users pay to the regulator in order to guarantee discrimination. The value of this information rent tends to increase with increasing convergence in water use attitudes between users belonging to different farm typologies justifying or not the adoption of differential strategies instead of the actual ones.

Formalising the problem of the regulator, a new constraint is added in equation 4:

$$IC:\quad \pi_i(x_i^{irr*}) - T_i^k \geq \pi_i(x_j^{irr*}) - T_j^k \quad \forall\ k \in \{a,b\},\ \ i \in n,\ \ j \in n-i \tag{5}$$

IC is the incentive constraint required to avoid the risk of adverse selection. As it gets binding this constraint allows for a second best solution, resulting in higher tariffs for some farmers than in the case without adverse selection. This rent extraction undermines the overall benefit given by the maximisation of the differences between profits and tariffs. Differential strategies are practicable if the payment of this contribution leads to higher levels of benefit than is the case for the actual pricing scheme.

The current tariff scheme does not consider farm heterogeneity. However, currently the RIB has the ability to offer contracts that induce farmers to qualify themselves. This would theoretically allow the regulator to classify users in different farm typologies on the basis of their intrinsic characteristics, enabling the differentiation of tariffs according to differences in water use attitudes. Here, the risk of adverse selection could be prevented by means of incentives that would induce users to reveal to which farm typology they belong, hence fostering the adoption of more consistent pricing schemes.

Monitoring

The problem of moral hazard is at question, given the need to control those water users opting for the incentive tariff. Indeed, the incentive tariff requires users to self-report the quota of irrigated farmland and the regulator to verify if users comply with their statements. This is done by means of monitoring activities. The theory of incentives allows for the investigation of the regulator's ability to identify monitoring strategies by way of indirect signals that positively influence the level of compliance with minimum costs (Laffont and Martimor 2002). Here, the regulator's monitoring strategy adopted in order to check for compliance, namely, direct monitoring, is not being questioned, but rather the way in which monitoring costs are allocated among users. The cost of direct monitoring and the way to share it among users conditions the application of the incentive tariff, threatening the optional strategy adopted by the RIB.

Monitoring activities, linked to the incentive tariff, limit the risk of false reporting. This results in added costs for water users. The need to bear monitoring costs in order to avoid the risk of moral hazard requires identifying an efficient way of sharing costs among farmers. Usually, the regulator should charge costs to those farmers that choose the incentive tariff. At present, contrary to what would be expected, the regulator charges monitoring costs to each farmer, regardless of the corresponding tariff attitudes. Both monitoring cost allocation scenarios are analysed in order to verify under what conditions the actual policy is more or less efficient than the expected one.

Hence, the regulator could only charge users whose behaviour generates the requirement of monitoring activities (*m1*) or allocate monitoring costs among all users (*m2*), independently of their tariff preferences. In the first case, equation 5 would change as follows:

$$\Pi_i^k(x_i^{irr*}) = \max[\pi_i(x_i^{irr_a}) - T_i^a, \quad \pi_i(x_i^{irr_b}) - (T_i^b(x_i^{irr_b}) + m)] \tag{7}$$

Unlike equation 5, a new parameter, m, affects the producer's decision making process resulting in growth in the threshold given in equation 3:

$$Tr_i = x_i^{irr_b} + [\pi_i(x_i^{irr_a}) - \pi_i(x_i^{irr_b})] / T^b + m / T^b \tag{8}$$

As the regulator chooses to recover monitoring costs by charging them to all of the beneficiaries, independently of their tariff attitudes, assuming perfect information among farmers, each producer would agree to a specific tariff scheme given other producers' preferences:

$$\Pi_i^k(x_i^{irr*}) = \max[\pi_i(x_i^{irr_a}) - (T_i^a + \sum_{j=1}^{n} \beta_j),\ \ \pi_i(x_i^{irr_b}) - (T_i^b(x_i^{irr_b}) + \gamma_i * m + \sum_{j=1}^{n} \beta_j)]$$
$$\text{with} \quad T_i^{k*} = [T_i^a, T_i^b(x_i^{irr_b})], \quad \beta_j = 0 \ \ \text{if}\ T_j^{k*} = T_j^a, \ \ \beta_j = \gamma_j * m \ \ \text{if}\ T_j^{k*} = T_j^b(x_j^{irr_b}) \tag{9}$$

γ_i is the quota of the total harvested area under the jurisdiction of the RIB owned by farm typology i; β_j is the quota of monitoring costs that producers i should pay given the choices of other producers. Accordingly, the threshold is given by the following equation:

$$Tr_i = x_i^{irr_b} + [\pi_i(x_i^{irr_a}) - \pi_i(x_i^{irr_b})]/T^b + \gamma_i * m/T^b \tag{10}$$

The threshold is lower than the one in the previous problem (equation 8) enabling the adoption of the incentive tariff for a greater number of farmers at the expense of the no-incentive tariff adopters.

Results

The methodology has been implemented on a recent estimation of the profit and water use function of various farm typologies in the RIB's region (Viaggi et al. 2010). In order to facilitate the interpretation of the results, only two of the five farm typologies identified in the cited study are presented. In fact, farmers belonging to those typologies are the ones allowed to choose which tariff scheme to comply with as both of them have an average farm size greater than 15 ha. The first typology represents 3% of the RIB's farms and is characterised by an average farmland area of 33.85 ha, 37% of which is covered by orchards; the second represents 1% of RIB's farms and is characterised by an average farmland area of 75.07 ha, 16% of which covered by orchards. The two farm typologies cover 21% of the area managed by the RIB. The values of the results are contingent on the characteristics of each typology and the relative representativeness. However, tendencies with respect to differences in values between the ideal design and the current one go beyond the case study contingency.

Optional Tariff

In this section, by analysing producer behaviour in light of the optional tariff scheme adopted by the RIB, a comparative scenario is designed to verify the practicability of the optional tariff instead of the application of either the *no incentive* or the *incentive tariff*, widely adopted in numerous European contexts.

The two types of marginal profit functions with respect to the irrigated farmland (respectively $\pi_i^{'}(x_i^{irr*})$ and $\pi_j^{'}(x_j^{irr*})$), that matches with the relative demand function of irrigated farmland, are represented by the skewed lines in Fig. 1. Here, the horizontal lines represent the marginal function of the *incentive tariff* (T^b) and the value of the *non incentive tariff* (T^a) imposed by the RIB. The slope of the profit functions is conditioned by the crops grown by each farm typology. The dotted vertical lines represent the tariff thresholds obtained in equation 3 for each type (respectively Tr_i and Tr_j) and the ratio between the unit values of the *non incentive tariff* and the *incentive tariff* (T_a/T_b). The dotted lines cross the x-axis to a point given by the sum of the optimal quota of irrigated farmland and the incentive tariff (x_i^{irr*}) plus the ratio between the gross profit differences of the tariff options, that correspond to the triangular area AGH, and the level of T^b, that correspond to the respective heights. Thus the threshold can be rewritten as follows: $Tr_i = x_i^{irr_b} + (x_i^{irr_a} - x_i^{irr_b})/2$.

The tariff ratio given in Fig. 1 is over the j_{th} farm typology threshold (b) and below the i_{th} threshold (a). Hence, farmers of type i will opt for the incentive tariff, as the DEFG area is higher than the ABCD area, while farmers of type j will choose the non incentive tariff, as the DEFG area is lower than the ABCD area.

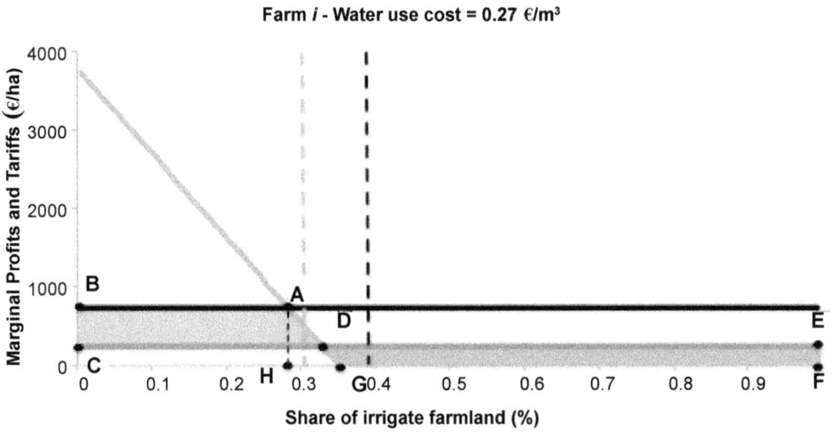

Figure 1. Tariff attitude of the two farm typologies given a specific water use cost.

As a result, only in the i_{th} typology does the irrigated farmland decrease by a quota corresponding to the difference between the intersection of the marginal profit function with the x-axis and the x-axis projection of the intersection between the marginal profit function and the incentive tariff level.

The private problem helps to understand how the public decision-maker should direct its pricing strategies in light of the new challenges imposed by the WFD, which, *de facto*, decentralise management functions for water resources by further empowering local authorities.

Figure 2 compares the average benefits of the current regulator tariff scheme, the *Optional Tariff*, with the benefit that derives respectively from the enforcement of the *incentive tariff* and the *no incentive tariff* for increasing water use cost levels (square dots and triangles dots in Fig. 2). Thus, the optional tariff, in which farmers are allowed to chose which tariff scheme to comply with, guarantees greater or equal benefits to the sole implementation of the two alternatives as the two lines are always lower or equal to 0.

For low levels of water use costs (from 0.00 €/m³ to 0.05 €/m³) the differences between the optional tariff and the no-incentive tariff is very small (see square dots in Fig. 2). Benefit differences between the two alternatives tend to grow at higher levels of water use costs because of increasing differences in surplus losses due to the presence of incentive

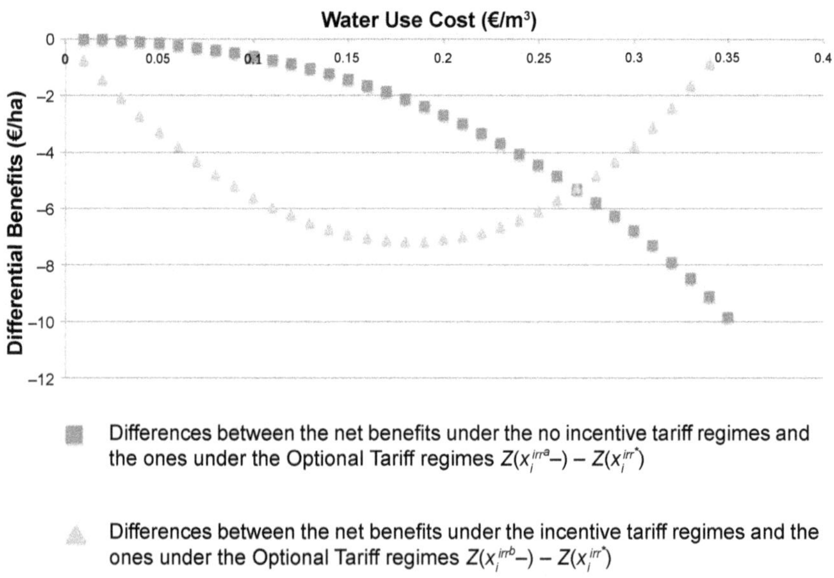

Figure 2. Variation in net benefits under both the no-incentive and the incentive tariff with respect to the optional tariff for increasing water use cost levels.

tariff adopters within the optional scheme. Here, only one of the two farm typologies is allowed to choose the *incentive tariff* up to a level of water use costs over which all farmers are allowed to join the incentive tariff scheme. From this point onwards (0.35 €/m^3) the optional tariff scheme matches with the incentive ones and the difference is equal to 0 (see triangles dots on Fig. 2).

From 0.05 €/m^3 to 0.35 €/m^3 both the no-incentive tariff and the incentive tariff allow, on average, to lower the level of benefits with respect to the optional tariff. This is mainly due to the fact that the optional tariff guarantees a partial discrimination among users, allowing the regulator to design tariffs that are closer to the real water use attitude of its beneficiaries. With no discrimination, one of the two types is forced to pay a tariff that exceeds the corresponding water use cost level (in accordance with the FCR constraint in equation 4). Thus, in between a given range of water use cost levels, whose entity increases with increasing farm heterogeneity, the optional tariff enables some farmers to choose the incentive tariff whilst others choose the no-incentive tariff, hence guaranteeing higher benefits than what would exist in the case of no discrimination. This is true independently if the optional tariff is compared to the incentive tariff or to the no-incentive tariff.

Differential Tariffs

The fixed Optional Tariff scheme discussed above reflects the current tariff policies enforced within the case study area. Here, assuming that the regulator knows the characteristics of each farm typology that falls under his jurisdiction, it would be possible to enforce different tariff options for each type. Thus, each farmer would contribute at a tariff value closer to the actual cost of irrigation water, hence minimising the impact of water use costs on the tariff. Under such conditions, whatever the level of the water use cost, all farmers are allowed to choose the incentive tariff since it is more efficient (lower surplus loss) than the alternative options. However, in order to avoid the risk of adverse selection, the regulator should design a menu of contracts that incentivises each farmer to choose the tariff option designed for the typology to which he belongs. This incentive results in additional costs for some producers whilst avoiding the risk that some other farmers opt for the tariff scheme intended for them. This additional cost is the "price" that some producers should pay in order to guarantee user discrimination.

Figure 3 shows that in order to avoid the risk of adverse selection, in some circumstances (when the incentive constraint is binding) the regulator must charge higher tariffs to one of the two farm typologies with respect to what would happen under conditions of perfect information.

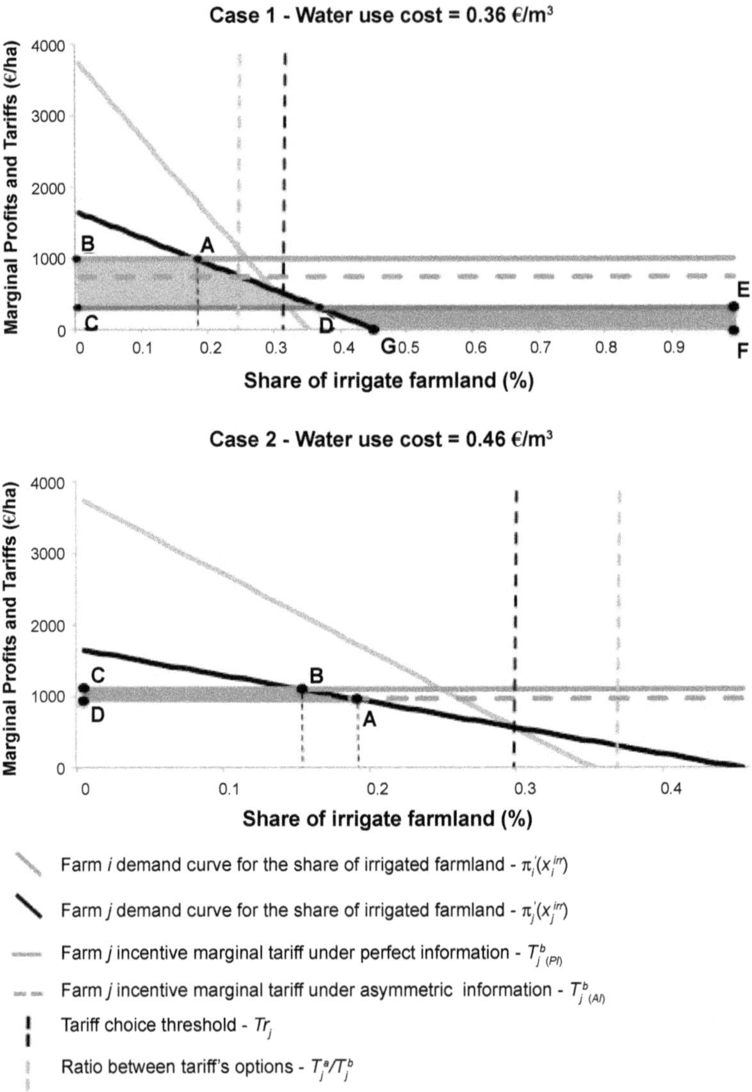

Figure 3. Tariff variation of the two-farm typology given the incentive constraint for the differential tariff strategy.

This additional cost is minimised thanks to the regulator's ability to influence farmers' tariff attitudes by inducing them to shift from one tariff option to the other. An example of this is given in Fig. 3, case 1, where the level of rent extraction is higher than the differences between the net profit under the no-incentive tariff scheme and the net profit under the incentive tariff scheme. This allows the regulator to adjust tariff option levels in order to induce farmer j to choose the no-incentive tariff. In order to avoid the risk of adverse selection, the regulator increases farmer j's incentive tariff to a level equal to the one reached under non-users discrimination. This condition does not nullify the regulator's attempt to differentiate tariffs thanks to the discriminatory power of the optional tariff.

The dotted horizontal lines in Fig. 3, case 1, represent the incentive tariff level that farmer *j* should correspond to under conditions of perfect information. The sloped lines represent the irrigated farmland demand functions for both farms.

In order to avoid the risk of adverse selection, the regulator adjusts tariff option levels by lowering the ratio (dotted vertical grey line) under the tariff threshold (dotted vertical black line). Accordingly the ABCD area is greater than the area DEFG. This means that the reduction in net profit is greater under the incentive tariff regimes with respect to the no-incentive tariff regimes. In these circumstances, farmer *j* is allowed to switch from the incentive tariff to the no-incentive tariff.

In Fig. 3, case 2, adverse selection is overcome by increasing farmer *j*'s incentive tariff levels up to a new level below the incentive tariff level of farmer *i*. Unlike the previous case, the ratio between the alternative tariffs is still above the threshold without conditioning farmer *j*'s attitudes. Hence, farm j suffers a surplus loss due to the risk of adverse selection equal to the ABCD area.

The intersection between $\pi'_i(x_i^{irr*})$ and $\pi'_j(x_j^{irr*})$ shown in Fig. 3 affects the regulator's incentive strategies. That, depending on the water use cost level, conditions the feasibility of the Differential Tariff instead of the current strategy.

By comparing the Differential Tariff scheme under asymmetric information (with the Incentive Constraint) and under conditions of perfect information (without the Incentive Constraint), the overall benefit will always be equal to or higher than zero (Fig. 4). The incentive constraint ceases to be binding when the two lines overlap. This happens for both low and high levels of water use costs as farmers belonging to the two farm typologies show high differences in water use attitudes. These differences tend to shrink for the intermediate level of water use costs. As a result, the regulator raises the tariff for the less efficient group of producers in order to guarantee discrimination that will affect the overall benefit. Equality between the IC and the OT scenarios is achieved at the level of zero water use costs and over the water use cost level at which farmer j ceases to irrigate. Differential benefits between the IC and the OT scenarios change according to the water use cost levels. Thus, water use cost levels are the parameters against which the regulator should guide tariff policies.

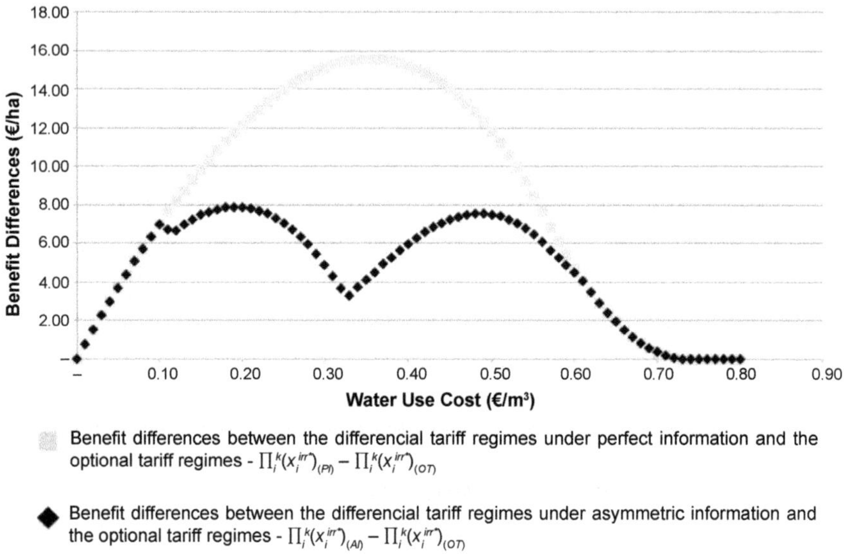

Figure 4. Profit variation of the Differential Tariff scheme with and without the incentive constraints with respect to the Optional Tariff scheme with increasing water use costs.

Monitoring

The ability to distinguish between different farm typologies enables the detection of incentive strategies, with the purpose of fostering the adoption of tariff schemes that are more efficient than the current one. On the other hand, the absence of indirect information connected to water use precludes the possibility of inducing compliance with the stated quota of irrigated farmland for farmers adopting the *incentive tariff*. Accordingly, in order to avoid the risk of moral hazard, the regulator bears monitoring costs. The magnitude of those costs could inhibit the adoption of the incentive tariff, hence undermining the feasibility of the actual tariff scheme. Assuming that monitoring activities prevent opportunistic behaviour, the question then is how to share the relevant costs between beneficiaries. As previously mentioned, the regulator shares the additional costs among all users (*m2*) rather than charging it to those users that opt for the incentive tariff (*m1*).

By comparing equations 7 and 9, it is expected that the *m2* strategy, unlike *m1*, induces the adoption of the incentive tariff for a higher value of monitoring costs as the tariff threshold switches only for a fraction of the additional cost. As the slope of the two lines shown in Fig. 5 changes, first one (see the A1, A2 dotted vertical lines in Fig. 5) and then the other farm typology (see the B1 and B2 dotted vertical lines in Fig. 5) switches

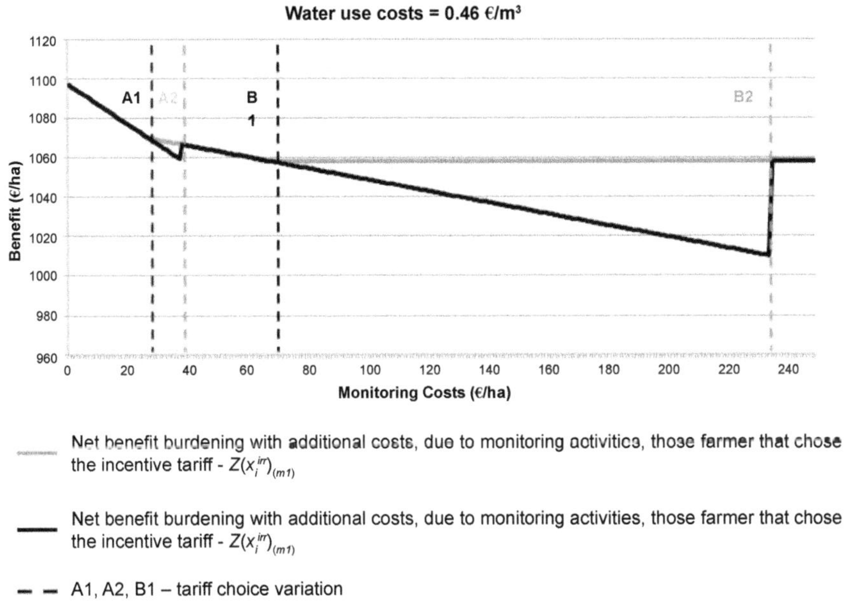

Figure 5. Benefit trend with increasing monitoring costs for both *m1* and *m2* scenarios.

from the incentive tariff to the alternative option. B1 and B2, respectively for the *m1* and the *m2* scenarios, mark the monitoring cost value beyond which the incentive tariff is no longer feasible for a given water use cost level. In between the A1-A2 range and the B1-B2 range, *m2* is lower than *m1*. This means that under the current monitoring cost allocation scenario (*m2*), the benefit gained by some farmers who opt for the incentive tariffs is followed by higher decreases in benefits for other farmers with respect to what was the case in the ideal scenario (*m1*). Under such circumstances, a pareto optimal solution is given only when the lines in Fig. 5 overlap (below A1, in between A2 and B1, and behind B2).

Conclusion

The current European policy guidelines on water governance should drive local populations towards an environmentally friendly self-management of water resources. This is a challenge, in particular for the agricultural sector, in light of the fact that open canals constitute the most common delivery system for irrigation water in the world. This undermines the ability to control uses and calls for the need to identify economic instruments able to deal with the actual policy framework under conditions of asymmetric information. The present study addresses the issue of asymmetric information by analysing the pricing strategies of a water authority in northern Italy, in an area served by surface water. In light of the actual regulatory framework, the case study pricing scheme is compared with other current and ideal pricing designs in order to assess its feasibility.

Specifically, the implementation of the full cost recovery principle and the incentive pricing principle of the European Water Framework Directive for unmetered water may result in an inefficient allocation of water use costs among beneficiaries and an inadequate control of water use that ultimately jeopardise the regulator's targets. In economic theory the effects of asymmetric information are tackled by way of two fundamental concepts: adverse selection and moral hazard (Laffont and Martimor 2002). For the management of irrigation water, adverse selection occurs when the regulator is unable to identify different kinds of users, while moral hazard occurs when the regulator is unable to control uses. The study deals with the issue of adverse selection by identifying appropriate incentive strategies, while moral hazard is approached through monitoring activities.

The study starts from the description of the regulator's pricing mechanism which relies on a discrimination strategy, the Optional Tariff, that induces farmers to partially reveal their use attitudes. This strategy is compared with the ones commonly adopted in the absence of water metering (Bogaert 2012). Assuming that the regulator acts in compliance with the WFD, up to a certain level of water use cost value, the Optional Tariff results in a lower benefit reduction with respect to the alternative ones. This validates the RIB's tariff policies.

Beside the existing RIB's tariff strategies the incentive theory provides an alternative design option for discrimination. These issues are addressed in the second part of the study. Here, assuming that the RIB knows the type of users in the area under its jurisdiction, the current RIB strategies are compared with an Optional Tariff designed for each farm typology. The RIB is able to discriminate users according to the farm typology they belong to at the cost of an information rent. The extent of this information rent, that goes to the detriment of some producers, varies with the water use cost level, more or less justifying the adoption of the ideal tariff instead of the current one. Here, in light of the FCR principle, water use cost allocations are questioned on the basis of the degree of benefit among the RIB's beneficiaries. The study explores the insight of adverse selection, seeks to contribute to the improvement of the dedicated literature (Besanko and Sappington 1987; Smith 1995; Viaggi et al. 2010) and reinforces the finding that more sophisticated tariff strategies do not always lead to better solutions.

The last part of the study addresses another issue generated by the incentive tariff: moral hazard. Here, the risk of false reporting undermines the ability to influence water uses and, consequently, to meet the WFD criterion of incentive pricing. Opportunistic behaviours are inhibited by monitoring activities. This generates costs that should be shared among beneficiaries who opt for the incentive tariff. However, the regulator shares the additional costs among all users rather than placing an extra burden on the farmers who generated the additional monitoring costs. Such behaviour can be explained by the tendency to encourage a sober use of irrigation water by its beneficiaries. Otherwise, there is no justification to spread the additional monitoring costs, arising from the adoption of the incentive tariff, among all users. However, whatever the allocative strategy, monitoring costs may be too high, making the incentive tariff unfeasible. Moral hazard is not properly considered as the present study analysed the way to allocate monitoring cost among users without questioning the viability of the regulator monitoring strategies. Currently, the case study allows just a few producers to choose the incentive tariff. This comes at the detriment of the WFD Incentive Pricing Principles (IPP) against which the regulator should discourage the overexploitation of irrigation water. However, this is the case since the incentive tariff requires the farmer to self-report the quota of irrigated farmland, obliging the RIB to undertake controls to monitor compliance. Costs arising from direct monitoring are extremely onerous and severely limit the regulator's ability to check for uses. This explains the regulator's decision to limit the enforcement of the incentive tariff to only a few farmers. Thus, in light of the WFD principles, further investigation is deemed necessary in order to verify the existence of costless indirect signals which could positively influence the level of compliance.

References

Bandaragoda, D.J. 1998. Design and Practice of Water Allocation Rules: Lessons from Warabandi in Pakistan's Punjab. Research Report #17, IIMI, Colombo, Sri Lanka.

Bazzani, G.M., Di Pasquale, S., Gallerani, V. and Viaggi, D. 2004. Irrigated agriculture in Italy and water regulation under the European Union water framework directive. Water Resources Research 40(7): 1–8.

Besanko, D. and Sappington, D. 1987. Designing Regulatory Policy with Limited Information. Harwood Academic Publisher, New York.

Bogaert, S. 2012. The role of water pricing and water allocation in agriculture in delivery sustainable water use in Europe, European Commission, February 2012, Brussels.

Bowen, R.L. and Young, R.A. 1986. Appraising alternatives for allocating and cost recovery for irrigation water in Egypt. Agricultural Economics 1(1): 35–52.

Carlton, D.W. and Perloff, J.M. 1994. Modern Industrial Organization, 2005. Boston: Addison Wesley.

Coletta, A. 2010. Le strategie di tariffazione per l'uso di risorse comuni: il caso dell'acqua consortile per irrigazione. Rivista di Economia Agraria LXV(4): 619–635.

Dayton-Johnson, J. 2000. Choosing rules to govern the commons: a model with evidence from Mexico. Journal of Economic Behaviour and Organization 42: 19–41.

EC. 2012. Blueprint to Safeguard Europe's Water Resources. COM (2012) 673 final. Brussels.

Galioto, F., Raggi, M. and Viaggi, D. 2012. Potenzialità applicative della teoria dei contratti nella gestione delle risorse idriche in agricoltura: un caso di studio nell'area del Consorzio di Bonifica della Romagna Occidentale. Rivista di Economia e Diritto Agroalimentare XVII(3): 359–373.

Gallerani, V., Raggi, M. and Viaggi, D. 2005. Pricing irrigation water under asymmetric information and cost recovery constraints. Water Science and Technology: Water Supply 5(6): 189–196.

INEA 2011. Quadro Normativosull' Usodell' Acqua in Ambito Agricolo, INEA edition.

Johansson, R.C., Tsur, J., Roe, T.L., Doukkali, R. and Dinar, A. 2002. Pricing irrigation water. A literature review. World Bank working paper 2449.

Laffont, J.J. and Martimor, D. 2002. The Theory of Incentives. Princeton University Press. Oxford.

Schlagler, E. and Ostrom, E. 1992. Property rights regimes and natural resources: a conceptual analysis. Land Economics 68(3): 249–262.

Smith, R.B.W. 1995. The conservation reserved program as a least-cost land retirement mechanism. American Journal of Agricultural Economics 77: 93–105.

Smith, R.B.W. and Tsur, T. 1997. Asymmetric information and the pricing of natural resources: understanding the case of unmetered water. Land Economics 73(3): 392–403.

Viaggi, D., Raggi, M., Bartolini, F. and Gallerani, V. 2010. Designing contracts for irrigation water under asymmetric information: are simple pricing mechanisms enough? Agricoltural Water Management 97: 1326–1332.

Viaggi, D., Raggi, M. and Gallerani, V. 2011. Asimmetria informativa e contribuzione irrigua nell'ottica della direttiva quadro. Economia & Diritto Agroalimentare XVI: 55–69.

Zilberman, D. 1997. Incentives and economics in water resource management. Proceedings of the Second Toulouse Conference on Environment and Resource Economics. Toulouse, France, May 14–16.

SECTION 3

Water Management and Effects from Water Framework Directive, Agriculture and Energy Policy

CHAPTER

9

A Spatial Econometric Approach to Assess the Impact of RDPs Agri-Environmental Measures on the Gross Nitrogen Balance: The Case Study of Emilia-Romagna

Valentina Marconi,[1,a,*] *Meri Raggi*[2] and *Davide Viaggi*[1,b]

Introduction

Nutrient pollution is one of the main causes of impairment of water quality. Recent studies estimated that nitrogen pollution-related damage in EU27 ranges between 70 and 320 billion Euro each year, equivalent to 150–750 euro/capita (Brink et al. 2011).

[1] Department of Agricultural Sciences, Viale Fanin, 50, University of Bologna.
[a] Email: v.marconi@unibo.it
[b] Email: davide.viaggi@unibo.it
[2] Department of Statistical Sciences, Via delle Belle Arti, 41, University of Bologna.
Email: meri.raggi@unibo.it
* Corresponding author

Agriculture is the main source of nitrogen loading (EEA 2012) and is the sector with the largest remaining emission reduction potential (Sutton et al. 2011). A nutrient deficiency can reduce soil fertility and crop yields, but a nutrient surplus, in excess of crop and forage needs can lead to nutrient release in the environment and, potentially, to water contamination.

Surpluses of nitrogen and phosphate are forecast to grow in the next decade, while those of potash are likely to remain more or less stable (FAO 2008). This trend is expected in spite of the fact that the price of nitrogen-based fertilizers has almost doubled from year 2000 to 2010, when the yearly increase was on average equal to 10% in the US (USDA 2012). The value of main agricultural inputs, such as fertilizers and fuel, represent more than one third of the value of agricultural income in Italy (Fanfani and Gutierrez 2011).

The European Union efforts to reduce nutrient over enrichment of waters were put into action since the early '90s with the enactment of the Nitrates Directive (1991), which aims to protect water quality across Europe by preventing nitrates from agricultural sources polluting ground and surface waters and by promoting the use of good farming practices.

The Common Agricultural Policy also contributes to the mitigation of pollution of waters by nitrates, mainly through Rural Development measures. The intervention strategy is based on direct support to farmers who will voluntary apply agri-environmental measures in order to reduce nitrates pollution, such as organic farming, low input farming, cover crops, efficient management of livestock waste and extensive farming.

The EU demands an evaluation of Rural Development Plans (RDPs) from the national authorities of the European member states.

The EU Commission has provided a set of indicators and evaluation criteria to be adopted for the RDPs assessment (CMEF 2006). Within the designed monitoring framework, the gross nutrient balance (baseline indicator 20), provides an estimate of the potential surplus of nitrogen and phosphorous on agricultural land (kg/ha). Nutrient balance indicators, as an environmental driving force, link to the state (or concentration) of nutrients in water bodies.

This study aims to evaluate the role of RDPs implementation in preventing water quality deterioration due to nutrient runoff and infiltration. The analysis focuses on the impact of Agri-Environmental Schemes on the reduction of nitrogen surplus through the implementation of specific sub-measures such as organic farming, set aside, etc.

In order to provide a quantitative evaluation, we applied a spatial regression model. The model aims to explain the variation of nitrogen surplus between two monitoring years: 2000 and 2010. In order to build a suitable dependent variable, we carried out a calculation of the Gross

Nitrogen Balance (a proxy for nitrogen surplus) at the municipality scale in Emilia-Romagna region, for both monitoring years.

The study area is characterized by the presence of an intensive agricultural and livestock farming system. Gross Nitrogen Balance was calculated following the OECD and EUROSTAT method.

The chapter is structured as follows: the next section describes background and study area, then methodology, results and discussion are presented.

Background and Study Area

The Nitrates Directive (91/676/CEE) was first implemented in Italy in 1999 (D.Lgs. 152/99). The establishment of action programs, to be implemented by farmers within NVZs on a compulsory basis, and the constitution of a monitoring and reporting system took place in 2007. In the study area, 29.9% of the land is classified as Nitrate Vulnerable Zones (NVZs), which represent the 17% of national NVZs. NVZs include the hydrological vulnerable area of Ferrara province, indicated in light green in Fig. 1.

The Common Agricultural Policy contributes to the mitigation of water pollution by nitrates through RDPs (agri-environment measures, support for investments in storage of manure, and training), cross-compliance

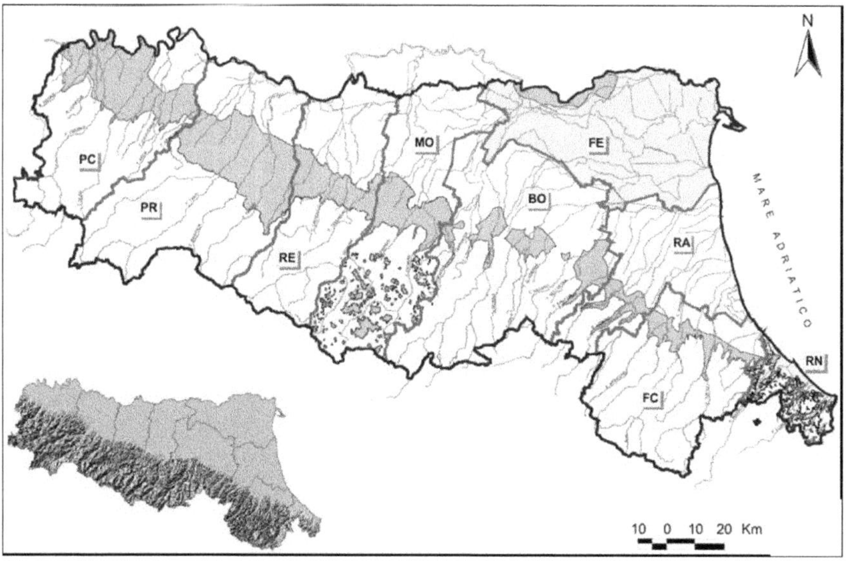

Figure 1. Grey lines are the borders of the 8 provinces, blue lines indicate rivers, green areas correspond to NVZs, light green area in Fe province represents vulnerable areas for water contamination (source ARPA). The picture in the left corner is a simplified physical map of the region.

(including the Nitrates Directive, establishment of buffer strips along water courses), and operational programs for fruit and vegetables.

Within the RDPs, the highest budget is assigned to measure (214), which deals with agri-environmental schemes and includes the potentially most effective actions for the mitigation of nitrate release in waters. Analysis will focus on those sub-measures that may have a direct influence on the Gross Nitrogen Balance: 214/1 (integrated farming), 2 (organic farming: restrains the use of nitrogen based fertilizers), 8 (extensive meadows: conversion of intensive crops into pasture, fodder crops and grasslands), 9 (protection of natural, semi-natural and agricultural landscape: restrains the extension of intensive cultivation), 10 (set aside of arable crops for environmental purpose: reduces the extension of intensive cultivation). For all these sub-measures, farmers are the direct beneficiaries of the payments.

It is worth pointing out that the regional budget for measure 214 exceeded the amount requested by applicants, therefore all the regular applications for measure 214 were funded by the RDP 2007–2013. Sub-measure 1 could be applied only in preferential areas, whereas the applications for the remaining sub-measures were given absolute priority if belonging to a preferential area. Preferential areas include the NVZs and protected areas at local and national level (e.g., Natura 2000).

The uptake of the sub-measures was provided by the regional government for each municipality as the extension of enrolled surfaces on the extension of the Utilized Agricultural Area of the municipality. Data are referred to the whole measure (214) and to its corresponding sub-measures in terms of number of participants on the total number of farms.

A detailed dataset of crop mix and livestock amounts for years 2000 and 2010, at the municipality scale, is available from the Italian Statistical Institute (ISTAT).

Emilia-Romagna is characterized by the presence of an intensive agricultural and livestock farming system, 12% of the total fertilizer sale in Italy is concentrated in this region, which represents the third topmost national area for fertilizer distribution. Organic farming represented only 2.4% of the UAA (Utilized Agricultural Area) in year 2000 and slightly increased up to 3.3% in year 2010.

According to the last two agricultural censuses, the percentage of utilized agricultural area (UAA) on the total extension of the region decreased from 50% in year 2000 to 47% in year 2010.

The cropping pattern of Emilia-Romagna is dominated by arable production (Fig. 2, left chart), which represents 72% of the agricultural utilized land, while fruit trees (orchards, vineyards and olive groves) and grassland (including permanent grazing) constitute 16% and 12% of the UAA, respectively. On the whole, the variations in percent UAA per crop between year 2000 and 2010 are negligible. However, it is worth mentioning

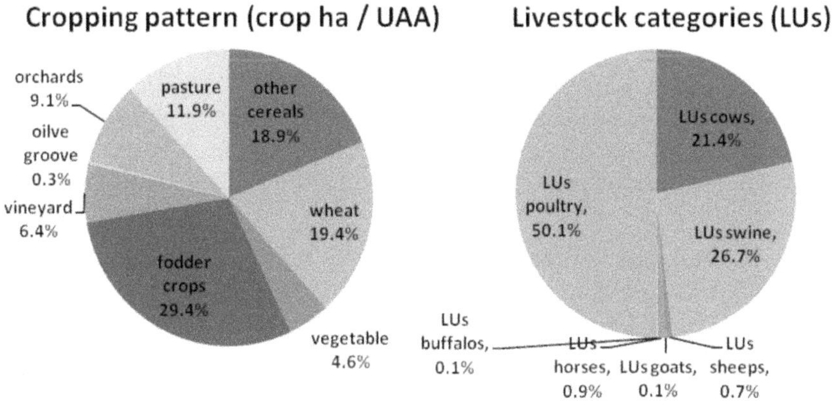

Figure 2. Main features of the agricultural farming system in Emilia-Romagna: cropping pattern (left chart), livestock categories (right chart).

that the share of arable crops on the total UAA has increased by 4% over the considered decade, whereas areas dedicated to grassland have decreased by 1% over the same time period.

Together with the other regions located in the Po plain (Veneto, Lombardia and Piemonte), Emilia-Romagna owns 70% of the national livestock. Half of the regional livestock is constitute by poultry (50% of LUs), whereas cows and swine represent 21% and 27%, respectively, of the total amount of LUs (right chart of Fig. 2).

The average livestock density, expressed as LUs per hectare of UAA, in Emilia-Romagna was equal to 1.5 in year 2000 and has decreased by 5% in 2010. On the whole, the variations in shares of the livestock categories between year 2000 and 2010 are negligible. The average number of LUs in livestock has also decreased from 39.21 in year 2000 to 30.51 in year 2010.

The total farm extension in Emilia-Romagna reached 1467237 ha in year 2000 and decreased to 1361153 ha in year 2010 (–8%). In the same period, the number of farms decreased by 33%. In the study area, about 20% of the farms are smaller than 5 ha and 40% of the farms have an extension greater than 50 ha. According to the last two agricultural censuses, in year 2000, 41% of the farms in Emilia-Romagna included livestock, this percentage has increased up to 71% in 10 years.

On the whole, dairy farms and swine livestock are mainly located in the western part of the region (especially Parma, Modena and Reggio Emilia provinces), whereas large poultry livestock are widespread in the southern provinces of Rimini and Forlì-Cesena. Orchards and vegetables are commonly cultivated in the eastern part of the region (especially Ravenna and Forlì-Cesena provinces). Cereals and fodder crops production characterizes the northern provinces.

Methodology

According to the objective of assessing the impact of AES in reducing nitrogen surplus, we applied a spatial econometric regression model. In this model, we set as dependent variable, the difference in the value of the indicator Gross Nitrogen Balance between two census years: 2000 and 2010. The selected explanatory variables can be grouped in three main categories:

1. Policy related variables: uptake of the AES (participation to sub-measures no. 1,2,8,9,10 of measure 214), allocated budget for measure 214, location in priority areas (such as LFA, Less Favoured Areas, and NVZs).
2. Structural variables: farm size, farm specialization, general landscape characteristics (location in plain or mountains, density of inhabitants).
3. Farmers' characteristics variables: age (younger than 40 years, between 40 and 55, older than 55 years) and education of the farmers (university or bachelor degree, or basic education).

All the variables included in the models are listed in Table 1 and divided into the three main categories. Variables included in the grouped named "Farmers' characteristics" are all related to year 2000, before the beginning of the last RDP, as well as a part of variables included in the remaining two groups: percentage of small (UAA <5 ha) and medium (UAA between 5 and 30 ha) farms, population density, Gross Nitrogen Balance in year 2000, dummy variable indicating the location in NVZs (Y NVZ) and in LFA (Y LFA).

All the variables describing the difference between year 2010 and 2000 of factors potentially affecting the variation of Gross Nitrogen Balance in the decade are indicated with a "D" in the table below. Negative values of these variables indicate a decreasing from 2000 to 2010 and vice versa in the case of positive values.

All these categories of regressors are likely to influence the optimal fertilization rate adopted by farmers and therefore trends in nitrogen surplus distribution.

The choice of the optimal fertilization rate, together with the structural and policy variables are all part of an "investment problem", which is simplified as follows:

$$\pi = \pi\ (N,\ P,\ Pol,\ x)$$

Where π is the profit, depending on the optimal fertilization rate (N), the vector of prices (including price of fertilizer, P), policy (Pol) and structural aspects of the farms (x).

Table 1. Independent variables included in the regression models.

FARMERS' CHARACTERISTICS	POLICY VARIABLES	STRUCTURAL VARIABLES
INDIVIDUAL COMPANY/ TOTAL COMPANIES	BUDGET MEASURE 214 (€)/UAA (ha)	D UUA/TAA
UNIVERSITY DEGREE	214-1 EXTENSION/UUA	D AVERAGE FARM SIZE (ha)
HIGH SCHOOL DIPLOMA	214-2 EXTENSION/UUA	D NUMBER OF FARMS
AGE BETWEEN 40–54	214-8 EXTENSION/UUA	D LIVESTOCK/TOT FARMS
AGE > 55 YEARS	214-9 EXTENSION/UUA	D LIVESTOCK UNITS/ LIVESTOCK
	214-10 EXTENSION/UUA	D LIVESTOCK UNIT/UUA (lsu ha)
	Y NVZ MUNICIPALITIES	D ORG. LAND/TAA
	Y LFA MUNICIPALITIES	D ARABLE/UAA
		D ORCHARDS/UAA
		UUA BETWEEN 5–30 (HA)
		UUA LOWER THAN 5 (HA)
		N SURPLUS (kg)/UAA (ha)_2000
		POPULATION DENSITY
		Y HILL
		Y MOUNTAIN

In order to determine if there is a spatial dependence, the variables must be related to a spatial weight matrix (w_{ij}), which provides the structure of the spatial relationship among observations.

Data and spatial weight matrix were geo-coded at the municipality scale. Thus, the empirical model was based on aggregated information rather than modeling individual farms.

Three spatial weight matrixes were calculated with the software GEODA based on queen contiguity criterion: municipalities are neighbors if they share a common edge. The three matrixes, named queen 1, queen 2 and queen 3 hereafter, are respectively characterized by increasing level of contiguity. In the first order, the neighbours of a municipality are only the adjoining ones, while the third order of municipalities will include a very large part of the total land as municipalities sizes are homogeneous and the shape of the region is regular.

We applied the Moran's *I* test statistic as spatial dependence test (Moran 1950):

$$I = e'We / e'e$$

where e=(y-y)/sd(y). Global Moran's *I* is the slope of line fit to scatter of We, e. This index is a measure of autocorrelation at the scale of the entire study area.

We performed a LISA (Local Indicators of Spatial Association; Anselin 1995 and 2003) analysis in order to test the data for local spatial autocorrelation, which is a measure of the local clustering of the data. LISA analysis led to the construction of cluster maps, which are tested for significance.

Spatial dependence was added to the regression (Ordinary Least Square) in two ways: spatial lag and spatial error.

The spatial lag model includes a spatially lagged dependent variable (*Wy)*, which is a weighted average of its neighbors' values. The spatial lag model reduced form equation is:

$$(I - \rho W)y = x\beta + e$$

where ρ is the spatial dependence parameter.

The spatial error model includes spatially correlated errors due to unobservable features or omitted variables associated with location.

The spatial error regression reduced form equation is:

$$(I - \lambda W)y = (I - \lambda W)'x\beta + u$$

where λ is the spatial dependence parameter.

As previously stated, nitrogen surplus can be estimated through the calculation of the indicator Gross Nitrogen Balance.

The OECD and EUROSTAT Gross Nitrogen Balance method, modified for calculations at sub-national level (OECD 2007), consists of an analytic INPUT-OUTPUT model (Fig. 3). All the model terms were calculated for each municipality of the region.

The inorganic fertilizers input has been calculated taking into account local application rates for each of the 8 crop partition elements considered: cereals (excluding wheat), wheat, vegetables, fruits, vineyards, olive grove, fodder crops, grassland. According to the OECD method, the estimated inorganic fertilizers input must be referred to the amount of inorganic fertilizers sold (or distributed at regional level), through the application of an "adjustment factor". Nitrogen deriving from livestock manure is another input to the OECD model, which requires the quantity of nitrogen

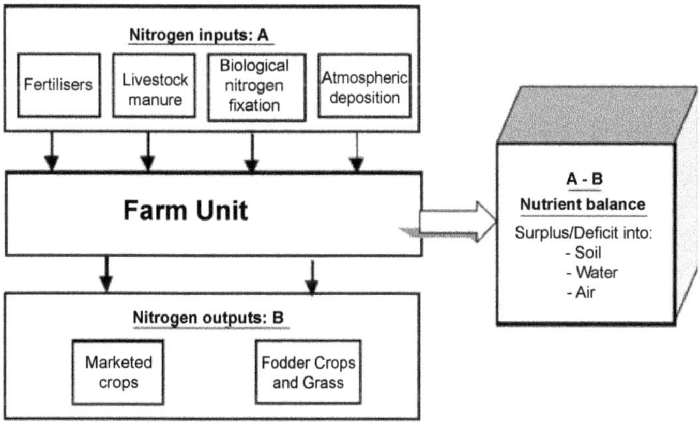

Source: OECD (2001), *Environmental Indicators for Agriculture—Volume 3: Methods and Results*, Publications Service Paris, France.

Figure 3. Calculation of the gross nitrogen balance at the regional (sub-national) level.

potentially available from local livestock. For each livestock category, the nitrogen quantity in the manure was estimated multiplying the number of heads by a specific manure coefficient.

Secondary nitrogen inputs included in the analysis are nitrogen biological fixation (symbiotic and non-symbiotic) and atmospheric deposition. Biological fixation (non-symbiotic) of nitrogen was calculated for arable, permanent crops, and grassland through the application of specific fixation rate. Symbiotic fixation was estimated for leguminous crops considering an average fixation rate of a variety of species.

Atmospheric deposition of nitrogen was estimated considering regional average values (N deposition in kg/ha) differentiated for plain and hilly areas.

Nitrogen output is the uptake by crops and grassland. This quantity depends on crop yields and on N uptake rates, which were estimated for each of the 8 considered categories.

Local fertilization rate, crop yield, crop nitrogen uptake, nitrogen biological fixation (grassland and leguminous crops) and atmospheric deposition were provided by the regional institutions dealing with agricultural (RER) and environmental issues (ARPA). Biological fixation rates of arable and permanent crops were calculated according to OECD indications (national average). Information on the nitrogen content of livestock manure was taken from APAT (National Environmental Agency).

As an environmental indicator, the balances only reveal the *potential* for nitrogen pollution.

Results

Gross nitrogen balance

The gross nitrogen balance is in the range between –50 and 50 kg/ha in most of the municipalities in both the monitoring years. The average N surplus has decreased from a positive average value of ca 26.6 kg/ha in year 2000 to a negative average value equal to –10.5 kg/ha in 2010 (Table 1). N surplus ranges from –103 to 436 in year 2000, and from –105 to 983 in year 2010, when the indicator variability increases (standard deviation is 82 in year 2000 and 86 in year 2010). Extremely high values are recorded in small municipalities where large livestock is located. The average variation of N surplus (given as difference between the 2010 and the 2000 value) in the considered time shift is –36.4 kg/ha, which indicates an overall decreasing of this value. The greatest N surplus decreasing is equal to 319 kg/ha, whereas the greatest rise registered is of 999 kg/ha (Table 2). The N surplus variations between year 2000 and 2010 are displayed in the map of Fig. 4, which points out

Table 2. Descriptive statistics of: N surplus in 2000, in 2010, and N surplus variation between these years.

	mean	dev st	min	Max
N surplus (kg)/UAA (ha) 2000	26.6	82.0	–103	436
N surplus (kg)/UAA (ha) 2010	–10.5	86.0	–105	983
N surplus variation 2000–2010	–36.4	75.5	–319	999

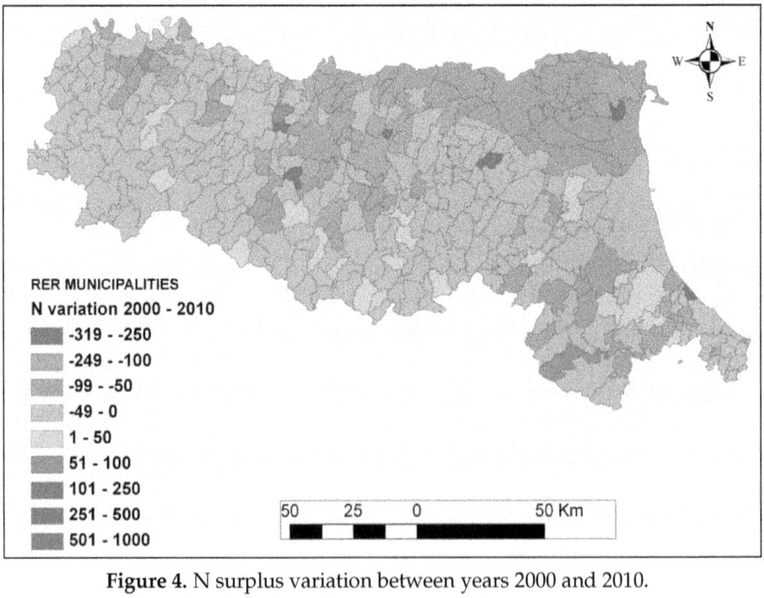

Figure 4. N surplus variation between years 2000 and 2010.

how the maximum variations (in both positive and negative directions) are located especially in the central part (Reggio-Emilia and Modena provinces) of the region and in the northeastern plain area (Ferrara).

The variations of the inputs and the outputs of the model for the calculation of the Gross Nitrogen Balance from year 2000 to 2010 are listed in Table 3. It is possible to observe that the quantity of nitrogen introduced into the agricultural system has decreased by 21%, whereas the total output of nitrogen is almost not changed (+ 3%).

The two main inputs to the model, the N deriving from inorganic fertilizers and N deriving from livestock manure have both decreased by –34% and –14%, respectively, from year 2000 to 2010.

Table 3. Gross Nitrogen Balance; input and output of the model for year 2000, 2010 and variation in the time shift.

	2000	2010	Difference %
TOTAL INPUT (N kg)	198109174.4	156885597.2	–21%
FERTILIZERS (N kg)	93733700.0	61957900.0	–34%
LIVESTOCK (N kg)	64500548.5	55684008.6	–14%
BIOLOGICAL FIXATION (N kg)	21527028.2	21477388.6	–0.2%
ATMOSPHERIC DEPOSITION (N kg)	18347897.7	17766300.0	–3%
OUTPUT (N kg)	160403499.2	164547315.0	3%
UAA (ha)	1115379.8	1064213.0	–5%
TOT INPUT/UAA (N kg/ha)	177.6	147.4	–17%
LIVESTOCK DENSITY (UBA/ha)	1.6	1.5	–5%

LISA

In this part of the chapter, the results of Moran scatter plots and the results of LISA cluster maps are presented.

The global Moran's I values for the three variables are listed in Table 4. The Moran's I values were calculated increasing the order of contiguity of the same queen contiguity matrix from 1 to 3 (queen 2 and queen 3, in the table).

In all the three spatial weight hypotheses, the global Moran's I value for N surplus in year 2000 is greater than zero showing that there is a positive spatial association in the N surplus distribution. Increasing the order of queen contiguity, the autocorrelation is reduced from 0.51 (queen 1) to 0.31 (queen 3). The same trend can be observed for the N surplus in year 2010; the global Moran's I values are lower and vary from a maximum of

Table 4. Global Moran's I of N surplus in 2000, N surplus in 2010 and of N surplus variation between these years, calculated using three spatial weight matrixes (queen 1, queen 2 and queen 3).

MORAN'S I	queen 1	queen 2	queen 3
N surplus 2000	0.53	0.45	0.31
N surplus 2010	0.21	0.17	0.12
N surplus variation 2000–2010	0.08	0.07	0.04

0.21 (queen 1) to a minimum of 0.12 (queen 3), which indicates that there is still a slightly positive autocorrelation of the variable. Finally, the Moran's I registered for the N surplus variation in the decade is close to zero under all the three spatial weight matrixes, indicating that there is no spatial dependence. In Figs. 5–7 the Moran scatterplots of the three variables for queen contiguity level are displayed.

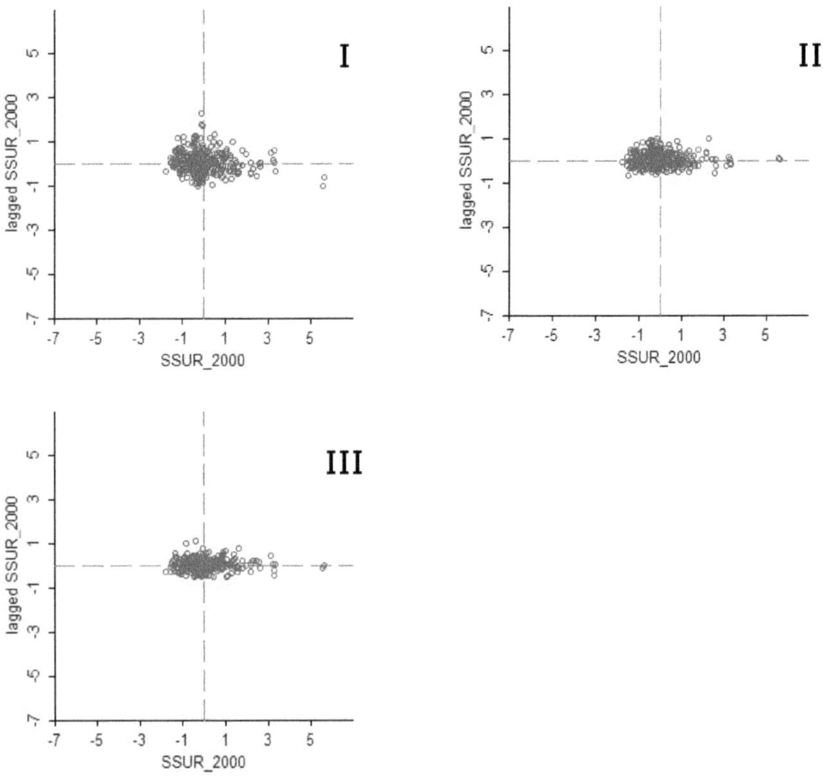

Figure 5. Moran scatterplots of N surplus in year 2000 under the hypotheses of three increasing contiguity levels: queen 1 (I), queen 2 (II) and queen 3 (III).

In each scatterplot the x-axis is the deviation from the mean for the observed value, while the y-axis is the average value of the deviation from the mean of the neighboring observations.

In Fig. 5, representing the Moran scatterplots of N surplus in year 2000, the values are scattered both in the upper right quadrant and in the bottom left quadrant. In Fig. 6 (N surplus in year 2010), most of the municipalities show values close to the origin of the axes), as well as in Fig. 7, which refers to the N surplus variation in the decade.

The location in the four quadrants represents a regime of spatial association: the high and right quadrant indicates the presence of hot spot clustering (high-high correlation), the low and left quadrant indicates cold spot clustering (low-low correlation). The remaining quadrants indicate low-high (high and left quadrant) and high-low correlations (low and right

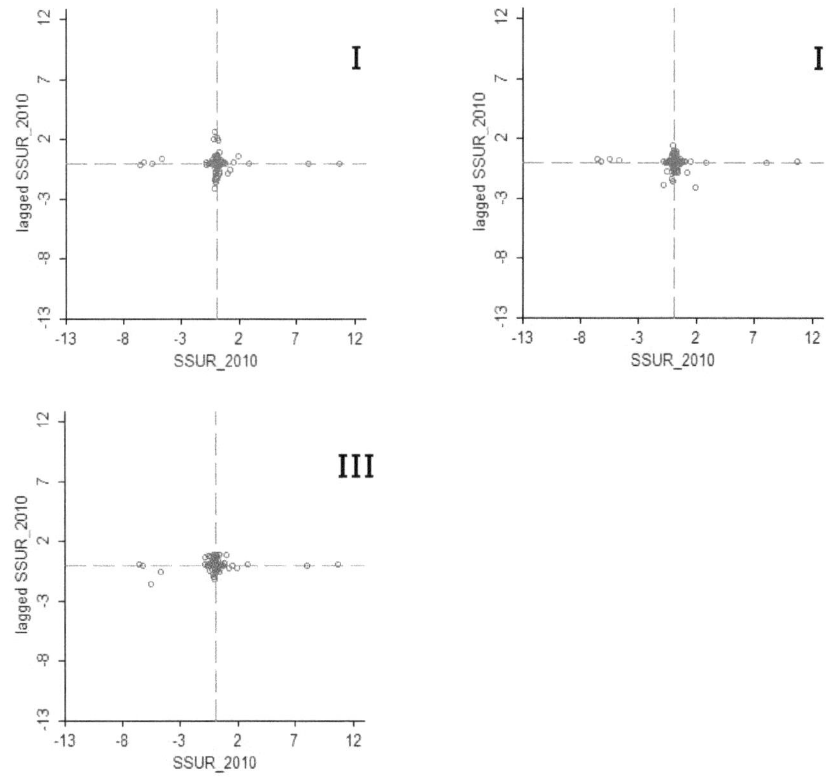

Figure 6. Moran scatterplots of N surplus in year 2010 under the hypotheses of three increasing contiguity levels: queen 1 (I), queen 2 (II) and queen 3 (III).

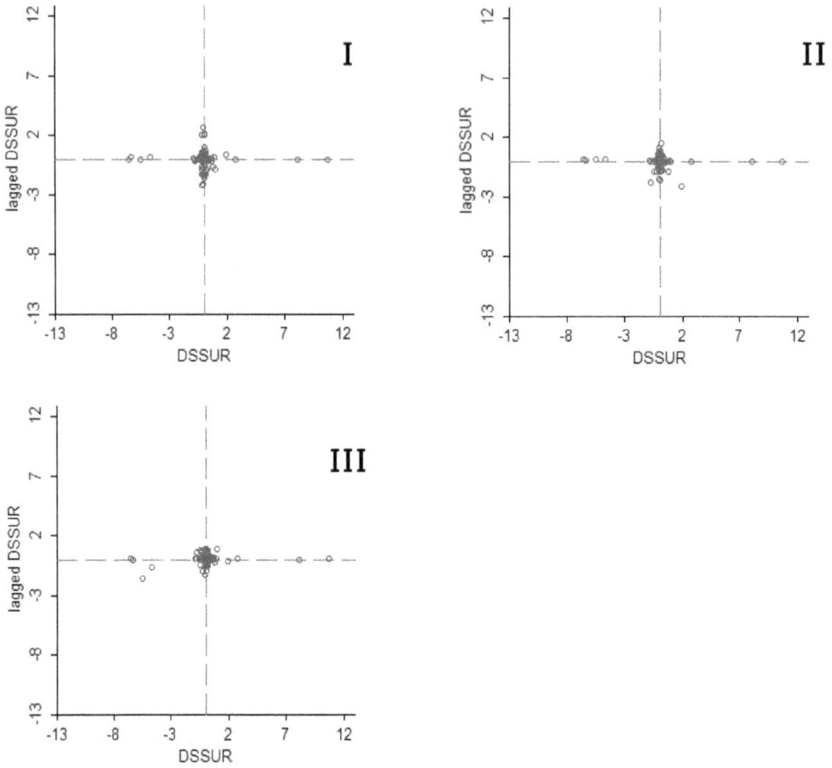

Figure 7. Moran scatterplots of N surplus variation in the decade 2000–2010, under the hypotheses of three increasing contiguity levels: queen 1 (I), queen 2 (II) and queen 3 (III).

quadrant). Clustering is displayed in LISA cluster maps (Figs. 10–12). The red colors represent a hot spot cluster, while the blue represents the cold spot cluster. The pink represents high-low correlations and the sky-blue, the low-high correlations. The white municipalities correspond to not statistically significant cluster.

All the painted municipalities are statistically significant at least at 0.05 level.

Figure 8 confirms that several municipalities located in the south-western border of the region (mountainous and hilly area) have the lower N surplus values and this also applies to the neighbors without the presence of spatial outliers values (high-low or low-high correlations). Hot spot clustering, mostly evidenced at the second order of contiguity, can be found on the eastern border (Forlì-Cesena province) and in northern part of the region (Reggio-Emilia and part of the Modena provinces). At increasing

Figure 8. LISA of the N surplus in year 2000 using queen contiguity matrix of the first (A), second (B) and third (C) level.

orders of contiguity, both these hot spot clusters become surrounded by spatial outliers (sky blue municipalities): they represent a group of municipalities with lower N surplus in areas dominated by municipalities with higher N surplus.

LISA maps of N surplus of year 2010 shows three distinct cold spot clusters (Fig. 9): the mountainous-hilly area of the southwestern border

Figure 9. LISA of the N surplus in year 2010 using queen contiguity matrix of the first (A), second (B) and third (C) level.

(Piacenza and Parma provinces), the mountainous-hilly area of Bologna province (central part of the region), the Ferrara province (in the northeastern corner). The hot spots clusters are found again in Reggio-Emilia and Forlì-Cesena provinces. Both the clustering spot categories have smaller extent and at the third order of contiguity, at which the number of spatial outliers increases (especially low-high clusters, in sky blue).

LISA maps of N surplus variation between years 2010 and 2000 show that clustering of low values (cold spot) is located at the north western border of the region (Fig. 10), especially at the first order of contiguity. Small scale neighborhood effects characterize some high value associations. Spatial outliers are common at the second and third level of contiguity.

Spatial regression model

In this section, the results of the regression models, ordinary least square (OLS), spatial lag and spatial error, are presented (5). All the models were run under the hypothesis of the first order of contiguity (queen 1).

The R-squared varies from 0.76 in the OLS estimation to 0.77 in the spatial models. Spatial dependence is indicated in the spatial lag model by the Rho coefficient of the lagged dependent variable, which is significant at 5%, and by the coefficient Lambda in the spatial error model, which is significant at 1% (respectively Rho and Lambda in Table 5).

The uptake of agri-environmental sub-measures is not significant in any of the performed models.

The variation of the number of farms in the municipality during the considered time shift (D Number of farms in Table 5) is significant at 5% in all the three models and has a negative coefficient: –0.11 in the OLS model and –0.02 in the spatial models.

Significant variables with positive coefficient are the variation of the percentage of arable crops (cereals and vegetable) on the UAA (significant at 1%), the percentage of farmers aged between 40 and 54 years in year 2000 (significant at 5%) and the variation of livestock density in the decade (significant at 1%).

In the spatial models, beta coefficients of the significant variables are relatively low in the case of D Number of Farms and Age between 40–54 (less than –1 and less than 1, respectively), whereas the variables D Arable/UAA and D Livestock Units/UAA have beta coefficients with magnitude of about 100 and 18, respectively (spatial lag and spatial error mode, Table 5).

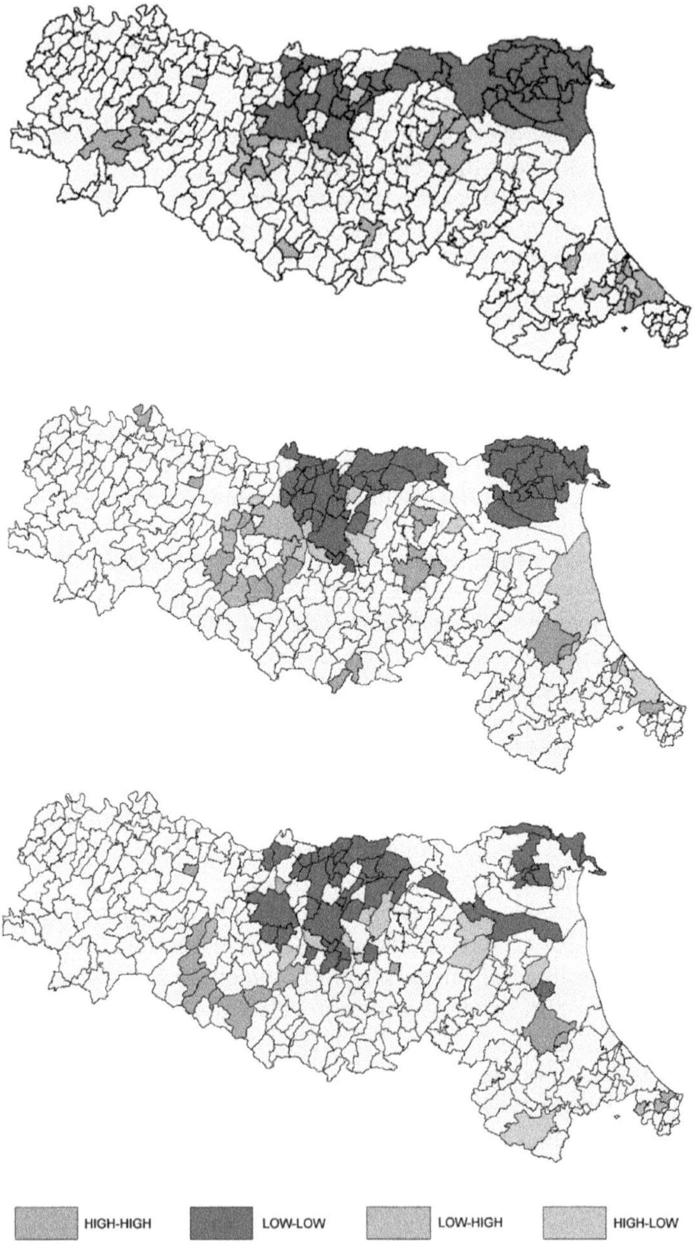

Figure 10. LISA of the N surplus variation in the decade 2000–2010 using queen contiguity matrix of the first (A), second (B) and third (C) level.

Table 5. Results of the regression models with Gross Nitrogen Balance variation between 2010 and 2000 as dependent variable.

DEPENDENT VARIABLE	Gross Nitrogen Balance variation between 2010 and 2000					
MODEL	**OLS**		**Spatial Lag**		**Spatial Error**	
R Square		**0.76**		**0.77**		**0.77**
Standard Error of Regression		38.81		36.50		36.04
	Beta	**p Value**	**Beta**	**p Value**	**Beta**	**p Value**
(Constant)		0.086	-62.37	0.069	-60.50	0.080
N SURPLUS (kg)/UAA (ha)_2000	**-0.08**	**0.029**	**-0.07**	**0.036**	**-0.08**	**0.023**
Y HILL	0.04	0.275	2.79	0.650	1.49	0.824
Y MOUNTAIN	0.06	0.221	6.09	0.466	7.31	0.415
POPULATION DENSITY	0.02	0.641	0.00	0.669	0.00	0.815
D UAA/TAA	-0.02	0.519	-22.64	0.408	-19.39	0.473
D AVERAGE FARM SIZE (ha)	-0.04	0.228	-0.22	0.148	-0.17	0.261
D NUMBER OF FARMS	**-0.11**	**0.015**	**-0.02**	**0.012**	**-0.02**	**0.023**
D LIVESTOCK/TOT FARMS	0.01	0.805	0.08	0.880	0.12	0.816
D LIVESTOCK UNITS/ LIVESTOCK	0.00	0.953	0.00	0.745	0.00	0.660
D LIVESTOCK UNIT/UAA (LSU/ha)	**0.82**	**0.000**	**17.69**	**0.000**	**17.68**	**0.000**
D ORG. LAND/TAA	-0.03	0.455	-11.93	0.263	-11.43	0.274
D ARABLE/UAA	**-0.12**	**0.000**	**109.40**	**0.000**	**98.98**	**0.002**
D ORCHARDS/UAA	0.01	0.826	69.28	0.198	69.94	0.201
BUDGET MEASURE 214 (€)/UAA (ha)	-0.08	0.065	0.00	0.065	0.00	0.081
214-1 EXTENSION/UAA	-0.01	0.661	-0.68	0.467	-0.67	0.490
214-2 EXTENSION/UAA	0.01	0.833	0.29	0.427	0.37	0.322
214-8 EXTENSION/UAA	-0.01	0.862	0.02	0.950	-0.03	0.935
214-9 EXTENSION/UAA	0.02	0.580	6.65	0.350	6.30	0.374
214-10 EXTENSION/UAA	0.01	0.797	-0.49	0.840	0.71	0.777
Y NVZ MUNICIPALITIES	-0.01	0.669	-2.17	0.644	-1.92	0.709
Y LFA MUNICIPALITIES	0.05	0.113	6.45	0.170	7.67	0.102
INDIVIDUAL COMPANY/TOT COMP.	0.03	0.296	0.45	0.090	0.37	0.169
UAA >5 (ha)	0.02	0.723	0.08	0.747	0.08	0.734
UAA BETWEEN 5–30 (HA)	-0.05	0.406	-0.40	0.172	-0.44	0.136
UNIVERSITY DEGREE	-0.01	0.766	-0.33	0.627	-0.05	0.938
HIGH SCHOOL DIPLOMA	-0.01	0.786	-0.07	0.593	-0.09	0.473
AGE BETWEEN 40–54	**0.07**	**0.033**	**0.94**	**0.030**	**0.88**	**0.037**
AGE >55	0.00	0.956	-0.16	0.557	-0.15	0.582
RHO			**0.14**	**0.012**		
LAMBDA					**0.28**	**0.000**

Discussion

Moran's statistics indicate a positive spatial dependence in the distribution of the N surplus (Table 4) in both years 2000 and 2010, meaning that there is a significant spatial association. As we expected, in both years the cold spot clustering is located in the mountainous and hilly area of the region, whereas the hot spot lies in the plain (Figs. 8 and 9). Thus, the location in a favorable landscape may be the cause itself for the spreading of an intensive farming system (and vice versa in the case of mountainous areas). However, LISA analysis allows a formal and more precise identification of the significant clustering: in year 2000, the cold spot partially overlaps the mountain belt, with the exception of Rimini province (Fig. 8). In the case of N surplus in year 2010 (Fig. 9), the fragmented location of the cold spots suggests that landscape is not the only driving factor in the distribution of low-low spatial associations. Within the plain, the hot spots clustering are mainly located in the provinces with higher livestock production (Modena and Reggio-Emilia, in Fig. 1), which could be a consequence of the sensitivity of the indicator to this parameter.

Global Moran's I indicates that the overall spatial autocorrelation is not significant for the variation of the Gross Nitrogen Balance from 2000 to 2010 (Table 4). However, LISA maps show local significant clustering (Fig. 10): a cold spot clustering is located in the north-eastern side of the region (plain areas mainly belonging to the Ferrara province). In this case, the cold spot indicates a significant association of municipalities characterized by the decrease of the indicator from year 2000 to 2010. Ferrara province is characterized by intensive arable and orchards production, the occurrence of high participation rate to sub-measure 1 (integrated production) and the location in NVZs for its entire extension (Fig. 1). However, policy variables do not result in significant explanatory variables for the variation of Gross Nitrogen Balance between year 2010 and 2000 (Table 5). According to the regression models, the indicator Gross Nitrogen Balance is not dependent upon any of the agri-environmental measures of the RDPs. All the considered sub-measures aim to promote extensive cropping systems instead of intensive production. Thus, measure 214 potentially affects the first input to the Gross Nitrogen Balance (inorganic fertilizers). However, the action of measure 214 does not directly influence the livestock regime, to which the indicator is sensitive. Moreover, the aggregate scale at which both the indicator and the uptake are considered may affect the capability of the model to capture the correlation between the two variables, especially in terms of spatial dependence. Although, in regional land use modeling (Isik 2004; Holloway and Lapar 2007) it is often stated that the uptake of AES is shared at the aggregated level.

The percentage of arable crops on the total extension of the UAA is highly significant. As we expected, the variation of the Gross Nitrogen Balance is positively dependent upon this variable, whose influence on the indicator is measured by a beta coefficient of about 100 in the models accounting for spatial dependence. The indicator is, obviously, also dependent upon the variation of the livestock density, which is a proxy for the input of organic nitrogen in the agricultural system. Regression models not including this variable result in a poor general fit, as testified by decreasing R squared (about 0.15) and increasing Aikake Info Criterion. Besides the Ferrara province, Gross Nitrogen Balance has also decreased in Modena and Reggio-Emilia (Figs. 1 and 4), which is likely to be due to a downturn in the production of swine (–20%) and cows (–10%), widespread in these areas. Caution should be applied when considering the magnitude of the beta coefficient for the livestock density, as it is expressed as LUs/ha, whereas the dependent variable is given in kg/ha. The average coefficient applied to convert the livestock number (in LUs) into nitrogen kg is of about 0.46, which means that the real magnitude of the variable is about 8. This is still a remarkable value, suggesting that a reduction of 1 unit of the livestock density will reduce the indicator variation 8 times (in the same time shift).

Within the group of variables related to farmers' characteristics (which refer to year 2000), age between 40–54 is the only significant variable. The positive beta coefficient indicates that the Gross Nitrogen Balance has increased in the municipality where there was a higher percentage of farmers with age included in this range before the implementation of RDPs 2007–2013, with respect to farmers with age lower than 40 years and greater than 55.

According to the model results, reduction of Gross Nitrogen Balance is also slightly favoured by the increase of the number of farms in the municipalities, suggesting that where small farms are widespread, there is an overall decrease in the inputs of nutrients.

Both spatial models yield improvement to the original OLS model, by increasing general model fit (increased R squared and decreased Standard Error of regression; in Table 5). Furthermore, the spatial coefficient Rho and Lambda are positive and significant, revealing the presence of spatially correlated errors (due to unobservable features or omitted variables associated with location, Lambda), and measuring the average influence on observations by their neighboring observations (Rho). As a result, the spatial error model appears to be the most improved model among all.

Conclusion

Gross Nitrogen Balance in Emilia-Romagna (26.6 kg/ha in year 2000 and 10.5 kg/ha in year 2010) is well below the EEA's nutrient surplus calculations, which estimated for Europe an average nitrogen surplus of 55 kg per ha in year 2000 (EEA 2005) and 30 kg/ha in year 2010 (ETC 2010). The Gross Nitrogen Balance in Emilia-Romagna has generally decreased (–36 kg/ha on average) from year 2000 to year 2010. Results indicate that the variation of the indicator is positively dependent upon the variation in the share of arable crops and to the variation of livestock density. These two explanatory variables directly link to the application rate of inorganic fertilizers and the presence of livestock. Inorganic fertilizers show the greatest reduction: –34% from 2000 to 2010. This decrease is probably linked to the rise of fertilizers price, which could not be taken into account in our model.

Moran statistics applied to the Gross Nitrogen Balance of year 2000 and 2010 points out that the influence of the neighbors enhances the nitrogen surplus in intensive farming systems and is effective in reducing the nitrogen surplus in extensive farming areas. High-high association are common in the plain and low-low in the mountain/hill; however, the clustering distribution suggests that landscape is not the only driving factor.

All the models explain more than 75% of the variability of the difference between Gross Nitrogen Balance between year 2010 and 2000. The indicators do not appear to be dependent upon policy variables: agri-environmental schemes of RDPs (sub-measures 214/1,2,8,9,10). The indicators reflect the livestock density distribution and, as a matter of fact, the sub-measures of the RDP's agri-environmental scheme are not designed for preventing the excess of nitrogen deriving from livestock. Similar to what is observed for AES, the occurrence of NVZs is also not significant.

However, other omitted variables and spatial factors should be further investigated, as suggested by the significance of the Lambda coefficient in the spatial error model. The availability of better estimation of changes in nitrogen surplus, such as the calculation at the farm scale, would be an important component to allow for a more robust use of spatial econometrics in RDP evaluation related to N surplus reduction.

References

Anselin, L. 1995. Local Indicators of Spatial Association—LISA, Geographical Analysis 27: 93–115.

Anselin, L. 2003. GeoDa™ 0.9 User's Guide. Luc Anselin and The Regents of the University of Illinois, USA.

Brink, C., van Grinsven, H., Jacobsen, B.H., Rabl, A., Gren, I., Holland, M., Klimont, Z., Hicks, K., Brouwer, R., Dickens, R., Willems, J., Termansen, M., Velthof, G., Alkemade, R., van Oorschot, M. and Webb, J. 2011. Costs and benefits of nitrogen in the environment. pp.

513–541. *In*: Sutton, M., Howard, C., Erisman, J., Billen, G., Bleeker, A., Greenfelt, P., van Grinsven, H. and Grizetti, B. (eds.). The European Nitrogen Assessment—Sources, Effects and Policy Perspectives. Cambridge University Press, UK.
[CMEF] Common Monitoring Evaluation Framework. 2006. Rural Development 2007–2013. Handbook on Common Monitoring and Evaluation Framework—Guidance document, Directorate General for Agriculture and Rural Development, European Commission, 16 p.
[EEA] European Environment Agency. 2005. EEA European Environment 2005; State and Outlook. EEA, Kongens Nytorv, Copenhagen, 584 p.
[EEA] European Environment Agency. 2012. European Waters. Assessment of Status and Pressures. EEA Report no. 8/2012, Kongens Nytorv, Copenhagen, 100 p.
[ETC] European Topic Centre on Water. 2010. Freshwater Eutrophication Background Report. ETC Water Technical Report no. 2/2010, Litevská, Praha, 84 p.
[FAO] Food and Agriculture Organization of the United Nations. 2008. Current world fertilizer trends and outlook to 2011/12, FAO, via delle Terme di Caracalla, Rome, 57 p.
Fanfani, R. and Gutierrez, L. 2011. L'ottovolante dei prezzi agricoli internazionali: up-down-up and what next? Economia e diritto agroalimentare XVI(2): 169–181.
Holloway, G. and Lapar, M.L. 2007. How big is your neighbourhood? Spatial implications of market participation among Filipino smallholders. Journal of Agricultural Economics 58(1): 37–60.
Isik, M. 2004. Environmental regulation and the spatial structure of the U.S. dairy sector. American Journal of Agricultural Economics 86: 949–962.
Moran, P.A.P. 1950. Notes on continuous stochastic phenomena. Biometrika 37(1): 17–23.
[OECD] Organisation for Economic Cooperation and Development. 2007. Gross Nitrogen Balance Handbook, OECD jointly published with EUROSTAT, Paris, 24 p.
Sutton, M., Howard, C., Erisman, J., Billen, G., Bleeker, A., Greenfelt, P., van Grinsven, H. and Grizetti, B. (eds.). 2011. The European Nitrogen Assessment—Sources, Effects and Policy Perspectives. Cambridge University Press, UK.
[USDA] United States Department of Agriculture. 2012. Fertilizer Use and Price—Overview, USDA Economic Research Service, Washington DC: http://www.ers.usda.gov/data-products/fertilizer-use-and-price.aspx.

CHAPTER

10

The Disproportionality Principle in the WFD: How to *Actually* Apply it?

Julia Martin-Ortega,[1,*] *Dimitris Skuras*,[2] *Angel Perni*,[3] *Silje Holen*[4] and *Demetris Psaltopoulos*[2]

Introduction

In October 2000, Directive 2000/60/EC, establishing a framework for action in the field of water policy, also known as the Water Framework Directive (WFD), entered into force leading to a substantial reform of water policy in Europe. The main aims of the WFD include stopping deterioration, improving the state of aquatic ecosystems and promoting the sustainable use of water by achieving 'good ecological status' (GES) in all water bodies by 2015. In the context of its implementation, economic instruments (e.g., water pricing), methods (e.g., cost-effectiveness analysis) and principles (e.g., the polluter-pays-principle) are prescribed to reach these objectives. Economic considerations also play a role to justify exemptions, with

[1] Social, Economic and Geographical Sciences Group, The James Hutton Institute (United Kingdom).
[2] University of Patras (Greece).
[3] University of Murcia (Spain).
[4] Norwegian Institute for Water Research (Norway).
* Corresponding author: julia.martinortega@hutton.ac.uk

Article 4 stating that the WFD allows the derogation of environmental objectives if meeting them has *disproportionately high costs*. In this context, the following types of derogation can be granted (Finnegan 2009): i) an extension of the deadline for reaching GES with up to two planning cycles if a water body requires additional measures which involve disproportionate costs to meet the deadline (if GES can be achieved after 2015); and ii) the setting of a less ambitious environmental target, i.e., reaching an *acceptable ecological state*. The Directive further specifies *disproportionately high costs*: i) in relation to the financial ability to meet them, or ii) compared to the benefits of meeting the objective. The first case is related to affordability. This argument only warrants a postponement of the deadline and does not constitute a justification for setting a less ambitious target. In this case, the WFD allows Member States certain flexibility in adopting a sector-specific and progressive approach depending on the affected sectors' ability to pay (Postle et al. 2004). The second argument, which is the main focus of this chapter, requires proving that the costs of the Programme of Measures (PoM) which achieve GES are higher than the associated expected benefits (European Commission 2003).

Although the disproportionality principle is crucial to the WFD, the Directive itself includes neither an accurate definition of this concept, nor operational criteria for establishing exemptions. It hence remains a Member-State-specific political decision. Yet, this decision needs to be informed by economic analysis.

In order to clarify relevant economic aspects of the WFD, a Common Implementation Strategy guidance document was released in 2003 (European Commission 2003). According to this document, decisions on disproportionality may be based on more or less sophisticated criteria depending on the type of exemption. In the case of postponing the GES objective, simple financial criteria could be used to assess disproportionality at first place. In later stages and in cases of lowering the GES objective, disproportionality analysis may require a thorough investigation which includes both qualitative and quantitative indicators of various cost and benefit categories. This vague guidance has led to inconsistent and often unclear ways of approaching this issue across Europe. Even more, proportionality assessments have not yet been widely undertaken and there is a great level of uncertainty and lack of clarity on their implementation. Thus, the practical interpretation of the terms "disproportionately costly" remains disputed. For instance, in proportion to what are costs considered as disproportionate, and what is the threshold value for disproportionality?

From an economic perspective, it is seems straightforward that Cost–Benefit Analysis (CBA) is the obvious tool for disproportionality analysis (Martin-Ortega 2012). But while Cost-Effectiveness Analysis (CEA) has been widely adopted by most national guidelines in Europe for the assessment of costs and prioritization of measures (Balana et al. 2011), and the estimation of the environmental benefits of the WFD under different contexts has received a significant attention by the literature (Birol et al. 2006; Martin-Ortega et al. 2011; Alcon et al. 2012; Perni et al. 2012), the way these two should be joined up in a CBA to assess disproportionality has received much less attention at both the research and policy levels (Vinten et al. 2012). Moreover, a purely monetary assessment of costs and benefits might not provide all the information needed for making an appropriate policy decision in this context. Attention is also required to include issues such as distributional effects and equity considerations. Further, such an assessment requires establishing decision-criteria regarding spatial and temporal scales and the consideration of local specificities and stakeholder views that have not been sufficiently discussed up to date.

The literature (both peer-reviewed and grey—e.g., policy documents) on the conceptual and practical implementation of the disproportionality principle of the WFD is not only scarce but also very fragmented, often covering partial aspects (Martin-Ortega 2012). There are still a significant number of unresolved issues and different approaches to this rather crucial aspect of the WFD. Thus, poor guidance and deficient applications can lead to an important risk of ecological targets not being met on the basis of improperly undertaken economic analysis if, for example, non-market benefits are not included or properly discounted. Similarly, disregarding issues such as distributional effects can impose excessive burden to specific sectors causing social inequalities.

This paper aims to fill this gap by: i) providing a comprehensive discussion of the key concepts underpinning WFD's (dis)proportionality from a theoretical point of view, ii) reviewing existing applications both at the policy and academic level and identifying their contributions and limitations, and iii) providing guiding principles for addressing the key challenges in practice. This discussion is driven by the need for matching economic scientific prescriptions and policy needs.

The Conceptual Basis of Disproportionality in the WFD

Defining costs and benefits

The identification of costs and benefits is a rather complicated exercise, as there is no consensus as to what should be included in the relevant estimations. In Table 1 the main costs and benefits that have been signalled by the guiding documents of the WFD (WATECO 2003) and the subsequent literature (e.g., Postle et al. 2004; Hanley and Black 2006; Brouwer et al. 2009; Balana et al. 2012; Martin-Ortega 2012) are presented. Although some of

Table 1. Main costs and benefits of the WFD.

Type	Definition	Examples/References
COSTS		
Direct financial costs	The financial costs of measures aimed at improving water quality. Expenditures from the actors involved in the PoM implementation.	These include all direct, indirect, maintenance and operating costs of the measures (European Commission 2003). For example, increased waste water treatment costs.
Associated non-water environmental costs	Environmental costs related to impacts of water projects different from the water-related ones.	Increased transport, emissions or noise, during the construction of some measures (Postle et al. 2004).
Cost derived from wider economic effects	Knock-on effects of the costs in one sector to other sectors and indirect social costs. These may be substantial if the measures are of a large scale and affect substantially the prices of products or inputs.	Decreased expenditure and associated jobs in a local economy, "multiplier effects", and reduced competition (Balana et al. 2011).
Costs to regulating authorities	Costs of monitoring and enforcing compliance.	Definition of ecological indicators to assess ecological status. Establishment of networks of water monitoring to collect information on ecological status over time (Balana et al. 2011)
Contemporaneous and intertemporal costs	This is a type of "residual" environmental cost once mitigation measures have been applied and refer to opportunity costs of water exploitation beyond its natural rate of recharge that deprives other users from abstracting desired water quantities now (contemporaneous) or in the future (intertemporal) (Skuras et al. 2010).	Groundwater overexploitation may affect different water uses such as agriculture or recreation in the present and/or the future. Therefore, resource costs have to be incorporated in the decision making (Martinez-Paz and Perni 2011).

Table 1. contd....

Table 1. contd.

Type	Definition	Examples/References
BENEFITS		
Market benefits from increased water quality	Commercial benefits that may increase due to water quality improvement or generated by some measures.	Increased benefits in commercial fisheries due to the increase in fish populations (Hanley and Black 2006).
Non-market benefits from increased water quality	Welfare gain resulting from the improvement of water quality from the current to GES (Brouwer et al. 2009).	Improvements in aquatic biodiversity or recreation (Ramajo-Hernandez and Saz-Salazar 2012).
Associated non-water environmental benefits	Benefits beyond the water-related ones, i.e., wider benefits in terms of their impact on other ecosystem goods and services.	Tree planting in buffer strips may enhance biodiversity and landscapes, and sequester carbon (Borin et al. 2010).
Benefits from reduced contemporaneous costs	Benefits from reducing the opportunity costs imposed by the current use of water.	Improving water quality at source may reduce drinking water costs, e.g., in water treatments or bottled water (EEA 2012).
Benefits from reduced intertemporal costs	Benefits from reducing the opportunity costs imposed by using the water today and due to over-abstraction deprive future users desiring to abstract the demanded quantities.	Maintaining aquifer table levels allow future generation to use groundwater (Stone and Webster Consultants 2004; Postle et al. 2004).

them may be difficult to quantify, they should be at least acknowledged and qualitatively described in order to guide the decision making process. In the following section, we discuss key issues regarding their estimation.

Measuring costs and benefits

Estimating the costs of improving the good ecological status

The WFD prescribes the elaboration of a PoM, to be included in the River Basin Management Plans. These measures have first to be selected according to cost-effective criteria, i.e., measures which can achieve the target at the least cost are the ones to be selected (European Commission 2003), and then their costs compared with the benefits they provide. The literature on how to estimate such costs is very extensive and a detailed review is out of the scope of this chapter (for that purpose, we recommend Balana et al. 2011, 2012). Here we just include a brief general discussion.

Financial costs associated with changes in private business as a direct outcome of policy implementation are relatively easy to identify and quantify, and usually represent the main focus of compliance. Indirect wider economic impacts are generally less tangible and more difficult to estimate, but general equilibrium models can be applied to this end (Brouwer et al. 2008).

Private financial costs include all types of cost incurred by private actors, firms or farmers, in order to comply with mitigation measures. For example, a change in irrigation practices from sprinkler to drip irrigation will incur cost that will be covered by farmers. Wider economic effects arising from knock-on effects on other economic actors as a result of measures and/or achieving WFD objectives include increased or decreased expenditure and multiplier effects, job creation or displacement, regeneration and enhanced competition. For example, in the case of job displacement, higher public expenditure for unemployment benefits may incur. Non-water costs can also arise as a result of implementation of a measure or achievement of WFD objectives that are not directly related to the water environment. Such costs may indicatively incorporate increased waste and energy use due to new treatment processes, and increased noise, odours and congestion. Mitigation measures may incur increased cost to regulating authorities in the form of, for example, monitoring, inspection and control costs.

Mitigation measures may also produce opportunity costs, i.e., foregone benefits. Opportunity costs may be contemporaneous or inter-temporal. For example, water over-abstraction from the energy sector may deprive aquaculture or agriculture from abstracting or using the desired water quantities in the current period (contemporaneous opportunity cost). However, over abstraction or pollution, for example by agriculture in the current period, may deprive aquaculture or conservation from using the water resource in the future (inter-temporal opportunity cost).

As Balana et al. (2012) point out it is usual practice for cost-effectiveness analysis of environmental measures not to account for corresponding indirect social costs. The rationale of such a practice is based on the assumption of great flexibility in labour markets, which is in turn compatible with neoclassical economic analysis. Similarly, impacts on firm competitiveness are not usually included in the estimations of indirect compliance costs, especially in the case of EU environmental policies.

Assessing environmental benefits of good ecological status

The assessment of environmental benefits of the WFD requires the estimation of welfare gains derived from the improvement of the water bodies from the current ecological status to the GES (Brouwer et al. 2009).

Some of these benefits correspond to the market. For example, improved biota and aquatic ecosystem health can result in increased fish populations, and subsequent market benefits for the fishing sector. However, a significant part of benefits generated by improved water-quality status corresponds to ecosystem goods or services that are not traded in existing markets (Birol et al. 2006; Brouwer et al. 2008). Indicatively, increased water quality contributes to healthy habitats and enhanced biodiversity, valued by society for their scenic beauty and recreational potential, as well as other cultural and non-use values (UK NEA 2011). Moreover, increased water quality also has positive effects on human health and the availability of drinking water.

Environmental economists have long recognized the need for the valuation of non-market values and developed several valuation methods based on revealed and stated preferences (Bateman et al. 2002). These methods are based upon either observed behaviour (revealed preferences) towards some marketed good with a connection to the non-marketed good of interest, or stated preferences in surveys with respect to the non-marketed good. Revealed Preference methods values are "revealed" through studying consumers' choices and the resulting price changes in actual markets that can then be associated with changes in the provision of ecological status. These methods include the Travel Cost method, Hedonic Price analysis, Averting Behaviour, Production Function, Market prices, Replacement Costs and Mitigation Costs. In the case of Stated Preference methods, willingness-to-pay (WTP) (or willingness-to-accept—WTA—compensation) for changes in ecological status are "stated" by respondents in surveys based on structured questionnaires. These methods include Contingent Valuation and Choice Experiments.

The literature on the economic valuation of water non-market benefits is longstanding and very extensive. Moreover, the WFD has generated increased efforts to improve the assessment of water related benefits through Stated Preference techniques, such as Contingent Valuation and Choice Experiments (for a comprehensive review of different applications, see Martin-Ortega 2012).

Table 2 shows the variety of existing environmental valuation methods with a brief description of the approach they take and examples of applications in the water context. Limitations and difficulties associated with each method are also presented.

Beyond conventional Cost–Benefit Analysis: extended, dual and distributional CBA

The Cost–Benefit Analysis (CBA) methodology relies on the principle of economic efficiency as the best means to make fund-allocation decisions to alternative forms of public investment projects, through the application of the Kaldor-Hicks criterion.[1] Table 3 presents the most commonly used indicators in a conventional CBA, including Net Present Value (NPV), Internal Rate of Return (IRR), Benefit–Cost Ratio (BCR) and Payback Period (PB).

Traditionally in water management, CBA has focused only on market costs and benefits (the so-called *private* or *classic* CBA). However, as already mentioned, the achievement of GES is expected to generate substantial non-market benefits (Bateman et al. 2006a; Birol et al. 2006; Brouwer 2008). To be applicable in the context of the WFD, CBA indicators have to be based on a comprehensive inclusion of all costs and benefits, i.e., all the market and non-market values derived from achieving GES (European Commission 2003). Therefore, disproportionality assessments of the WFD require *extended* CBA in order to include environmental externalities that may affect the public through a long period of time (Almansa and Calatrava 2007).

There are a number of studies including non-market benefits into CBA of water management options. Birol et al. (2010) evaluate the artificial recharge of the Akrotiri aquifer in Cyprus with treated waste water, using Stated-Preference methods to capture the total economic benefits of the action. Almansa and Martínez-Paz (2011a,b) study an irrigation water

[1] The Kaldor-Hicks criterion is a less restrictive economic efficiency principle than the Pareto one. While the latter considers that an action is efficient if at least someone actually gains and no-one actually loses, the former assumes if the beneficiaries can (potentially) compensate those bearing the costs, the right economic decision would be to approve the evaluated action (Farrow 1998).

Table 2. Environmental valuation methods.

Valuation method	Approach	Applications	Examples	Limitations
Market prices	Observe prices directly in markets	Environmental goods and services that are traded in markets	The ecosystem's marginal contribution to, e.g., commercial fisheries	Market prices can be distorted, e.g., by subsidies. Environmental services often not traded in markets
Replacement cost	Estimate cost of replacing an environmental service with a man-made service	Ecosystem services that have a man-made equivalent that could be used and provides similar benefits to the environmental service	Water storage and filtration by wetlands	Over-estimates value if society is not prepared to pay for man-made replacement. Under-estimates value if man-made replacement does not provide all of the benefits of the environmental service
Damage cost avoided	Estimate damage avoided due to ecosystem service	Ecosystems that provide protection to houses or other assets	River flow control by wetlands (e.g., flood mitigation)	Difficult to relate damage levels to ecosystem quality
Net factor income	Revenue from sales of environment-related good minus cost of other inputs	Ecosystems that provide an input in the production of a marketed good	Filtration of water by wetlands; the ecosystems marginal contribution to, e.g., commercial fisheries	Over-estimates ecosystem values
Production function	Estimate value of ecosystem service as input in production of marketed good	Ecosystems that provide an input in the production of a marketed good	Filtration of water by wetlands; the ecosystems marginal contribution to, e.g., commercial fisheries	Technically difficult. High data requirements
Hedonic pricing	Estimate influence of env. characteristics on price of marketed goods	Environmental characteristics that vary across goods (usually houses)	Proximity to waste dumps, polluted watercourses	Technically difficult. High data requirements

Travel cost	Travel costs to access a resource indicate its value	Recreation sites	Protected areas, watercourses, bathing places and fishing spots	Technically difficult. High data requirements
Contingent valuation	Ask survey respondents directly for WTP for environmental service	Any environmental good or service	Species loss, good ecologic status, good bathing water quality	Expensive to implement
Choice modelling	Ask survey respondents to trade-off environmental and other goods to elict WTP	Any environmental good or service	Species loss, good ecologic status, good bathing water quality	Expensive to implement. Technically difficult
Value transfer	Use values estimated at other locations	Any environmental good or service	Any of the examples mentioned	Possible transfer errors. Can be as technically difficult as primary valuation

Source: Adapted from Pagiola et al. (2004).

Table 3. Indicators of Cost-Benefit Analysis.

	Indicator	Calculation	Decision criteria/comments
Conventional CBA	Net Present Value (NPV)	$NPV(r)=\sum_{t=0}^{t}\frac{B_t^m - C_t^m}{(1+r)^t}$	$NPV(r) \leq 0 \rightarrow$ *Reject* $NPV(r) > 0 \rightarrow$ *Accept*
	Internal Rate of Return (IRR)	$IRR = k$ *if* $NPV(k)=\sum_{t=0}^{t}\frac{B_t^m - C_t^m}{(1+k)^t}=0$	$IRR \leq r \rightarrow$ *Reject* $IRR > r \rightarrow$ *Accept*
	Benefit–Cost Ratio ($R_{B/C}$)	$R_{\frac{B}{C}} = \frac{\sum_{t=0}^{t}\frac{B_t^m}{(1+r)^t}}{\sum_{t=0}^{t}\frac{C_t^m}{(1+r)^t}}$	$R_{\frac{B}{C}} \leq 1 \rightarrow$ *Reject* $R_{\frac{B}{C}} > 1 \rightarrow$ *Accept*
	Payback period (PB)	The length of time required for cumulative benefits returns to equal the cumulative costs	$t \geq PB$ A shorter PB is desirable
Extended and dual CBA	NPV with environmental effects	$NPV(r,r^a)=\sum_{t=0}^{t}\frac{B_t^m - C_t^m}{(1+r)^t}+\sum_{t=0}^{t}\frac{B_t^a - C_t^a}{(1+r^a)^t}$ For extended approach: $r = r^a$ For dual approach: $r > r^a$	$NPV(r,r^a) \leq 0 \rightarrow$ *Reject* $NPV(r,r^a) > 0 \rightarrow$ *Accept*
	Critical Environmental Rate (CER)	$CER = k^a$ *if* $NPV(r,k^a)=\sum_{t=0}^{t}\frac{B_t^m - C_t^m}{(1+r)^t}+\sum_{t=0}^{t}\frac{B_t^a - C_t^a}{(1+k^a)^t}=0$	$CER \leq r^a \rightarrow$ *Reject* $CER > r^a \rightarrow$ *Accept*
	Intergenerational Transfer Amount (ITA)	$ITA = NPV(r, r^a) - NPV(r)$	A bigger ITA is desirable

Distributional CBA	Implicit distributional weights	Assuming $\sum_{i} a_i NPV_i > 0$ and only two actors involved in the decision-process, then: $0 = NPV = a_1 NPV_1 + NPV_2 \rightarrow a_1^* = \frac{NPV_1}{NPV_2}$	a_1^* can be used to determine what weight is justified from society or political point of view
	Explicit distributional weights	The weights are estimated as follows: $a_i = \left(\frac{\overline{Y}}{Y_i}\right)^{-e}$	Conventional decision criteria
	Parameters and terminology used: B^m and C^m = market benefits and costs; B^a and C^a = environmental benefits and costs; r = conventional discount rate; r^a = environmental discount rate; a_i = weights in Distributional CBA; e = elasticity of the marginal utility of income; $\overline{Y}$ = average income per capita in the evaluation area; Y_i = income of the ith group or sector		

Source: Authors' own elaboration based on Pearce et al. (2006) and Almansa and Martinez-Paz (2011a,b)

desalination plant to prevent the eutrophication of a coastal lagoon, including environmental benefits obtained via the Avoided Cost Method. Alcon et al. (2012) incorporate non-market environmental benefits obtained via a Contingent Valuation survey of the use of reclaimed waste water for irrigation in the south east of Spain. Vinten et al. (2012) incorporate non-market benefits obtained from a Choice Experiment into a CBA of diffuse pollution mitigation in Scottish lochs.

In general, these studies do not address the CBA of a full PoM and therefore do not strictly deal with WFD's disproportionality analysis, but rather with the assessment of the economic efficiency of individual measures. An exception in this respect is Vinten et al. (2012) who assess a full set of measures to mitigate phosphorus diffuse pollution (although not necessarily full achievement of good ecological status); and Birol et al. (2010), who assess a complete aquifer management plan.

Once the costs and benefits of certain actions (e.g., those included in PoM) are quantified for a period of time, they must be summed in order to estimate evaluation indicators which in turn, guide the decision process. In order to compare costs and benefits that are not simultaneous in time, a discount rate must be applied to represent higher society's preference for consumption in the present than in the future. Thus, the higher the discount rates are, the lower importance is attributed to costs and benefits in the future. In relation to environmental goods and services, this raises theoretical and ethical considerations about whether it is appropriate to attribute lower importance to costs and benefits of future generations in relation to current ones. To address this issue, part of the literature proposes to apply different discount rates depending on the nature of costs and benefits (Almansa and Martinez-Paz 2011a). It has been argued that lower discount rates should be applied to non-market values due to sustainability and intergenerational solidarity reasons (Roumboutsos 2010; Almansa et al. 2012). This has been addressed by a dual CBA approach which applies different discount rates to different kind of benefits (Kula and Evans 2011). This provides decision-makers with new indicators which improve the information obtained by CBA, such as the Critical Environmental Rate (CER) and the Intergenerational Transfer Amount (ITA), used for instance in Almansa and Calatrava (2007) (Table 3).

In this context, which discount rate to choose specifically for market and non-market costs and benefits is another critical decision in CBA, as different options can acquire special relevance in deciding postponement or lowering of WFD requirements. The European Commission suggests a standard social discount rate between 3.5 and 5.5 per cent (European Commission 2008). However, there is a debate about which is the most appropriate discounting scheme in the context of the WFD. For instance, in their CBA of the WFD in Scotland, Hanley and Black (2006) use a discount rate of 6%

and suggest that the social value of the WFD would be higher assuming the discount rate of 3.5% adopted by the UK Treasury. This discount rate is used by Vinten et al. (2012) in their study of disproportionality of water quality improvement of Scottish lochs. Meyerhoff and Dehnhardt (2007) use an even lower discount rate of 3%.

Regardless of the rate selected, several guidelines recommend that several discount rates should be checked in order to determine possible bias in the evaluation of investments with long time horizons (European Commission 2008; MARM 2008). In this respect, Birol et al. (2010) select constant rates of 3% and 6%, but also a declining discount rate (4.5% during the first decade of evaluation and declined gradually for the next 190 years until about 2%). Almansa and Martínez-Paz (2011a) conclude that using lower discount rates for environmental effects is generally endorsed by experts and propose a set of discount rates for different ranges of time horizons based on the results of an international Delphi survey.

Finally, disproportionality analysis relying only on the Kaldor-Hicks criterion can have undesirable social implications. Whether the cost of achieving a certain environmental target is disproportionate or not also depends on the social desirability of the benefits and costs distribution among different socio-economic actors. This is particularly important in the context of the WFD, where in large parts of Europe, the costs of improving water quality are going to be mostly borne by land managers in rural areas, while benefits are likely to be higher for urban residents (Bateman 2011), leading to distributional effects and equity problems. Therefore, the answer to disproportionality cannot be found by analysing a single criterion, but through simultaneously examining a range of criteria assessing efficiency and distributional aspects of proposed actions.

Some CBA approaches incorporate these distributional effects (see Table 3). The OECD guide on CBA in environmental policy (Pearce et al. 2006) presents a procedure for calculating relative weights for different actors in a CBA as a way of addressing distributional effects. Different weights associated to different actors can lead to a change in the final decision about a certain action with respect to non-distributional CBA. Two approaches to dealing with distributional effects are proposed: implicit and explicit weights. Implicit weights are the result of the estimation of the NPV obtained for each of the societal actors affected by a policy or programme. These weights provide information on what is the relative importance that a certain sector should have in society for the final decision (based on CBA) to change. Explicit weights, on the contrary, are directly allocated by the analyst and represent a given weight to each of the involved societal actors, which influences the final outcome of the CBA. These weights can be established in relation to variables such as elasticity of the marginal utility income, average income per capita and income of a group sector. However, both these approaches are not free of theoretical controversy and practical difficulties (Wegner and Pascual 2011). In fact, "we know too little about what values these weights should take" (Atkinson and Mourato 2008).

Spatial and temporal scales

The WFD does not specify at what scale the disproportionality analysis has to be carried out, but, in practice, there have been different space-scale decisions and suggestions. For example, River Basin Management Plans have to be approved at the basin scale by planners, while measures that allow achieving GES are implemented at sub-catchment or water body level. The WATECO guidance document (European Commission 2003) does explicitly mention that while cost-effectiveness analysis of the PoM is best performed at the river basin level, derogations can be justified at the water body level (note that they use the word "can", not "should"). At the same time, WATECO also mentions that the relevant scales for the calculation of benefits are "river basin, sub-river basin, sector and sub-sector". Stemplewski et al. (2008) advocate for the sub-basin level and below (water body), as the appropriate working area for the integrated socio-economic planning of the WFD. Their argument is based on the need to balance the two following issues: a) larger planning areas can lead to economies of scale in planning organization and processes, and b) the larger a planning area is, the more difficult it is to get a sufficient detailed comprehension of the local conditions and systems.

The Collaborative Research Programme On River Basin Management Planning Economics in the UK (Jacobs 2007), states that disproportionality analysis should be undertaken based on the minimum amount of evidence (i.e., lowest level of detail) required to make a decision within acceptable limits of risk and uncertainty, bearing in mind the full consequences of the decision.

The choice of spatial scale can have an important influence on the final output of disproportionality analysis, as the value of the environmental change is spatially dependent (i.e., the public does not value changes in the environment irrespective of where they occur) (Bateman et al. 2002; Hein et al. 2006). More specifically, water status improvement preferences and values are expected to be determined, at least partly, by the spatial distribution of beneficiaries throughout the river basin (Bateman et al. 2006b). Distance decay (i.e., the decline in value with increasing distance from a certain site) is a well-established concept in the relevant literature (Bateman et al. 2006b). Other approaches have been utilised in order to account for spatial heterogeneity of preferences (Brouwer et al. 2010; Schaafsma 2010), leading to the conclusion that the economic jurisdiction (i.e., the area incorporating all those who hold economic values regarding a project; Bateman et al.

2006b) does not necessarily coincide with political jurisdiction, nor with natural (hydrological) boundaries. Therefore, estimating benefits at lower scales (e.g., water body) can potentially lead to the exclusion of a great number of beneficiaries of water quality improvements, and therefore into downward-biased benefit estimations.

Regarding the costs, it is widely expected that the WFD will give rise to substantial additional costs which are necessary to cover both new administrative tasks (e.g., monitoring and administration) and water protection measures. In some examples, however, implementing measures mainly in one or two upstream municipalities could be the most cost-effective way of improving the water quality of the rivers and lakes of a river basin. This will however require new mechanisms for allocating costs across administrative boundaries which offer adequate incentives for the most cost-effective measures for the whole river basin.

The temporal scale specification (i.e., the period of time over which cost and benefits need to be compared) is also a difficult issue. The WFD imposes its own planning cycles (e.g., 2015, 2021 or 2027), which could be assumed as the time-horizon of the evaluation process. For example, this is what Vinten et al. (2012) apply to the disproportionality analysis of mitigating diffuse pollution in Scottish lakes. However, the costs of measures and the benefits of improving GES do not necessarily occur in accordance with these deadlines, because costs can extend beyond them. In some WFD CBA applications the time-horizon has been equated to the physical or economic life of the measures in the PoM (Nocker et al. 2007). This is, for example, the case of Martinez-Paz et al. (2013) and Birol et al. 2010, who apply a 50 and 200 years time horizon respectively. Moreover, once environmental features have been restored, benefits do not disappear, and therefore contribute to longer-term (i.e., beyond WFD deadlines) welfare improvements. Furthermore, some decisions may be very sensitive to the time-horizon. For example, if a PoM requires high initial investments, very short time horizons may result in "non-real" disproportionate situations, because benefits cannot compensate costs in such a time frame. Thus, the decision may be to lower instead of postponing the environmental objectives. Hence, economic analysts must describe the time-related assumptions of CBA to supply complete information.

Review of Applications of the WFD's Disproportionality Principle in Europe

In this section, approaches to the disproportionality analysis applied in a number of European countries are presented and discussed. These include applications by decision-makers or regulators, using grey literature, and also academic applications published in the scientific literature. Görlach and

Pielen (2007) undertook an early revision, which is significantly expanded and updated here. A summary of the review is presented in the Annex, including reference to which of the CBA approaches described in Section 2 are used, and whether the concerned applications address distributional effects, and at which spatial and temporal scales the analysis is undertaken. The Annex also includes references to scientific literature addressing in part the issue of disproportionality in the WFD, but not associated to any particular Member State. It should be noted, however, that carrying out an exhaustive inventory of all disproportionality applications in Europe and/or in the literature is out of the scope of the chapter, which, in turn, attempts to provide a flavour of the approaches that have been taken in the different countries.

In the United Kingdom, a number of early guidance documents were produced to advise policy makers on how to undertake economic analysis within the WFD context. Among those, Postle et al. (2004) conceptually discuss the issue of disproportionality for developing a methodology to be applied in England by the Department for Environment, Food and Rural Affairs (DEFRA). While acknowledging that the decision on disproportionality is ultimately a political one, they advocate that CBA should be the main basis for decisions. The distributional assessment would then serve to accommodate the concerns and objections of affected sectors. Further, DEFRA undertook an Overall Impact Assessment for the WFD which aimed at comparing the overall costs and benefits of implementing the WFD in England and Wales. The assessment includes two options of measures (compliance by 2015 and by 2027), but warns that these should only be considered for illustrative purposes.

Despite the wealth of studies on the assessment of environmental benefits of the WFD in the UK (indicatively, Bateman et al. 2006a; Hanley et al. 2006a; Glenk et al. 2011), aimed, in principle, at feeding into the discussion on disproportionality, the actual applications of CBA in this context are very few. The approach of the Scottish Environment Protection Agency (SEPA) largely relies on CEA alone (SEPA 2005). This decision was based on the outputs of the Impact Assessment of the River Basin Management Plan (Scottish Government 2008), which include a qualitative assessment of benefits and led the regulator to assume that mitigation is usually proportionate, unless costs seem particularly high or if such a concern is raised. In principle, this implies that not all environmental benefits are estimated quantitatively. Some relevant academic applications include Hanley and Black (2006) who undertook a CBA of the implementation of the WFD in Scotland at water body and national scales, including mainly market effects. Non-market ones are only included by the authors when previous studies for transferring benefits are available. The authors conclude that the implementation of the WFD can generate more benefits than costs, but at the same time warn that GES could impose disproportionate costs in some river stretches, so that it may be necessary to designate some water bodies as high-modified in order to maximize social net benefits. Also, Vinten et al. (2012) undertake a CBA to assess (dis)proportionality of the costs of improving the water status of Scottish lochs at the national scale. In their study, benefits from GES improvement are obtained from a published Choice Experiment (Glenk et al. 2011). They conclude that improving 67 per cent of loch area to GES by 2015 is a proportionate objective.

In Germany, Meyerhoff and Dehnhardt (2007) combine CBA with two valuation techniques, Contingent Valuation and Replacement Costs methods, to evaluate the suitability of riparian wetlands as a measure to achieve the GES. Another study is that by Held and Krull (2008), who analyse disproportionality and compensations in the hydropower sector under the WFD context. However, there is a general scepticism with regards to the potential use of valuation studies in policy making in Germany. A different approach is proposed by Ammermüller et al. (2008a,b) and Klauer et al. (2007). These guidance papers establish a sequential procedure with different criteria to eliminate those water bodies in which achieving GES is not disproportionate.

In France, the discussion on disproportionate costs is mainly influenced by the water agencies, which are responsible for the WFD implementation. In Seine-Normandie, proportionality of cost is assessed in proportion to the current level of expenditure. If the annual costs of all measures needed to achieve GES do not exceed current expenditure for water management by more than 20 percent, they are not considered as disproportionate. If they do exceed this limit, a more detailed analysis is required, based on the comparison of costs and market and non-market benefits of the measures (Skuras et al. 2010). Another approach, suggested by Artois-Picardie Water Agency, consists of assessing disproportionate costs on the basis of household incomes and the cost of water supply (Laurans 2006).

The Environment Ministry of Spain proposes two arguments for establishing whether the costs of compliance are disproportionate or not (MARM 2008). The first one consists of checking whether water users are able to afford these costs or not, in which case an alternative funding-mechanism should be contemplated. The second criterion, and the one of interest to this discussion, is based on estimating potential benefits expected from water quality changes, such as reduced cost in water provisioning, reduced risk of floods and droughts, or new opportunities for economic activities, amongst others. While acknowledging the existence of non-market benefits, the Spanish legislation advocates for a valuation approach based the cost of the measures necessary to prevent, restore or mitigate environmental damage. This cost-based approach assumes that the value of the environmental good is at least as great as the cost to protect or restore it. This is the approach taken by Molinos-Senante et al. (2010, 2011a,b), who conduct a CBA in which environmental benefits of wastewater treatment are quantified by calculating the shadow prices of undesirable outputs from the process. However, this approximation to the value of ecosystem services is questionable in the context of the WFD, since the estimation of the benefits is precisely needed for a comparison with the costs. Also, Martinez-Paz et al. (2013) adopt a different approach to the estimation of the non-market benefits of the WFD, through using a Contingent Valuation method. They incorporate these benefits in CBA models (classic, extended and dual) and conclude that non-market benefits estimation combined with dual discounting is the more appropriate approach to assess disproportionate costs. Alcon et al. (2012) also apply a dual CBA approach to evaluate the feasibility of supplying reclaimed water for irrigation under scarcity conditions.

In the Netherlands, differences between financial and economic impacts of the WFD have been considered. Brouwer (2008) compares the cost of implementing the WFD with the benefits estimated by means of Stated Preference methods. His analysis shows that WFD may involve disproportionate costs at national level. Brouwer et al. (2008) also found for the Netherlands that the macroeconomic costs of a 10, 20 and 5 percent reduction of the emission levels in the year 2000 of ten priority substances in the EU WFD vary between 0.2 and 9.4 percent of Net National Income (NNI). Furthermore, they show that a large share of total economic costs is borne by important sources of pollution like commercial shipping, the chemical and metal industry. Van der Veeren (2010) also discusses several CBA applications of WFD's effects in this country, and concludes that economic analysis contributes to the enhancement of the transparency of decision-making and, therefore, to the achievement of socially accepted PoMs.

In Denmark the implementation of the WFD is pursued through the Law of Environmental Aims. A guidance document on the overall technical and economic assessments of the WFD in Denmark (Miljøstyrelsen 2006) describes and discusses methods which can be applied to reveal whether costs of achieving good ecological status are disproportionate. It is suggested that an increase on the economic burden put on the public in terms of increased water and sewage charges could be taken as an indicator of disproportionate costs (e.g., increases above 50 percent). Another suggestion is to look at total costs of measures, by setting a fixed limit on total costs and deciding that all individual measures exceeding this limit should be further assessed in depth. A third option deals with the comparison of the size of the planned investment with the current level of investments for the funding sector. If the costs for the planned PoM constitute a large percentage of total current level of investments, this could be a basis for more in-depth analysis of whether costs are disproportionate. The guidelines for the compilation of PoM (Miljøminesteriet 2010) also include a chapter describing exemptions from environmental objectives. These guidelines follow a stepwise CBA approach which should be implemented at sub-district level, using the costs from the cost-effectiveness analyses compared to at least a qualitative assessment of benefits.

Disproportionate costs are generally stated in the Swedish River Basin Management Plans as a reason for exemptions caused by eutrophication, physical changes and environmental toxins. Exemptions and reasons for these are only explained qualitatively, without any calculations. The Valuation Study Database for Environmental Change in Sweden (ValueBaseSWE) was developed at the Beijer Institute of Ecological Economics within the framework of a project funded by the Swedish Environmental Protection Agency. This database is the result of a survey of empirical economic valuation studies on environmental change in Sweden (Sundberg and Söderqvist 2004). The need for and difficulties in describing changes in ecosystem services in monetary terms, generated the background for a report written by Kinell et al. (2010), commissioned by the Swedish Environmental Protection Agency. This report suggests monetary standard values for environmental change in Sweden but emphasizes that the valuation studies carried out in Sweden today are far too few to meet the needs for valuing environmental changes and impacts on ecosystem services.

In Finland, a discussion of economic analysis and disproportionality of costs has been carried out in connection to environmental permit procedures and discussions concerning the definition of Best Available Technologies (BAT). Legislation on Environmental Impact Assessment (EIA) and Strategic Environmental Assessment (SEA) includes considerable economic impacts but the role of economic analysis including the evaluation of disproportionality of costs is unclear. As yet, the Finnish authorities have clarified that the Cost–Benefit Analysis necessary to assess disproportionality was not possible in the first cycle of the WFD, due to lack of data, notably regarding benefits (European Commission 2012).

In Norway, Magnussen and Holen (2011) carried out a report on economic aspects of the WFD in a set of European Union countries and made a proposal for the use of economic analysis in Norwegian water management. Two possible approaches are proposed in this report. The first approach is a stepwise reference value approach based on experiences from the German WFD management which are proposed as a screening methodology for identifying disproportional costs. If measures with possibly disproportionate costs are identified, the sub-basin authorities must do a thorough CBA comparing the costs with benefit assessments and user interests in the area. The second approach proposed is a traditional CBA approach. In this report, the authors propose the further development and adaptation of experiences from Germany to Norwegian conditions and that the water regions perform a detailed analysis based on evaluating the costs from the Cost-Effectiveness Analysis compared to comparable benefit assessments and an assessment of user interests in the area. Bayesian network models have also been used in Norway to conduct Cost-Effectiveness and Cost–Benefit Analysis under uncertainty, responding to the economic analysis requirements of the EU Water Framework Directive (Barton et al. 2008). Barton et al. (2009) also assessed the economic benefits of good ecological status in selected lakes in Norway by using Contingent Valuation and Choice Experiment methods.

Proposed Guiding Principles for the Analysis of Disproportionality of the WFD

The above discussion has shown that specific guidelines on how to *actually* apply disproportionality in the context of the WFD are still missing and homogenous criteria for all European countries have not been established. This section aims at contributing to filling this gap by suggesting a series of critical steps that need to be observed in the application of the assessment of disproportionality. Applications would need to be adapted to national and local circumstances, but through following a series of common principles and guidelines across Europe, comparative analyses are more likely to be feasible.

Our proposal is based on the principle that a co-constructed understanding of the interactions between water ecosystem services and society through the integration of different disciplines and strands of knowledge is needed to design economically efficient and socially acceptable programmes to improve the water status. This transdisciplinary approach requires the integration of hydro-chemical and economic models, accompanied by stakeholder consultation processes, through participatory approaches.

Stakeholder consultation in this context is coherent with the public participation principle of the WFD[2] and facilitating a deliberative process in the assessment of whether costs of measures are disproportionate could be

[2] Although the expression "public participation" does not appear in the Directive, three forms of public participation with an increasing level of involvement are mentioned: i) information supply; ii) consultation; and iii) active involvement. According to the Directive, the first two are to be ensured, the latter should be encouraged (European Commission 2003). Public participatory approaches include all methods which directly involve affected stakeholders and allow them to participate in an on-going process or discussion (Owens 2000; Davies 2001).

another relevant form of public participation (Wright and Fritsch 2011). More generally, as in the case of other complex socio-ecological problems, factors affecting water quality are dynamic, complex, multiscalar and inherently uncertain and therefore require "transparent decision-making that embraces a diversity of knowledge and values" (Reed 2008). Stakeholder participation is a key mechanism to elicit these different forms of knowledge (Parker 2012).

On the basis of this general principle, a step-wise general procedure for the assessment of disproportionality of the WFD is presented next.

Step 1: Collaborative scoping of problems and solutions

The identification of problems regarding compliance with the WFD and its resolution is strictly part of the elaboration of the so called PoM. This is a necessary requirement of the Directive that should, in principle, precede the disproportionality analysis. For the sake of completeness and coherence with the above-described principle, we will simply mention here that a collaborative scoping (between scientists, practitioners, policy makers and local stakeholders) of the main pressures affecting water bodies, as well as measures potentially able to address the problem, seems to be a good basis for the elaboration of the PoM. Rowe and Frewer (2000) identify and review a wide range of methods that can be used to consult (e.g., consultation documents, opinion polls and referendums, focus groups and surveys) or participate (e.g., citizen's juries, consensus conferences, task-forces and public meetings with voting) with stakeholders. Tippett et al. (2007) also provides a useful review of participatory process designs, and a wide range of tools and methods. A practical example on the collaborative scoping of solutions to be incorporated in the WFD's PoM can be found in Martin-Ortega et al. (2012a,b) and Perni and Martinez-Paz (2013), who use focus groups and face-to-face interviews respectively.

Step 2: Establishing arguments for the disproportionality analysis

The starting point should be to establish clear arguments why a disproportionality analysis is needed with particular reference to the economic actors that are affected or serious objections already expressed by stakeholders. Economic analysis, and particularly one including extensive primary data collection, can be expensive. There is no need to launch a full disproportionality analysis when it is clear that the proposed measures are obviously not disproportionate or in case where no stakeholders have already objected to them or are likely to do so (Jacobs 2007). When there is doubt as to whether proposed measures are disproportionate, stakeholders have objected to them or their effectiveness is uncertain, then it should be carried out.

A-priori reasons for why DA is needed are summarized below:

- There is a clearly and narrowly-defined economic activity or a group of economic agents that should undertake the cost of compliance.
- If stakeholders argue for any of the following reasons (examples):
    - The cost to comply (abatement cost) is very high for one industry and lower for another (e.g., phosphate abatement is lower in agricultural production and significantly higher in animal husbandry activities).
    - There are obvious technological obstacles to adopt mitigation measures and thus, economic agents will be forced to close down or re-locate.
    - There are differential costs among agents in the same economic activity due to, for example, scale, or production technology, market orientation, etc. For example, it is well documented that there is an optimum size for Municipal Wastewater Treatment Plans that results in penalizing very small and very large communities.
    - Any other reason that should be brought up by stakeholders and is seriously documented.

Step 3: Identifying the right spatial and temporal scales for analysis

The review of applications shows that EU Member States are proceeding in different ways in this respect (see Table in the Annex), and there are arguments to support different spatial scales. In principle, we advocate for the sub-catchment level as the middle-way between a broader catchment scale, where local specificities might be lost, and a lower (water body or even field) level, which might produce more costly accurate estimates, but that can entail huge assessment costs. More importantly, estimating benefits at lower scales can lead to the exclusion of a great number of beneficiaries of water quality improvements, and therefore into downward-biased benefit estimations.

However, if stakeholder consultation demonstrates that there are significant benefits at a wider scale (which is likely to be the case, for example, in areas of recreational interest attracting people beyond the sub-catchment or even catchment level), these should be incorporated, at least in the discussion of the results.

Therefore, rather than establishing rigid criteria about spatial scales it would possibly be better to adopt an approach based on the analysis of socio-hydrological systems, i.e., taking into account the specificities of the relationship between water bodies and society in each case (again, supported by stakeholder consultation).

Regarding time scale used in the disproportionality analysis we detect that not only Member States, but also scientists differ in their interpretations (see Table in the Annex). From our review, we identify the three main criteria that can drive the selection of the time horizon for disproportionality analysis:

- The "legal criterion", based on the WFD prescription of achieving GES by 2015 and the subsequent planning cycles (2021 and 2027). This is the approach taken by Vinten et al. (2012) in Scotland.
- The "technical criterion" for those cases in which the PoM heavily relies on technological and/or 'hard' solutions (e.g., construction of wastewater treatment plants or transformation of the irrigation systems). In these cases, the lifetime of the measures can set a reasonable criterion. This is the case of Martinez-Paz et al. (2013), who establish 50 years as the lifetime of measures such as restoration of degraded watercourses or wells to extract polluted groundwater that affects a wetland.
- The "policy criterion" based on the capacity of the governments and public authorities to allocate water improvements and costs of compliance to the public and private sectors, taking into account affordability issues in public investment as well as among societal actors (Laurans 2006; MARM 2008).

Only the legal criterion can ensure comparability across different river basins and countries. However, it can occur that compliance by 2015 (or even by some of the further planning cycles) is deemed to be unrealistic in some context and therefore, disproportionality analysis becomes merely an intellectual or academic exercise. Therefore our proposal is to undertake an assessment at different time scales, using a combination of technical and policy criteria adapted to local circumstances in consultation with policy makers and practitioners, plus an assessment using the legal deadlines for comparability purposes.

Step 4: Assessing costs and benefits

As already noted, a Cost-Effectiveness Analysis of the PoM is a pre-requisite to any disproportionality analysis. CEA is used to select the least costly combination of measures that can ensure the achievement of the GES. Hydrochemical models can then be used to assess if cost-effective measures achieve compliance[3] (Balana et al. 2011). However, a critical challenge here is the combination of the outputs of these hydrochemical models with cost-optimization approaches, as there seems to be a trade-off between modelling power and accuracy, attributed to the fact that the two approaches often have a different focus. In more detail, monitoring data used by hydrochemical models can be at a different scale (e.g., catchment level) compared to that used by CEA analysts (e.g., sub-catchment level or/and land management level); in such a context, small-scale interventions such as buffer strips are unlikely to be effectively modelled. Within this context, one can also acknowledge the difficulty of obtaining information on the link between measure-specific pollution reduction (e.g., reduction of fertiliser application) and ecology (e.g., reduction of phosphorus levels).

Independently of the CEA approach, it should be pointed out that it must be *linked* to the disproportionality analysis, not only *related*. As we mentioned above, both CEA and disproportionality analysis should be devised in a common framework where spatial and temporal scales are compatible.

In relation to benefits, Section 2 has outlined the key methods available for the estimation of the benefits related to water quality improvements. We advocate for including monetary assessments of non-market benefits, since they amount to a very significant share of the total benefits, and not incorporating them can lead to biased decisions. Estimating non-market benefits is a very challenging task, but significant progress has been made in the recent literature from which WFD applications can benefit (see Martin-Ortega 2012 for a review).

Although the application of Stated Preference methods to the assessment of WFD benefits is widely accepted, an important problem associated with the costs of undertaking primary analysis (Stated Preference methods are based on the use of costly surveys and can entail sophisticated statistical requirements) remains. Financial and human resource constraints specific to such an analysis have led researchers to investigate the benefit transfer (BT) method and assess its capacity to serve as an alternative (Navrud and Ready 2007; Johnston and Thomassin 2010; Bateman et al. 2011). The benefit transfer method estimates the value of a change of an environmental good at a target 'policy' site drawing on existing values derived from research undertaken at another 'study' site (Hanley et al. 2006b). BT has a particular appeal to policy makers as it is an inexpensive way to quickly obtain values

[3] Hydrochemical models could, in principle, be used to simulate catchment-scale effectiveness of measures for improving water quality, and results could be incorporated into a cost optimization model, which allows the selection and ranking of the most economically efficient combination of mitigation measures. An example for the Thames catchment in England is Whitehead et al. (2013).

for various policy purposes compared to the commissioning of a new study at the policy site. However, BT may come at the cost of reduced accuracy and precision of benefit estimates. Policy makers hence face a trade-off between accuracy and precision of benefit estimates, which would arguably be greatest for a new study at the policy site, and the cost associated with obtaining the benefit information (Brookshire and Neill 1992).

Due to the high importance placed on the BT technique and the laborious process of matching and adapting similar studies, certain databases have been developed. The most important are the ones developed for the European Commission (DG Environment) and the UK. Other countries in Europe maintain regional databases (e.g., Sweden—ValueBaseSWE, Greece—GEVAD). The Greek Environmental Valuation Database (GEVAD) that is maintained by the National Technical University of Athens reviews 317 studies from 49 countries concerning 15 general and 167 specific environmental goods using 38 different valuation methods. The Swedish ValueBaseSWE (Sundberg and Söderqvist 2004), which was updated by Kinell et al. (2010), suggests monetary standard values for environmental changes in Sweden.

An additional general challenge that affects this whole process is uncertainty. Uncertainty affects costs, estimations and effectiveness (and therefore compliance) assessments (Brouwer and De Blois 2008). Several approaches are available for dealing with cost uncertainty, from simple sensitivity analysis to more sophisticated Bayesian approaches or Monte Carlo simulations (Balana et al. 2011; Berbel et al. 2011). Nevertheless, evidence of the ecological impacts of mitigation and restoration measures is still emerging and a categorical demonstration of benefits of restoration in terms of final ecosystem services currently represents a major challenge, in general and specifically, for the assessment of WFD disproportionality. Addressing this challenge requires, in the longer run, interdisciplinary research to deepen our understanding of the specific bio-physical processes underpinning water interventions and ecosystem services delivery and the way these are valued by the public. Yet, there is an urgent need to incorporate these values into the WFD decision-making process. As an immediate way forward, we suggest developing valuation scenarios based on best available evidence of the changes associated with restoration, and including an element of uncertainty in ecosystems provision. This has been done before in the valuation literature, for example, in the context of atmospheric contamination (see Wielgus et al. 2009 for a review), climate change mitigation (Glenk and Colombo 2010) and water supply under scarcity conditions (Rigby et al. 2010; Mesa-Jurado et al. 2012).

Step 5: Assessing economic efficiency

The decision about whether the costs disproportionately exceed the benefits should be based on the use of pre-established indicators. For the CBA indicators to be applicable in the context of the WFD, they have to be based on a comprehensive inclusion of all costs and benefits, i.e., all the market and non-market values derived from achieving the GES. As we explained in Section 2, the new advances in CBA provide a set of indicators available to be used in this context. Notably, the extended dual CBA, in which non-market and market cost and benefits are discounted at different rates, is the one we advocate.

Step 6: Distributional effects and equity considerations

In Section 2, we described two existing approaches for including distributional effects in CBA, based on the allocation of implicit or explicit weights to different societal actors affected by a policy or programme. These approaches have been applied in the economic literature (Somanathan 2006; Anthoff et al. 2009; Prasanthi 2010). However, their specific application to the WFD has not yet been tested. Also, our review of country applications (see Annex) shows that this issue has hardly been addressed. In the WFD context, the main challenges for applying distributional CBA are to quantify the costs and benefits attributable to each societal actor and determine the actual weights on the basis of judgements about society's preferences (for example, using multi-criteria approaches). Notwithstanding, the implicit approach can provide the set of weights that could tip the balance between recommending to implement certain PoM or not.

Where this quantitative approach might not be possible, distributional effects and equity issues should still be discussed qualitatively. For this, a mapping of the main cost bearers and beneficiaries of the improvement of the ecological status can still be made and a 'narrative' of the benefits and costs flows across the different groups can be developed (again in consultation with local stakeholders).

Step 7: Flagging the wider benefits

Many interventions to ensure cost-effective compliance with policy requirements may also generate wider benefits in terms of their impact on other ecosystem goods and services. For example, the implementation of buffer strips and restoration of riparian area to mitigate diffuse pollution can contribute to timber production, carbon storage and soil fertility (Borin et al. 2010; Qiu and Dosskey 2012) as well as enhance biodiversity protection and act as green corridors (Le Maitre et al. 2007). Similarly, constructed wetlands can also provide flood control and carbon storage (Everard et al. 2012). Moreover, improved water ecological status can also generate cultural, recreational, health and other less tangible shared social values (UK NEA 2011). These wider benefits should be acknowledged in any holistic analysis of interventions to maintain or improve water quality.

These wider benefits are to be considered as additional benefits provided by measures to improve water quality and have been largely ignored until now (none of the studies reviewed here and presented in the Annex account for them, with the exception of Meyerhoff and Dehnhardt (2007), which does include a discussion in relation to the multiple benefits provided by wetlands). This is not surprising, since the quantification of the range of services provided by ecosystems and the accurate estimation of associated values (avoiding issues such as double counting, Ojea et al. 2012) is one of the most important, as well as challenging, tasks that environmental economics are currently facing (MA 2005; Fisher et al. 2009; UK NEA 2011). It cannot be expected that disproportionality analysis in the context of the WFD can deal with such issues in the time frame required for compliance. It is therefore not reasonable to expect a quantification of the wider benefits that water quality improvement can provide for its inclusion on any CBA. However, when these benefits can be expected to be significant, they should be acknowledged and used in the final interpretation of the CBA results. Local stakeholders can be consulted in this respect for the identification of potential benefits, such as for example, knock-on effects on the local economy due to increased recreation potential of a certain area. However, we consider that lay knowledge in this particular context may not be sufficient. For this reason, multidisciplinary cooperation by means of expert consultations in a case-by-case basis is advisable. Examples of expert and stakeholder consultation to identify the wider benefits of measures to mitigate diffuse pollution can be found in Martin-Ortega et al. (2013a,b), who use an expert survey and interactive consultation display and a stakeholder workshop.

Conclusion

The implementation of the disproportionality principle in the context of the WFD remains a political decision, but it needs to be informed by economic analysis. In this context, this chapter has discussed the key theoretical and regulatory concepts associated with WFD disproportionality, critically reviewing existing research applications across Europe.

Our discussion highlights two important key issues associated with the disproportionality principle and its actual application. First, despite the flexibility granted by the Directive, any inference on disproportionality should be based on the use of concrete definitions, criteria and (thus) economic and financial indicators. However, any assessment suggestions and proposals should in advance refrain from being applicable at a very high cost. In other words, analysts should avoid proposing assessment methods which are disproportional themselves, and rather non-affordable. Second, specified methodologies should be designed and implemented in a manner that facilitates their capacity to adjust to different contexts, and thus, ensures their realization. Within this framework, the participation of local stakeholders in the assessment process and especially in the definition of what is disproportionate seems to be a rather critical condition.

Taking the above key issues as well as the significant variation in terms of the approaches to this rather crucial aspect of the WFD into account, this chapter has attempted to provide specific guiding principles on how to *actually* apply disproportionality and homogeneous criteria which could be adopted in a European context. These guiding principles can help the application of disproportionality assessment and avoid risks such as non-compliance to specific ecological targets and excessively unequal distributional effects. Also, the adoption of common guidelines would enable the adjustment of applications in accordance to national and local conditions and thus, facilitate the wider application of disproportionality assessment. Further, this rather "disciplined" flexibility in applications can enhance both consistency and accommodate comparative analysis of problems and solutions across Europe.

The proposal specified here is largely based on a combination of co-constructed understanding of interactions between water ecosystem services and society from different strands of knowledge and rigorously defined analytical methodological guidelines. The active engagement of stakeholders throughout the whole WFD-planning process is considered as key to ensuring the design of economically efficient and socially acceptable water-quality improvement action. Hence, co-constructed procedures advocated here are specific to crucial steps of disproportionality assessment, including scoping of problems and solutions, establishing arguments on the need for disproportionality analysis, identifying temporal and spatial scales for analysis, assessing distributional effects and equity considerations and flagging wider benefits. Indicatively, existing weaknesses with regards to evidence on the ecological impacts of mitigation and restoration measures mean that a categorical demonstration of benefits of restoration in terms of final ecosystem services currently represents a major challenge for the assessment of disproportionality. To this end, stakeholder engagement can facilitate the definition and analysis of "option-pathways" specific to different uncertainty scenarios.

Finally, this paper has marginally dealt with rather important issues which can surely affect any efforts to assess disproportionality in the application of the WFD. Indicatively, these could include the specification of clear criteria and specific thresholds on the determination of affordability at both the public and private sector levels. Also, climate change effects can considerably influence the majority of proposed steps, as they introduce an additional layer of uncertainty specific to the above-mentioned interactions.

Acknowledgements

This work was carried out as part of the REFRESH FP7 Project: Adaptive Strategies to Mitigate the Impacts of Climate Change on European Freshwater Ecosystems, funded by EU Seventh Framework Program (Grant Agreement 244121): www.refresh.ucl.ac.uk. The work of Dr. Martin-Ortega was co-financed by the Scottish Government Research Programme (Theme 2, WP2.3 Effectiveness of measures to manage water quality); and the work of Angel Perni was financed with a PhD scholarship from "Fundación Séneca" (Spain). All REFRESH partners, and specially those of *WP6: Cost-effective mitigation adaptation and restoration strategies*, are to be thanked for their active participation in discussions that have nourished this paper. Only the authors are responsible for the content of this chapter. Authors are grateful to Irene Raya for her assistance.

References

Alcon, F., Martin-Ortega, J., Berbel, J. and de Miguel, M.D. 2012. Environmental benefits of reclaimed water: an economic assessment in the context of the Water Framework Directive. Water Policy 14(1): 148–159.

Almansa, C. and Calatrava, J. 2007. Reconciling sustainability in cost-benefit analysis: a methodological proposal. Ecological Economics 60(4): 712–725.

Almansa, C. and Martinez-Paz, J.M. 2011a. Intergenerational equity and dual discounting. Environmental and Development Economics 16: 685–707.

Almansa, C. and Martinez-Paz, J.M. 2011b. What weight should be assigned to future environmental impacts? A probabilistic cost benefit analysis using recent advances on discounting. Science of the Total Environment 409(7): 1305–1314.

Almansa, C., Calatrava, J. and Martinez-Paz, J.M. 2012. Extending the framework of the economic evaluation of erosion control actions in Mediterranean basins. Land Use Policy 29: 294–308.

Ammermüller, B., Klauer, B. and Bräuer, I. 2008a. Kosten-Nutzen-Abwägung zur Feststellung von Ausnahmen aufgrund unverhältnismäßiger Kosten. Im Kontext der EG-Wasserrahmenrichtlinie. Handbuch. Ecologic Institut für Internationale und Europäische Umweltpolitik.

Ammermüller, B., Klauer, B. and Bräuer, I. 2008b. Entwicklung einer Methodik zur nicht-monetären Kosten-Nutzen-Abwägung im Umsetzungsprozess der EG-Wasserrahmenrichtlinie. Ecologic Institutfür Internationale und Europäische Umweltpolitik.

Anthoff, D., Hepburn, C. and Tol, R.S.J. 2009. Equity weighting and the marginal damage costs of climate change. Ecological Economics 68: 836–849.

Atkinson, G. and Mourato, S. 2008. Environmental cost-benefit analysis. Annual Review of Environment and Resources 33: 317–344. DOI: 10.1146/annurev.environ.33.020107.112927.

Balana, B.B., Vinten, A. and Slee, B. 2011. A review on cost-effectiveness analysis of agri-environmental measures related to the EU WFD: key issues, methods, and applications. Ecological Economics 70: 1021–1031.

Balana, B.B., Lago, M., Baggaley, N., Castellazzi, M., Sample, J., Stutter, M., Slee, B. and Vinten, A. 2012. Integrating economic and biophysical data in assessing cost-effectiveness of buffer strip placement. Journal of Environmental Quality 41(2): 380–388.

Barton, D.N., Saloranta, T., Moe, S.J., Eggestad, H.O. and Kuikka, S. 2008. Bayesian belief networks as a meta-modelling tool in integrated river basin management—Pros and cons in evaluating nutrient abatement decisions under uncertainty in a Norwegian river basin. Ecological Economics 66(1): 91–104

Barton, D.N., Navrud, S., Lande, N. and Bugge, A.M. 2009. Assessing Economic Benefits of Good Ecological Status in Lakes under the EU Water Framework Directive. Case Study Report. Norway. NIVA-report no. 5732–2009, Oslo, Norway.

Bateman, I.J. 2011. Catchment Hydrology, Resources, Economics and Management: Integrated Modelling of WFDImpacts upon Rural Land Use and Farm Incomes. End of Award Report. http://www.cserge.ac.uk/sites/default/files/110616%20-%20ChREAM%20Project%20Report%20-%20Revised.pdf (accessed April 2013).

Bateman, I.J., Carson, R.T., Day, B., Hanemann, M., Hanley, N., Hett, T., Jones-Lee, M., Loomes, G., Mourato, S., Özdemiroglu, E., Pearce, D.W., Sugden, R. and Swanson, J. 2002. Economic Valuation with Stated Preference Techniques: A Manual. Edward Elgar Publishing, Cheltenham.

Bateman, I.J., Brouwer, R., Davies, H., Day, B.H., Deflandre, A., Di Falco, S., Georgiou, S., Hadley, D., Hutchins, M., Jones, A.P., Kay, D., Leeks, G., Lewis, M., Lovett, A.A., Neal, C., Posen, P., Rigby, D. and Turner, R.K. 2006a. Analysing the agricultural costs and non-market benefits of implementing the Water Framework Directive. Journal of Agricultural Economics 57(2): 221–237.

Bateman, I.J., Day, B.H., Georgiou, S. and Lake, I. 2006b. The aggregation of environmental benefit values: welfare measures, distance decay and total WTP. Ecological Economics 60(2): 450–460.

Bateman, I.J., Brouwer, R., Ferrini, S., Schaafsma, M., Barton, D.N., Dubgaard, A., Hasler, B., Hime, S., Liekens, I. and Navrud, S. 2011. Making benefit transfers work: deriving and testing principles for value transfers for similar and dissimilar sites. Environmental and Resource Economics 50(3): 365–387.

Berbel, J., Martin–Ortega, J. and Mesa, P. 2011. A cost-effectiveness analysis of water saving measures for the Water Framework Directive: the case of the Guadalquivirriver basin in Southern Spain. Water Resources Management 25: 623–640.

Birol, E., Karousakis, K. and Koundouri, P. 2006. Using economic valuation techniques to inform water resources management: a survey and critical appraisal of available techniques and an application. Science of the Total Environment 365(1-3): 105–122.

Birol, E., Koundouri, P. and Kountouris, Y. 2010. Assessing the economic viability of alternative water resources in water-scarce regions: combining economic valuation, cost-benefit analysis and discounting. Ecological Economics 69(4): 839–847.

Borin, M., Passoni, M., Thiene, M. and Tempesta, T. 2010. Multiple functions of buffer strips in farming areas. European Journal of Agronomy 32: 103–111.

Brookshire, D.S. and Neill, H.R. 1992. Benefit transfers: conceptual and empirical issues. Water Resources Research 28(3): 651–655.

Brouwer, R. 2008. The potential role of stated preference methods in the Water Framework Directive to assess disproportionate costs. Journal of Environmental Planning and Management 51(5): 597–614.

Brouwer, R. and De Blois, C. 2008. Integrated modelling of risk and uncertainty underlying the cost and effectiveness of water quality measures. Environmental Modelling and Software 23(7): 922–937.
Brouwer, R., Hofkes, M. and Linderhof, V. 2008. General equilibrium modelling of the direct and indirect economic impacts of water quality improvements in the Netherlands at national and river basin scale. Ecological Economics 66(1): 127–140.
Brouwer, R., Barton, D.N., Bateman, I.J., Brander, L., Georgiou, S., Martin-Ortega, J., Navrud, S., Pulido-Velazquez, M., Schaafsma, M. and Wagtendonk, A. 2009. Economic Valuation of Environmental and Resource Costs and Benefits in the Water Framework Directive: Technical Guidelines for Practitioners. Institute for Environmental Studies, VU University Amsterdam, The Netherlands.
Brouwer, R., Martin-Ortega, J. and Berbel, J. 2010. Spatial heterogeneity in choice experiments. Land Economics 3(86): 552–568.
Davies, A. 2001. What silence knows—planning, public participation and environmental values. Environmental Values 10: 77–102.
Department for Environment, Food and Rural Affairs (DEFRA). 2007. Overall Impact Assessment of the Water Framework Directive. www.defra.gov.uk/corporate/consult/river-basin/.
EEA. 2012. Towards Efficient Use of Water Resources in Europe. EEA report no. 1/2012. Environmental European Agency, Copenhagen.
European Commission. 2003. Common implementation strategy for the Water Framework Directive (2000/60/EC) Guidance Document No. 8 Public Participation in Relation to the Water Framework Directive Produced by Working Group 2.9—Public Participation.
European Commission. 2008. Guide to Cost–Benefit Analysis of Investment Projects—Structural Funds-ERDF, Cohesion Fund and ISPA, prepared for the Evaluation Unit. DG Regional Policy, Brussels.
European Commission. 2012. Report from the Commission to the European Parliament and the Council on the Implementation of the Water Framework Directive (2000/60/EC), River Basin Management Plans. Member State: Finland. Brussels, SWD (2012) 379 final.
Everard, M., Harrington, R. and McInnes, R.J. 2012. Facilitating implementation of landscape-scale water management: the integrated constructed wetland concept. Ecosystem Services 2: 27–37.
Farrow, S. 1998. Environmental equity and sustainability: rejecting the Kaldor-Hicks criteria. Ecological Economics 27(2): 183–188.
Finnegan, J. 2009. Economic aspects of preparing river basin management plans under the Water Framework Directive. Irish Journal of Agricultural Research 7: 11–18.
Fisher, B., Turner, K.R. and Morling, P. 2009. Defining and classifying ecosystem services for decision making. Ecological Economics 68(3): 643–653.
Glenk, K. and Colombo, S. 2010. How Sure Can You Be? A Framework for Considering Delivery Uncertainty in Benefit Assessments Based on Stated Preference Methods. Journal of Agricultural Economics 62(1): 25–46.
Glenk, K., Lago, M. and Moran, D. 2011. Public preferences for water quality improvements: implications for the implementation of the EC Water Framework Directive in Scotland. Water Policy 13(5): 645–662.
Görlach, B. and Pielen, B. 2007. Disproportionate costs in the EC Water Framework Directive—the concept and its practical implementation. Paper Presented at the Envecon 2007 Applied Environmental Economics Conference, London, 23 March 2007.
Hanley, N. and Black, A.R. 2006. Cost-benefit analysis and the Water Framework Directive in Scotland. Integrated Environmental Assessment and Management 2(2): 156–165.
Hanley, N., Wright, R.E. and Alvarez-Farizo, B. 2006a. Estimating the economic value of improvements in river ecology using choice experiments: an application to the water framework directive. Journal of Environmental Management 78(2): 183–193.

Hanley, N., Colombo, S., Tinch, D., Black, A. and Aftab, A. 2006b. Estimating the benefits of water quality improvements under the Water Framework Directive: are benefits transferable? European Review of Agricultural Economics 33(3): 391–413.
Hein, L., van Koppen, K., de Groot, R.S. and Van Ierland, E.C. 2006. Spatial scales, stakeholders and the valuation of ecosystem services. Ecological Economics 57(2): 209–228.
Held, S. and Krull, D. 2008. Disproportionality and compensations in the context of hydropower and Water Framework Directive. Wasserwirtschaft 98(9): 17–22.
JACOBS. 2007. Report on Guidance on the Evidence Required to Justify Disproportionate Cost Decisions Under the Water Framework Directive. Collaborative Research Programme on River Basin Management Planning Economics. Contract Number: CRP/SG Project 3.
Johnston, R.J. and Thomassin, P.J. 2010. Willingness to pay for water quality improvements in the United States and Canada: considering possibilities for international meta-analysis and benefits transfer. Agricultural and Resources Economics Review 39(1): 114–131.
Kinell, G., Söderqvist, T. and Hasselström, L. 2010. Default Monetary Values for Environmental Change. Report 6323. Swedish Environmental Protection Agency, April 2010.
Klauer, B., Mewes, M., Sigel, K., Unnerstall, H., Görlach, B., Bräuer, I., Pielen, B. and Holländer, R. 2007. Verhältnismäßigkeit der Maßnahmenkosten im Sinne der EG-Wasserrahmenrichtlinie: Komplementäre Kriterien zur Kosten-Nutzen-Analyse. Endbericht. Bund/Länderarbeitsgemeinschaft Wasser, Projekt Nr. AR 1.05.
Kula, E. and Evans, D. 2011. Dual discounting in cost-benefit analysis for environmental impacts. Environmental Impact Assessment 31(3): 180–186.
Laurans, Y. 2006. Implementing cost-effectiveness analysis: perspectives based on recent French pilot studies. Vortrag auf der Messe Wasser Berlin, 5. April 2006.
Le Maitre, D.C., Milton, S.J., Jarmain, C., Colvin, C.A., Saayman, I. and Vlok, J. 2007. Linking ecosystem services and water resources: landscape-scale hydrology of the Little Karoo. Frontiers in Ecology and the Environment 5(5): 261–270.
MA, Millennium Ecosystem Assessment. 2005. Ecosystems and Human Well-Being: Current State and Trends. Island Press, Washington, DC.
Magnussen, K. and Holen, S.N. 2011. Økonomiske aspekter ved vanndirektivet—Erfaringer fra EU og forslag til økonomiske analyser i norsk vannforvaltning. (Economic Aspects of the EU Water Framework Directive—Experience from the EU countries and proposal for the use of economic analysis in Norwegian water management.) SWECO rapportnr. 144811-1 (in Norwegian).
MARM. 2008. Instrucción de Planificación Hidrológica. Ministerio de Medio Ambiente Rural y Marino. Spain.
Martin-Ortega, J. 2012. Economic prescriptions and policy applications in the implementation of the Water Framework Directive. Environmental Science and Policy 24: 83–91.
Martin-Ortega, J., Giacomo, G. and Berbel, J. 2011. Environmental and resource costs under water scarcity conditions: an estimationin the context of the European Water Framework Directive. Water Resources Management 25(6): 1615–1633.
Martin-Ortega, J., McKee, A., Cooksley, S., Jackson-Blake, L. and Slee, B. 2012a. Workshop proceedings on collaborative scoping of solutions in the Dee catchment, UK. REFRESH Project Report—Adaptive strategies to Mitigate the Impacts of Climate Change on European Freshwater Ecosystems. Deliverable 6.5. www.refresh.ucl.ac.uk.
Martin-Ortega, J., Parker, H. and Old, G. 2012b. Workshop proceedings on collaborative scoping of solutions in the Thame catchment, UK. REFRESH Project Report—Adaptive strategies to Mitigate the Impacts of Climate Change on European Freshwater Ecosystems Deliverable 6.18. www.refresh.ucl.ac.uk.
Martin-Ortega, J., Perni, A., Parker, H. and Skeffington, R. 2013a. Proceedings of the workshop for informing the disproportionality analysis and flagging the wider benefits of improving water quality in the Thame sub-catchment, UK. REFRESH Project Report—Adaptive Strategies to Mitigate the Impacts of Climate Change on European Freshwater Ecosystems. Tasks 6.4 & 6.5 Interim Report. www.refresh.ucl.ac.uk.

Martin-Ortega, J., Perni, A., McKee, A. and Helliwell, R. 2013b. Proceedings of the workshop for informing the disproportionality analysis and flagging the wider benefits of improving water quality in the Loch of Skene and Leuchar burn, UK. REFRESH Project Report—Adaptive Strategies to Mitigate the Impacts of Climate Change on European Freshwater Ecosystems. Tasks 6.4 & 6.5 Interim Report. www.refresh.ucl.ac.uk.

Martinez-Paz, J.M. and Perni, A. 2011. Environmental cost of groundwater: a contingent valuation approach. International Journal of Environmental Research 5(3): 606–612.

Martinez-Paz, J.M., Perni, A. and Martinez-Carrasco, F. 2013. Assessment of the programme of measures for coastal lagoon environmental restoration using Cost-Benefit Analysis. European Planning Studies. DOI:10.1080/09654313.2012.722923.

Mesa-Jurado, A., Martin-Ortega, J., Berbel, J. and Ruto, E. 2012. The economic value of guaranteed water supply for irrigation under scarcity conditions. Agricultural Water Resources Management 113: 10–18.

Meyerhoff, J. and Dehnhardt, A. 2007. The European Water Framework Directive and economic valuation of wetlands: the restoration of floodplains along the River Elbe. European Environment 17: 18–36.

Miljøminesteriet. 2010. Retningslinjer for udarbeidelse af indsatsprogrammer. Version 4. By- og landskapsstyrelsen, januar 2010 (http://vandognatur.kontainer.com/files/folders/30088).

Miljøstyrelsen. 2006. Enhedsomkostninger og forureningsbegrænsning ved forskellige miljøforanstaltninger. Katalog til brug for overordnede tekniske og økonomiske vurderinger i forbindelse med Vandrammedirektivet. Miljøstyrelsen, marts. http://www.blst.dk/NR/rdonlyres/2770DB2D-0919-404E-9AD1-0F58AC16B074/0/KatalogVandrammedir.DOC.

Molinos-Senante, M., Hernandez-Sancho, F. and Sala-Garrido, R. 2010. Economic feasibility study for wastewater treatment: a cost-benefit analysis. Science of the Total Environment 408(20): 4396–4402.

Molinos-Senante, M., Hernandez-Sancho, F. and Sala-Garrido, R. 2011a. Assessing disproportionate costs to achieve good ecological status of water bodies in a Mediterranean river basin. Journal of Environmental Monitoring 13(8): 2091–2101.

Molinos-Senante, M., Hernandez-Sancho, F. and Sala-Garrido, R. 2011b. Cost-benefit analysis of water-reuse projects for environmental purposes: a case study for Spanish wastewater treatment plants. Journal of Environmental Management 92(12): 3091–3097.

Navrud, S. and Ready, R. 2007. Review of methods for value transfer. pp. 1–10. *In*: Navrud, S. and Ready, R. (eds.). Environmental Value Transfer: Issues and Methods. Springer, Dordrecht, The Netherlands.

Nocker, L., Broekx, S., Liekens, I., Görlach, B., Jantzen, J. and Campling, P. 2007. Costs and Benefits Associated with the Implementation of the Water Framework Directive, with a Special Focus on Agriculture: Final Report. DG Environment.

Ojea, E., Martin-Ortega, J. and Chiabai, A. 2012. Defining and classifying ecosystem services for economic valuation: the case of forest water services. Environmental Science and Policy 19(20): 1–15.

Owens, S. 2000. Engaging the public: information and deliberation in environmental policy. Environmental Planning A 32(7): 1141–1148.

Pagiola, S., von Ritter, K. and Bishop, J. 2004. Assessing the economic value of ecosystem conservation. World Bank Environment Department Paper No. 101. In collaboration with The Nature Conservancy and IUCN-The World Conservation Union.

Parker, H. 2012. Co-constructing knowledge for theWater Framework Directive: an analysis of the creation and integration of water quality knowledge in the Thame sub-catchment. Master Thesis. Oxford University, Oxford.

Pearce, D., Atkitson, G. and Mourato, S. 2006. Cost–Benefit Analysis and Environment: Recent Developments. OECD, Paris.

Perni, A. and Martínez-Paz, J.M. 2013. A participatory approach for selecting cost-effective measures in the WFD context: the Mar Menor (SE Spain). Science of the Total Environment 458: 303–311.

Perni, A., Martinez-Paz, J.M. and Martinez-Carrasco, F. 2012. Social preferences and economic valuation for water quality and river restoration: the Segura River, Spain. Water and Environment Journal 26(2): 274–284.

Postle, M., Fenn, T., Foottit, A. and Salado, R. 2004. CEA and Developing a Methodology for Assessing Disproportionate Costs. A Report Prepared for Defra. WAG. SE. and DOENI. Project: J467/WFD CEA. UK.

Prasanthi, U.A.D. 2010. Inequalities and externalities of power sector: a case of Broadlands hydropower project in Sri Lanka. Energy Policy 38: 726–734.

Qiu, Z. and Dosskey, M.G. 2012. Multiple function benefit–cost comparison of conservation buffer placement strategies. Landscape and Urban Planning 107(2): 89–99.

Ramajo-Hernandez, J. and Saz-Salazar, S. 2012. Estimating the non-market benefits of water quality improvement for a case study in Spain: a contingent valuation approach. Environmental Science and Policy 22: 47–59.

Reed, M. 2008. Stakeholder participation for environmental management: a literature review. Biological Conservation 141: 2417–2431.

Rigby, D., Alcon, F. and Burton, M. 2010. Supply uncertainty and the economic value of irrigation water. European Review of Agricultural Economics 37(1): 97–117.

Roumboutsos, A.B. 2010. Sustainability, social discount rates and the selection of project procurement method. International Advances in Economic Research 16: 165–174.

Rowe, G. and Frewer, L. 2000. Public participation methods: a framework for evaluation. Science, Technology & Human Values 25: 3–29.

Schaafsma, M. 2010. Spatial effects in stated preferences studies for environmental valuation, PhD Dissertation. Vrije Universiteit, Amsterdam.

Scottish Environment Protection Agency (SEPA). 2005. An Economic Analysis of Water Use in the Solwaye Tweed River Basin District. Summary Report.

Scottish Government. 2008. Impact Assessment of the River Basin Management Plan for Scotland River Basin District. Technical Report. Entec UK Limited.

Skuras, D., Kosola, M.L. and Kontolaimou, A. 2010. Assessing proportionality/ disproportionality of compliance costs: conceptual issues, methods, application. REFRESH WP6, Working Paper 2.

Somanathan, E. 2006. Valuing lives equally: distributional weights for welfare analysis. Economic Letters 90: 122–125.

Stemplewski, J., Krull, D., Wermter, P., Nafo, I.I., Palm, N. and Lange, C. 2008. Integrative socio-economic planning of measures in the context of the Water Framework Directive. Water and Environment Journal 22: 250–257.

Stone and Webster Consultants. 2004. Assessing Current Levels of Cost Recovery and Incentive Pricing. Final Report. DEFRA.

Sundberg, S. and Söderqvist, T. 2004. The Economic Value of Environmental Change in Sweden: A Survey of Studies. Swedish Environmental Protection Agency, report 5360, March 2004.

Tippett, J., Handley, J.F. and Ravetz, J. 2007. Meeting the challenges of sustainable development—a conceptual appraisal of a new methodology for participatory ecological planning. Progress in Planning 67: 9–98.

UK NEA—UK National Ecosystem Assessment. 2011. The UK National Ecosystem Assessment: Synthesis of the Key Findings. UNEP-WCMC, Cambridge.

Van der Veeren, R. 2010. Different cost-benefit analyses in The Netherlands for the European Water Framework Directive. Water Policy 12(5): 746–760.

Vinten, A.J.A., Martin-Ortega, J., Glenk, K., Booth, P., Balana, B.B., MacLeod, M. and Jones, M. 2012. Application of the WFD cost proportionality principle to diffuse pollution mitigation: a case study for Scottish Lochs. Journal of Environmental Management 97: 28–37.

WATECO. 2003. Common implementation strategy for the Water Framework Directive (2000/60/EC). Guidance Document No. 1. Economics and Environment—the Implementation Challenge of the WFD European Commission, Luxembourg.
Wegner, G. and Pascual, U. 2011. Cost-benefit analysis in the context of ecosystem services for human well-being: a multidisciplinary critique. Global Environmental Change 21: 492–504.
Whitehead, P., Crossman, J., Balana, B., Futter, M.N., Comber, S., Lin, L., Skuras, D., Wade, A.J., Bower, M.J. and Read, D.S. 2013. A Cost Effectiveness Analysis of Water security and Water Quality: Impacts of Climate and Land Use Change on the River Thames System. Philosophical Transactions A (in press).
Wielgus, J., Gerber, L.R., Sala, E. and Bennett, J. 2009. Including risk in stated-preference economic valuations: experiments on choices for marine recreation. Journal of Environmental Management 90: 3401–3409.
Wright, S.A.L. and Fritsch, O. 2011. Operationalising active involvement in the EU Water Framework Directive: why, when and how? Ecological Economics 70(12): 2268–2274.

Annex. Review of applications of the WFD's disproportionality principle in Europe.

Country	Description	CBA approach applied[4]	Distributional effects	Spatial scale	Time scale	Discount rate	Interaction with the public	References
United Kingdom	Analysis of the cost and non-market benefits in agriculture due to WFD	-	Spatial variation of the agricultural costs and non-market benefits of the WFD	Basin scale	-	-	Two-stage farm survey. Stated preferences methods	Bateman et al. (2006a)
	Guidance document that provides an overall theoretical framework for assessing disproportionate costs	-	-	Case-by-case basis	Guidance about time horizon is needed	-	Stakeholder participation should be encouraged	Postle et al. (2004)
	Overall impact assessment of costs and benefits of the WFD	-	Costs assessed by sector. Benefits to the general public	England & Wales	2015 – 2027 + Benefits at 95 years	3.5% for first 30 years and 3% for remaining years	A series of sector-based or themed working groups representing stakeholders	DEFRA (2007)

[4] We adopt the following terminology: conventional CBA (CBA); extended or environmental CBA (ECBA), that includes non-market values; and dual CBA (DCBA) that applies different discount rates depending on the nature of the costs and/or benefits (Almansa and Martinez-Paz 2011b).

	CBA of implementing WFD in Scotland	ECBA	Different spatial scale are used to quantify the impacts of a policy in different sectors	3 studies at water body level (Tummel & Dee Rivers and Forth Estuary) and 1 at national level (Scotland)	2002–2042	6%	Choice modelling	Hanley and Black (2006)
	Compare cost and benefits of improving Scottish lochs by 2015	ECBA	-	DA of improving Scottish Lochs at local and national level	2008–2015	3.5%	Down-scaling from a choice modelling conducted in Scotland	Vinten et al. (2012)
Germany	Study focused on remarking importance of wetlands' ecosystem services to achieve WFD	ECBA	-	Basin scale	20 years	3%	Contingent valuation	Meyerhoff and Dehnhardt (2007)
	Multi-step procedure to determine disproportionality	-	-	Water body	WFD deadlines	-	-	Ammermüller et al. (2008a,b; and Klauer et al. (2007)
Spain	Establishes the legal guidelines of disproportionate cost analysis for water planning in Spain	-	Criteria based on user availability to pay and public budgets increases	Case-by-case basis (user, water body or basin level)	WFD deadlines	Establishes that different discount rates should be used to identify possible biases in investments with long time horizons	Public participation regulated by water authorities	MARM (2008)

Annex. contd....

Annex. contd.

Country	Description	CBA approach applied	Distributional effects	Spatial scale	Time scale	Discount rate	Interaction with the public	References
	Assess proportionality of supplying reclaimed water in agriculture in a scarcity context	DCBA	-	Farm level	25 years	3.5%	Contingent valuation	Alcon et al. (2012)
	The benefits incorporated into the CBA are obtained by the distance function approach, which allows for estimating shadow prices of pollutants in wastewater treatments	ECBA	-	At local and river basin scale	15 or 20 years depending on the type of measures under evaluation	2%–3%	-	Molinos-Senante et al. (2010, 2011a,b)
	The authors use three different CBA schemes and recommend to incorporate use and non-use values together with dual discount rates for evaluating PoM	CBA, ECBA, DCBA	Intergenerational distributional effects are commented	Interrelated water bodies (sub-catchment)	Measures maximum lifetime (50 years)	$r_m = 2\%$ $r_a = 1\%$	Contingent valuation	Martinez-Paz et al. (2013)

France	Criteria to determine affordability and the need of introducing non-market benefits are addressed in this study	-	Comments	Household level for affordability issues and water body level for CBA	-	-	Emphasis on the need of interact with stakeholders	Laurans (2006)
Cyprus	This study shows the need of combining CBA, economic valuation and different discount schemes to assess the economic viability of alternative water resources	ECBA (using declining discount rates)	Benefits of different sectors (farmers and households) are identified and quantified separately	Water body (aquifer)	200 years	Test of 3%, 6% and a declining discount rate (from 4, 5 to 2%)	Choice experiment	Birol et al. (2010)
The Netherlands	This paper examines the issue of disproportionality of WFD implementation using public surveys	ECBA	Comments	National scale	Different time scales are used	4%	Contingent valuation	Brouwer (2008)
Denmark	Guidelines for the compilation of programs of measures have a chapter describing exemptions from environmental objectives, and a stepwise approach for the assessment of exemptions.	CBA—minimum a qualitative analysis of benefits.	General description	Sub-district	WFD deadlines	-	Qualitative benefit assessments need to be based on expert/stakeholder judgements	Miljøminesteriet (2010)

Annex. contd....

Annex. contd.

Country	Description	CBA approach applied	Distributional effects	Spatial scale	Time scale	Discount rate	Interaction with the public	References
	Unit costs and pollution reduction by various environmental measures. Catalogue for the overall technical and economic assessments of the Water Framework Directive, describing methods to assess whether costs of measures is disproportionate	CBA—minimum a qualitative analysis of benefits	General description	Sub-district	WFD deadlines	-	-	Miljøstyrelsen (2006)
Norway	Bayesian network models are used to conduct cost-effectiveness and benefit-cost analysis under uncertainty, responding to the economic analysis requirements of the EU WFD	CBA (Benefits related to bathing water quality)	-	Basin scale	-	-	Expert opinions	Barton et al. (2008)
	Willingness to pay per household per year to reach good ecological status is assessed	ECBA	-	The largest lakes in three different sub-catchments in the countries of Østfold and Akershus (Norway)	2008 → "good ecological status"	-	Contingent valuation and choice experiments	Barton et al. (2009)

	A proposal for the use of economic analysis in the Norwegian WFD management based on experiences in a set of European Union Countries	A reference value approach as a screening methodology for identifying dispropor-tional costs and traditional CBA based on evaluating the costs from the cost-effectiveness analysis compared to benefit assessments and assessment of user interests	Benefits of different sectors and user interests should be identified and assessed separately	Basin scale/ sub district. To develop a reference value, a median of costs of measures in a larger scale (water region) is calculated	WFD deadlines	-	Willingness to pay surveys, qualitative/ quantitative benefit assessments involving stakeholders and experts	Magnussen and Holen (2011)
Other papers that deal with the concept of dispropor-tionality and distribution-al effects	Review about DA approaches and applications in some Member States	-	Comments	Comments	Comments	-	Comments	Görlach and Pielen (2007)
	This study focus on how to integrate socio-economic aspects into PoM design	-	Comments	Depends on the type of decision (planning, derogation, etc.)	-	-	Active involvement of stakeholders	Stemplewski et al. (2008)

Annex. contd....

Annex. contd.

Country	Description	CBA approach applied	Distributional effects	Spatial scale	Time scale	Discount rate	Interaction with the public	References
	Description about WFD economic aspects, with special attention to the role of Cost-Effectiveness Analysis and DA	Comments	Comments	Comments	Comments	-	-	Finnegan (2009)
	This paper shows the need of including active involvement procedures to assess disproportionality	Comments	Comments	Case-by-case basis	-	-	Active involvement of stakeholders	Wright and Fritsch (2011)
	This paper reviews the advances and highlights the challenges of the WFD economic aspects such the issue of DA	Comments	Comments	Review about different approaches	Comments	-	"Co-constructing knowledge" with stakeholders	Martin-Ortega (2012)

CHAPTER

11

"Setting Up Young Farmers"—Impact of RDP Measures on Irrigated Agriculture in Greece

Bournaris Thomas,[a,*] *Moulogianni Christina*[b] and *Manos Basil*[c]

Introduction

EU gives many opportunities for its Member States through support measures co-funded under the Common Agricultural Policy. Every Member State implements a National Rural Development Plan (RDP) which must be based on EU Strategic Guidelines (EC 2005). The most important measure of RDP that targets age classes is "Setting up Young Farmers" (Measure 112). The "Setting up Young Farmers" measure has been included in RDPs in the majority of Member States in the last two programming periods, 2000–2006 and 2007–2013. The "Setting up Young Farmers" measure supports the entry of young persons into the agriculture sector and provides for a one-off grant to be paid to trained farmers between the ages of 18 and 40 who

Aristotle University of Thessaloniki, Department of Agricultural Economics, 54124 Thessaloniki, Greece.
[a] Email: tbournar@agro.auth.gr
[b] Email: kristin@agro.auth.gr
[c] Email: manosb@agro.auth.gr
* Corresponding author

have been set-up in farming for the first time (Department of Agriculture, Fisheries and Food of Ireland 2009).

This Rural Development Plan (RDP) affects people living in rural areas by contributing to farm incomes and by limiting farm income variability. It also has indirect impacts by supporting rural employment and maintaining the social fabric of rural areas, by changing the production plans of young farmers and by changing the farm structure with many environmental impacts. The "Setting up Young Farmers" measure also affects the rural economy and by promoting diversification enables local actors to unlock their potential and optimize the use of additional local resources. Finally, there are many impacts on the structural diversity of the farming systems, by improving the conditions for small farms and developing local markets, because in Europe, heterogeneous farm structures and production systems contribute to the attractiveness and identity of rural regions (EC/COM 2010).

Assessment of the "Setting up Young Farmers" regionally will help in re-addressing the CAP in the wider framework of EU policy objectives. In this context, this paper focusing on assessing the impacts of "Setting up Young Farmers" RDP measure in the irrigated agriculture of Greece. The study has been carried out in the region of Central Macedonia in Greece and is an important pilot process enabling the regional authorities to design, develop and implement IA for their regional policies. It tries to measure the impact of "Setting up Young Farmers" in irrigated agriculture measure ex-post so that policy makers will be able to measure the impact of their policies at the regional level. This study was made in the context of LIAISE project (Linking Impact Assessment Instruments to Sustainability Expertise-Network of Excellence-FP7 Environment).

The paper is organized as follows. In the following section, the methodology approach which includes Knowledge Brokerage Approach and MCDA model is provided. Section Three contains the results of the analysis and the last section concludes the paper.

Methodology Approach

Knowledge brokerage approach (KBA)

Knowledge brokerage (KB) is about knowing what knowledge exists, who owns that knowledge, and how that knowledge can be best exchanged among stakeholders and decision-makers. Knowledge brokering has gained increasing importance among all the strategies proposed in scientific literature to increase knowledge utilization during the last decade (Ziam et al. 2009). In the context of the LIAISE project, a coherent suite of test cases of different jurisdictions and policy fields was developed. In order to ensure consistency across test cases, LIAISE proposes a set of support modules, which are linked to the impact assessment process with a final goal to support future policies and design.

The IA Support Modules have a dual role: (1) they provide a research infrastructure in the form of temporary Support Modules during the implementation of Test Cases, and (2) they help structure future interaction processes between researchers and policy-makers, for example, helping facilitate use of long-term IA Toolbox developed by LIAISE. The modules also provide a framework for assessing the most appropriate KB strategy to use. They also include the crucial aspect of evaluation of the KB approach—how KB worked, what factors influenced it and how effective it was (Ward et al. 2009).

Table 1. LIAISE IA Support Modules used in "Setting up Young Farmers" study.

Support Modules	*Phases*	*Setting Up Young Farmers*
1) Test Case Formulation and Scheduling	Formulation Phase	Applicable
2) Identification of Test Case Team and Target Groups		Applicable
3) Policy Storylines	Scoping and Planning Phase	Not Applicable
4) Identifying Impact areas and scales		Partly Applicable
5) IA Scoping and Planning		Not Applicable
6) Tools Selection and Technical Specification	Instrumental Phase	Applicable
7) Indicators, Data requirements and sources		Applicable
8) Tool implementation: analysing impacts	Conceptual Learning Phase	Not Applicable
9) Reflection and evaluation		

Source: Modules for IA Support (LIAISE 2011).

The selection of the appropriate tools for this study was made according to the specific needs of policy makers (after interviews), the LIASE toolbox and the availability of data and tools already used for similar studies. Special attention was also given on Impact Scales and Impact areas. The main tools selected were Simple Tools and Multicriteria Decision Analysis:

- **Simple Tools.** Tools that can give answers when estimating impacts in a simple way, e.g., indicators linked to surveys and questionnaires in order to reflect young farmers' perceptions of rural areas where they live. In this context Economic, Environmental and Social Indicators were measured in order to assess the impacts of the measure. For measuring the impacts on irrigated agriculture the Water Use (m^3/ha) indicator was selected.

- **MCDA Model.** An MCDA model combines different criteria to a utility function under a set of constraints concerning different categories of land, labor, available capital, etc. The implementation of an MCDA model optimizes the Young Farmers farm plan in the prefectures taking into account the available resources (land, labour, capital). The MCDA methodology was implemented for the 2007–2013 programming period.

Methodology—Weighted goal programming and multicriteria analysis

In order to analyze how "Setting Up Young Farmers" measure may influence Greek irrigated agriculture, we have extended the Sumpsi et al. (1993, 1997) and Amador et al. (1998) methodologies for the analysis and simulation of agricultural systems based upon multicriteria techniques. These authors propose weighted goal programming as a methodology for the analysis of decision making. This methodology has been successfully implemented in real agricultural irrigated systems (Berbel and Rodriguez 1998; Bartolini et al. 2007b; Bartolini et al. 2007a; Gomez-Limon et al. 2002; Gomez-Limon and Riesgo 2004; Manos et al. 2008; Manos et al. 2006; Manos et al. 2007; Manos et al. 2010).

To this end, a Multicriteria Decision Model (MCDM) and specifically a Mathematical Programming Multicriteria Model is used in order to achieve better policy-making procedures and the simulation of the most realistic decision process. The MCDM model was chosen because of the variety of criteria taken into account by farmers when they plan their crop plans. It also assembles the multifunctionality of agriculture involving variables related with economic, social and environmental aspects. We used this methodology to estimate a utility function in order to simulate farmers' decision-making processes. Briefly, the methodology can be summarized as follows:

Variables

Each farmer has a set of variables X_i (crops), as described in the previous section. These are the decision variables that can assume any value belonging to the feasible set.

Objectives

This model will optimize at the same time, different criteria as profit maximization, fertilizer minimization, etc. At the preliminary stage, three objectives must be regarded as belonging to the farmer's decision-making process.

Profit maximization

Farmers wish to maximize profits, but calculation of profit requires the computation of some relatively difficult factors such as depreciation. Therefore, for convenience it is assumed that gross margin (GM) is a good estimator of profit, and maximization of profit is equivalent in the short run to maximization of gross margin.

The objective function included in the model is defined as follows:

$$MaxGM = \Sigma GM_i \times X_i \quad (1)$$

where GM is the total gross margin, X_i is crop i and GM_i is the gross margin of crop i.

Fertilizer minimization

Fertilizer minimization is a public objective. For this reason it is not considered in the decision process by farmers. The most obvious indicators are those related to the consumption of water and use of pesticides that are directly related to the pollution of water resources and appear more directly quantifiable at farm level. They are, nevertheless, not obviously subject to aggregation at higher level and their effects on the environment can be evaluated only after some elaboration of prediction models based on diffusion functions.

Fertilizer minimization is the main form for calculating the surpluses of nitrogen potentially dangerous for the environment. It would also be the main indicator of the impact of farming on the environment as groundwater quality is concerned.

In this way, all nitrogen reaching the cultivated soil is included as input. Similar indicators can be designed for other nutrients, such as phosphorus and potassium. For this reason, fertilizer is computed as the sum of fertilizers used for all crops (TF), and its objective function will be as follows:

$$MinTF = \Sigma F_i \times X_i \quad (2)$$

Minimization of labor inputs

The minimization of labor implies not only a reduction of input cost, but also an increase of leisure time and reduction of administration and management processes. The farmers usually show an aversion to hiring labor. An explanation of this behavior is that this parameter is connected with the complexity of crops, because the hired labor adds a degree of complexity to family farming.

For this reason, labor is calculated as the sum of labor for all farm activities (TL), therefore the objective function will be:

$$MinTL = \Sigma TL_i \times X_i \tag{3}$$

No other objectives are proposed in advance. We will assume that at the preliminary stage the three objectives mentioned above are enough to explain farmers' behavior.

Constraints

In order to analyze CAP's impacts we will use several constraints resulting from the implementation of the new CAP. The chosen constraints are the following: Total cultivation area, CAP constraints (subsidies, rights quotas), market and other constraints, rotational and agronomic considerations. All this information has been included in the model that forms the basis for the MCDM simulation.

Table 2. Objectives and constraints used for MCDA.

Objectives	**Constraints**
Profit maximization	CAP Single Farm Payments
$MaxGM = \Sigma GM_i \times X_i$	CAP Production Rights
	CAP Crop rotations
Fertilizer minimization	Land Total
$MinTF = \Sigma F_i \times X_i$	Land Irrigated
	Market Constraints
Labour minimization	Capital
$MinTL = \Sigma TL_i \times X_i$	Variable Costs
	Total Labor

Data requirements

The modeling approach suggested requires data collection from the specific prefecture. A sample of young farmers who have participated in the "Setting up Young Farmers" measure during 2007–2010 from the Region of Central

Macedonia in Greece was chosen. Secondary data also used was gathered from the Department of Agriculture and Veterinary of the region of Central Macedonia. Data were collected for crops, yields, prices, subsidies, income and variable costs (seeds, fertilizers, chemicals, machinery, labor and other costs (e.g., cost of water)), gross margin and fertilizer use.

Description of Case Study Area

The region of Central Macedonia is comprised of seven prefectures: Imathia, Thessaloniki, Kilkis, Pella, Pieria, Serres and Chalkidiki. It borders with the region of Western Macedonia to the West, the regions of Eastern Macedonia and Thrace to the east, the region of Thessaly to the south and the states of the former Yugoslavian Republic of Macedonia and Bulgaria to the north. Its geographical and strategic position has made it a crossroads for trade with the Balkan countries and Eastern Europe. The "Setting up Young Farmers" measure aims to (Ministry of Agriculture and Food 2007):

a. Achieve the transfer of land to young, trained farmers better able to meet the new challenges facing agriculture;
b. Offset the set-up costs faced by young people when establishing themselves in farming, and;
c. Provide assistance for the investments required on such holdings.

From 2000 to 2012, there were two programming periods where the "Setting up Young Farmers" measure was implemented in Central Macedonia, Greece. The first period was from 2000–2006 and the second is the programming period 2007–2013. In the first programming period there were four calls for participating in the measure (2001, 2002, 2003 and 2005). On the contrary, in the second programming period there was only one call in 2009. The sample of the case study was people who have participated in "Setting up Young Farmers" measure from year 2000 to year 2009. The analysis was made in two parts.

- The first part includes the analysis of the farm plans and the main technical and economic characteristics of the farmers, according to their farm plans submitted with their application, in comparison with the real situation in 2010. This part of the analysis was made using simple tools such as descriptive statistics and estimation of the main economic, social and environmental indicators.
- The second part of the analysis includes the implementation of an MCDA model. The farm plan is optimized for 2007–2013. In continuation we compare the values of the chosen indicators between the farm plans 1) submitted with their application, 2) the real (existent) situation in 2010, and 3) the optimum achieved by the MCDA model.

Results

In this section, we present the impact of the "Setting up Young Farmers" RDP measure, in irrigated agriculture in Greece by using the water use indicator. The following Tables 4, 5 and 6 belong to the first part of the analysis and show a comparison between the existent and real situation (2010) of the average farm of each prefecture for both programming periods. These results show how the "Setting up Young Farmers" measure changed the existent situation for the specific indicator. On the other hand in MCDA results Table 7, a comparison between real and optimum (MCDA) situation is given for the same indicator.

First part analysis

The analysis of the first part yielded some useful results for the young farmers' intentions as regards the irrigated land. The following table shows the results for the irrigated and non-irrigated land that young farmers used in both programming periods and the irrigated and non-irrigated land that they will use at the end of the "Setting up Young Farmers" program. The comparison between irrigated and future irrigated land, in both programming periods, shows that the young farmers intend to increase their irrigated land from 1.9 to 3.5 (total average ha) in 2000–2006 and from

Table 3. Average irrigated and non-irrigated agricultural land (existent and future).

	Land (ha)	Prefecture							
		Thessaloniki	Serres	Kilkis	Imathia	Pieria	Pella	Chalkidiki	Total
2000–2006	Irrigated Land (ha)	1.2	1.9	2.8	1.8	1.5	1.6	2.3	1.9
	Future Irrigated Land (ha)	3.8	4.8	4.0	4.0	2.4	2.6	2.6	3.5
	Non-irrigated Land (ha)	1.3	1.0	0.0	0.6	1.7	0.7	0.3	1.5
	Future Non-irrigated Land (ha)	0.2	0.7	0.0	1.5	3.1	2.3	0.9	1.4
	Total Land (ha)	2.5	2.9	2.8	2.4	3.2	2.3	2.6	3.4
	Future Total Land (ha)	4	5.5	4	5.5	5.5	4.9	3.5	4.9
2007–2013	Irrigated Land (ha)	1.9	3.4	2.2	4.3	2.4	2.7	2.1	2.9
	Future Irrigated Land (ha)	3.0	6.8	7.9	6.5	6.6	5.4	3.1	5.6
	Non-irrigated Land (ha)	2.5	0.1	1.5	0.0	0.0	0.0	3.2	0.8
	Future Non-irrigated Land (ha)	5.2	1.5	1.0	1.1	0.0	0.0	7.8	1.9
	Total Land (ha)	4.4	3.5	3.7	4.3	2.4	2.7	5.3	3.7
	Future Total Land (ha)	8.2	8.3	8.9	7.6	6.6	5.4	10.9	7.5

Table 4. Comparison between Existent and Future plans and Real Water Use (m^3/ha) for 2000–2006.

Prefecture	Existent Farm Plan	Future Farm Plan	Increase/Decrease % (existent-future)	Real (2010)	Increase/ Decrease % (existent-real)
Chalkidiki	1107.1	722	–34.78%	1769.1	59.80%
Imathia	4860.9	4779.7	–1.67%	4250.1	–12.60%
Kilkis	3382.9	3013.2	–10.93%	3802.5	12.40%
Pella	4367.3	4299.8	–1.55%	5098.2	16.70%
Pieria	2518	3142.6	24.80%	2768.3	9.90%
Serres	1451.9	2937.9	102.35%	2485.2	71.10%
Thessaloniki	2096.8	1769.5	–15.61%	1620.1	–22.70%

Table 5. Comparison between Existent and Future plans and Real Water Use (m^3/ha) for 2007–2013.

Prefecture	Existent Farm Plan	Future Farm Plan	Increase/ Decrease % (existent-future)	Real (2010)	Increase/ Decrease % (existent-real)
Chalkidiki	483.2	445.2	–7.87%	599.7	24.20%
Imathia	6041.1	6248	3.42%	5813.8	–3.80%
Kilkis	23.8	40.3	69.46%	41.4	74.00%
Pella	4955.4	5127	3.46%	4986	0.60%
Pieria	3540.7	3984	12.52%	2415.2	–31.80%
Serres	4108	4021.9	–2.14%	3554.9	–13.50%
Thessaloniki	3203.2	4076.1	27.25%	3409.8	6.50%

Table 6. Comparison between Real (2010) and MCDA Water Use (m^3/ha).

Prefecture	Real (2010)	MCDA	Increase/Decrease %
Chalkidiki	600	588	–2.02%
Imathia	5814	5779	–0.59%
Kilkis	1800	1832	1.75%
Pella	4986	5079	1.87%
Pieria	2415	2418	0.12%
Serres	3555	3469	–2.48%
Thessaloniki	3410	3683	8.01%

2.9 to 5.6 (total average ha) in 2007–2013. This is due to increasing their total land from 3.4 to 4.9 (total average ha) in 2000–2006 and from 3.7 to 7.5 (total average ha) in 2007–2013.

Table 7. Water Use Impact Comparison.

Prefecture	2000–2006	2007–2013	MCDA
Chalkidiki	↓	↓	↑
Imathia	↑	↑	↑
Kilkis	↓	↓	↓
Pella	↓	↓	↓
Pieria	↓	↑	↓
Serres	↓	↑	↑
Thessaloniki	↑	↓	↓

Programming period 2000–2006

Table 4 shows the comparison between the young farmers' existent and future crop plan and existent and real crop plan in the first programming period of 2000–2006 as regards water use. As we can see, in five prefectures, young farmers intended to reduce their water use at the end of the program; however, the results of the real situation in 2010 reveal that this goal was achieved only in two prefectures (Imathia and Thessaloniki). In the other five prefectures, young farmers had increased their water use. In two prefectures (Pieria and Serres), the future farm plans intended to increase the water use, but as we can see, the increase did not reach the limits that had been set.

Programming period 2007–2013

For the second programming period of 2007–2013, the comparison between young farmers' existent and future crop plan and existent and real crop plan shows that young farmers intended to reduce their water use in only two prefectures (Chalkidiki and Serres). From the comparison with the real situation in 2010, we can conclude that the farmers achieved better results by decreasing the water use in three prefectures (Imathia, Pieria and Serres).

Also, in four prefectures, they achieved an increase the water use but not as high as the limits that they had set for their future farm plans. In one prefecture (Chalkidiki), farmers did not' achieved to decrease water use.

Second part analysis

With the use of MCDA, we can achieve optimum production plans for the young farmers in each prefecture. These production plans are compared to the real ones (2010) in order to analyze the impact of the measure. The MCDA model was implemented only in the 2007–2013 programming period because this is an ongoing policy and the results can be found useful for the farmers.

MCDA

From the results of the MCDA analysis in Table 7, we can conclude that in three prefectures (Chalkidiki, Imathia and Serres) the MCDA model achieved decrease in water use from –0.59% to –2.02%. In four prefectures the model achieved an increase in water use from 0.08% to 1.87%.

Table 7 presents the impact of "Setting up Young Farmers" on water use indicators in the prefectures of Central Macedonia in Greece according to programming period and MCDA analysis. A green arrow indicates positive impact, and a red arrow, negative impact. We can conclude that the MCDA model achieved a decrease in water use (positive impact) in three prefectures (Chalkidiki, Imathia and Serres) and an increase in water use in four prefectures (Kilkis, Pella, Pieria and Thessaloniki); in the latter cases, however, the increase was near zero (no important change).

Conclusion

The "Setting up Young Farmers" RDP measure was implemented in the prefecture of Thessaloniki in Greece for two programming periods. The first period was 2000–2006 and the second, 2007–2013. In this study, an assessment of the impact of the "Setting up Young Farmers" RDP measure in the irrigated agriculture of Greece was made by measuring the water use indicator. The analysis was made in two parts. The first part included an analysis of the irrigated and non-irrigated agricultural land. Also included an analysis for irrigation water use according to the farmers' farm plans, which were submitted with their application. These two analysis were compared with the real situation (2010) for both of the programming periods. This analysis was made with the use of simple tools such as descriptive statistics and calculation of the water use indicator. The second part of the analysis included Multicriteria Decision Analysis. With the use of MCDA the farm plans of each prefecture of Central Macedonia, Greece were optimized, for the second programming period. With this methodology we compared the results of water use from the farmers' farm plans, both with the real situation in 2010 and with the optimum situation according to MCDA, only for the second programming period.

Our main conclusion is that the "Setting up Young Farmers" RDP measure achieved its goals to increase the irrigated land by increasing the total land of the young farmers, and to provide assistance for the investments required on such agricultural holdings. As regards water use, this measure achieved an increase in water use in many prefectures and decrease in others. The environmental impact was negative because of the water use increase in five prefectures for the first programming period (2000–2006) and four prefectures in the second programming period (2007–2013). By changing the young farmers' crop plans using MCDA analysis, we can conclude that MCDA achieved a decrease in water use (positive impact) in three prefectures (Chalkidiki, Imathia and Serres) and an increase in water use in four prefectures (Kilkis, Pella, Pieria and Thessaloniki), the increase however, being near zero (no important change).

References

Amador, F., Sumpsi, M.J. and Romero, C. 1998. A non-interactive methodology to assess farmers' utility functions: an application to large farms in Andalusia, Spain. European Review of Agricultural Economics 25: 92–109.

Bartolini, F., Bazzani, G.M., Gallerani, V., Raggi, M. and Viaggi, D. 2007a. The impact of water and agricultural policy scenarios on irrigated farming systems in Italy: an analysis based on farm level multi-attribute linear programming models. Agricultural Systems 1-3: 90–114.

Bartolini, F., Gallerani, V., Raggi, M. and Viaggi, D. 2007b. Implementing the water framework directive: contract design and the cost of measures to reduce nitrogen pollution from agriculture. Environmental Management 40: 567–577.

Berbel, J. and Rodriguez, A.O. 1998. An MCDM approach to production analysis: an application to irrigated farms in southern Spain. European Journal of Operational Research 107(1): 108–118.

Department of Agriculture Fisheries and Food of Ireland. 2009. Value for money review: the young farmers installation scheme. DAFFI Publications.

EC. 2005. Council Regulation (EC) No. 1698/2005.

EC/COM. 2010. The CAP towards 2020: meeting the food, natural resources and territorial challenges of the future. European Commission, Brussels.

Gomez-Limon, J.A. and Riesgo, L. 2004. Irrigation water pricing: differential impacts on irrigated farms. Agricultural Economics 31: 47–66.

Gomez-Limon, J.A., Arriaza, M. and Berbel, J. 2002. Conflicting implementation of agricultural and water policies in irrigated areas in the EU. Journal of Agricultural Economics 53(2): 259–281.

LIAISE. 2011. Modules for IA Support. Deliverable 6.2.

Manos, B., Bournaris, T., Kamruzzaman, M., Begum, A.A. and Papathanasiou, J. 2006. The regional impact of irrigation water pricing in Greece under alternative scenarios of European policy: a multicriteria analysis. Regional Studies 40: 1055–1068.

Manos, B., Begum, A.A., Kamruzzaman, M., Nakou, I. and Papathanasiou, J. 2007. Fertilizer price policy, the environment and farms behavior. Journal of Policy Modeling 29: 87–97.

Manos, B., Bournaris, T., Papathanasiou, J. and Chatzinikolaou, P. 2008. Evaluation of tobacco cultivation alternatives under the EU common agricultural policy (CAP). Journal of Policy Modeling 31: 225–238.

Manos, B., Papathanasiou, J., Bournaris, T. and Voudouris, K. 2010. A multicriteria model for planning agricultural regions within a context of groundwater rational management. Journal of Environmental Management 91: 1593–1600.

Ministry of Agriculture and Food. 2007. Greek Rural Development Plan 2007–2013. Athens.

Sumpsi, M.J., Amador, F. and Romero, C. 1993. A research on the Andalusia farmers' objectives: methodological aspects and policy implications. *In*: Aspects of the Common Agricultural Policy, VIIth EAAE Congress. Stresa, Italy.

Sumpsi, J.M., Amador, F. and Romero, C. 1997. On farmers' objectives: amulti-criteria approach. European Journal of Operational Research 96: 64–71.

Ward, V., House, A. and Hamer, S. 2009. Knowledge brokering: the missing link in the evidenceto action chain? Evidence & Policy 5: 267–279.

Ziam, S., Landry, R. and Amara, N. 2009. Knowledge brokers: a winning strategy for improving knowledge transfer and use in the field of health. International Review of Business Research Papers 5: 491–505.

CHAPTER

12

Irrigation Dams For Renewable Energy Production

A Case Study in an Agricultural Area in Greece

Thomas Patsialis, Ioannis Kougias, Jacques Ganoulis and *Nicolaos Theodossiou**

Introduction

The economic exploitation of existing agricultural infrastructure is a policy that has recently attracted the interest of the public sector, the local communities and investors, worldwide. In January 2013 the US Congress approved an "Hydropower Regulatory Efficiency Act" that promoted hydropower development in existing infrastructure. Agricultural infrastructure has, an important construction cost because its basic elements (irrigation networks, pumping stations and others) have a substantial

Division of Hydraulics and Environmental Engineering, Department of Civil Engineering, Faculty of Engineering, Aristotle University of Thessaloniki, 54124, Greece.
* Corresponding author: niktheod@civil.auth.gr

installation cost. In many cases the need for flow alterations of nearby creeks/rivers and the corresponding dam construction dramatically increases the total cost.

In many countries, a typical policy in remote, isolated and mountainous areas is to build small earth-fill dams. These dams, although they are relatively simple structures, play a significant role in the economic development of the local region. Their main contribution is to provide the timely and continuous supply of irrigation water, needed to meet the demands of crops and livestock. Moreover, they are used to prevent water flow into specific land regions and flood control. Typically, these constructions are of great importance for the inhabitants of the contiguous areas, since the agricultural production is related with the operation of the reservoir. The dam is connected to the irrigation network and supplies the needed water according to the irrigation schedule, the period of the year and the different types of crops.

Economic development of existing agricultural infrastructure

In the present chapter, a multi-purpose development of the existing agricultural infrastructure is presented. The proposed scheme is applied to irrigation dams in rural areas. These dams are typically constructed in the lower part of a river/creek, near the river basin outlet. In that position, the surface water that flows on the drainage basin is collected. This water satisfies the agricultural needs for water, especially during the months that the needs for irrigation water are notably increased.

The proposed scheme suggests the installation of a mini-scale hydroelectric station in the outflow of an existing reservoir, as an intervention that will exploit the hydropower potential. According to this idea, the surplus water of the reservoir will be used towards the production of electric energy, contributing to the local economy. At the same time the primary use of the reservoir, which is to cover irrigation needs, will not be affected and the agricultural sector's water will be fully covered.

The proposed modification in the reservoir's operation has an additional, important advantage. The construction of the dam and the corresponding reservoir are the most expensive parts of typical hydropower systems. Especially in mini-hydros, a need for dam construction might result to economically non-viable projects. Moreover dam construction can raise certain environmental concerns, even on small-scale weirs (Efstratiadis and Hadjibiros 2011). In the areas under consideration the reservoir has already been constructed in the site and the extension of the use of the reservoir towards energy production has a relatively low cost, without imposing additional environmental impact. Moreover, the sufficient precipitation in the studied mountainous areas during a long period of the year, allows the exclusive operation of the dam towards energy production. In that way, the economic benefits of the investment are substantially increased and so are the benefits to the local area.

The European strategy for renewable energy production

Increasing the non-polluting, renewable energy production is also an important aim due to environmental reasons. Developing small-scale hydroelectricity leads to "green" energy production and follows European Union's policy towards a European strategy for sustainable, secure energy supply. Small scale hydroelectric stations have always been regarded as a more environmentally friendly source of energy when compared to the large scale hydropower stations. The main reason for this is that they demand a small quantity of stored water or no quantity at all in case of run-of-the-river projects. Thus, small hydropower projects result in a minimum obstruction of the flow.

In the proposed scheme, since the dam has already been constructed in the area, the installation of the hydropower station leads to no other, additional impact to the creek's flow. In that way energy is produced without causing any additional environmental impact.

The importance of hydroelectric energy expands also to technical aspects. Hydropower is a very flexible technology for energy generation. In that sense small-scale hydro stations enhance Grid flexibility and address system variability by providing flexible reserves, reducing the load on the thermal fleet, and increasing overall system efficiency. Their quick response time and their ability to provide energy on a constant basis is an advantage compared to intermittent wind and photovoltaic renewable energy sources. These characteristics enable the use of hydroelectric technology towards the optimization of electricity production, meeting sudden fluctuations in demands. Despite being a mature technology, in comparison with other renewable energy sources, hydropower has still a significant potential. According to the European Commission and the Directive on Renewable Energy, the development of low-head or very low-head small hydro plants, is very promising.

Scheduling reservoir operations

Ensuring that the stored water will sufficiently cover both functions (irrigation and energy production) is an important part of the proposed approach. Obviously this stage requires a decision making procedure

(Kougias and Theodossiou 2012) that is closely related to the desires and needs of the local community. The different available options might be based either on technical hydrological-scientific research or on political decisions that answer local particularities and social needs.

Another important aspect includes the Environmental Flow (EF) regimes that reflect the volume of water that the ecosystem needs for its functionalities. Its determination is important in order to secure healthy river ecosystems and good ecological status of water bodies. Specifically, the ecological flow answers the following question: "Up to what extent can the flow of a river change in terms of natural hydrological features and water resources management for human use, maintaining at the same time important ecological features of the river?"

In the present chapter, the authors present in detail the data collection procedure that leads to a deeper knowledge of the river basin's characteristics. Moreover, they present a hydrological model that estimates the water balance in the river basin and the creek flow annual fluctuations. The software used for the hydrological modeling is MIKE SHE. Following these steps, it is quite possible to safely predict the available water that can be managed through the reservoir system. Moreover, it is possible to design an economically viable hydroelectric project that will operate in a profitable manner. This hydrological model also contributes to a detailed record of the flow regime that leads to increased flood-protection and control. Besides, the accurate estimation of future flows, which is of essential importance for the decision making and the optimum design regarding the hydroelectric station, offers a deep knowledge of the characteristics of the flow.

This sophisticated approach increases the safety of the local habitants and the studied area. That is also achieved by the integrated catchment management, implemented with the use of MIKE BASIN software. As it is demonstrated in the latter, catchment management software facilitates the decision making concerning the supply of water for the different functions of the reservoir. It defines different scenarios of water management in order to cover the demand in an optimum way. This knowledge results to a more efficient management of the river basin that improves the financial terms of the system, ensuring that the agricultural and production sector of the area will run in a continuous, unobstructed manner.

Small-Scale Irrigation Dams

Dams are classified according to their type, purpose and size. According to the recent US legislation (Texas Commission on Environmental Quality 2009), small scale dams have a height less than 15 m (40 ft) and a storage capacity between 60,000–1,250,000 m^3 (50–1,000 acre foot). The actual usable storage capacity of an irrigation reservoir must be greater than the net demand over a dry, irrigation season for the crops. Thus, the anticipated irrigation demand from a dam is linked to the yield of the catchment in any year.

As already mentioned, the studied type of dams is typically constructed in isolated and mountainous areas, in order to contribute to the agricultural production and enhance the local productivity and economy. Their height may vary according to the local topography; however they typically have a relatively low height of less than 15 m. Their volume capacity may also vary among a wide range. However, smaller reservoirs, as a rule, have a lower than 1.2 million cubic meters capacity.

Types of irrigation dams

Financial benefits from the cultivation of land in these areas are rarely large enough to allow for expensive, concrete structures to be built for impounding water. Thus, the typical irrigation dams are embankment dams. Besides, earth-fill dams have many advantages over equivalent concrete structures and when built on suitable sites and correctly designed and constructed, can be built using relatively unsophisticated design procedures and equipment (Stephens 2010). Once completed, embankment dams have generally cost less and are most appropriate for farm or other rural situations than a concrete wall.

Moreover, earth dams require minimal maintenance and the embankment is constructed from material excavated from the reservoir area. This provides a small increase in storage capacity and reduces costs. Dams classification for size can be based either on the height of the dam or the maximum storage capacity, Table 1 (Texas Commission on Environmental Quality 2009).

Table 1. Classification of Dams according to their size Texas Commission on Environmental Quality (2009).

Category of Dams	Impoundment Max. Storage (m^3)	Height (m)
Small	$18.5 \cdot 10^3 \leq S \leq 1.2 \cdot 10^6$	$2 \leq h \leq 13$
Intermediate	$1.2 \cdot 10^6 \leq S \leq 61.5 \cdot 10^6$	$13 \leq h \leq 30$
Large	$S \geq 61.5 \cdot 10^6$	$h \geq 30$

Reservoir control in rural-agricultural areas

Typically the earth-fill irrigation dams store water from a river or creek that has a fluctuant flow. The created reservoir is filled during the winter months and then the surplus water is released back to the river/creek. The stored water will cover the annual irrigation needs of the nearby cultivated areas, during the dry, irrigation period (May–September).

However, these mountainous areas, in many cases, receive a substantial amount of precipitation. Especially during the winter period, precipitation is sufficient for any agricultural activities and the operation of the reservoir for irrigation supply is paused until May. Moreover, the intense rainfall, snow and groundwater contribution result to additional runoff, during the winter period. This basin flow leads to an increased discharge in the river network and consequently to a relatively quick filling of the reservoir. After that, the flow of the creek cannot be stored and the surplus water is not exploited (Kollias 2009).

In that way the reservoir covers irrigation needs during the summer for a period of 4–5 months, when the need for agricultural water increases. As a result, during the rest of the year the reservoir is not used for any purpose. Considering the potential of the unexploited surplus water and the available hydraulic height in the dam, the authors propose the installation of mini-scale hydroelectric stations.

Hydropower towards renewable energy production

In such situations, reservoirs can be used for energy production during a long period of the year. Obviously, the hydropower installation has a construction and development cost. On the other hand, it brings a considerable income through directing the produced energy to the National Grid or towards local consumption. The proposed approach transforms the reservoir from a single purpose to a multi-purpose reservoir and the produced energy will contribute to the economical development of the local area through the utilization of existing infrastructure.

This idea is achieved by the installation of a turbine downstream the dam, through which the surplus water will flow. There is such a large extent of available turbines that their operational characteristics (Hydraulic Height—H, water flow—Q) can adapt to those of the existing irrigation dam, whatever they are (Papantonis 2001).

The turbines selected for this type of hydro plant must utilize a small–medium available hydraulic height between 5 m–15 m and a water flow between 100 l/s up to more than 1 m^3/sec. In many cases, Cross Flow turbines are selected, since they match the above criteria and their selection leads to the maximum capacity factor. If the available quantity of water is larger than 1.5 m^3/sec and the hydraulic height is not less than 15 m, Francis type turbines can also be effective. Finally, on situations where the available hydraulic height is rather small and an increased water flow is available, small Kaplan turbines can be installed. This can be documented from various sources. The British Hydropower Association (2005) suggests identical selection of turbines for small-scale hydropower projects. This suggestion is illustrated in Fig. 1.

Obviously the selection of turbines must be carefully made. The type of turbine has to be based on the hydrological data and needs to result in the maximum energy production. At the same time, the selection must balance the technical characteristics with the cost of the investment. In that direction the Power Capacity of the turbine must be balanced with the need to keep the installation's cost low and to an overall economically viable investment.

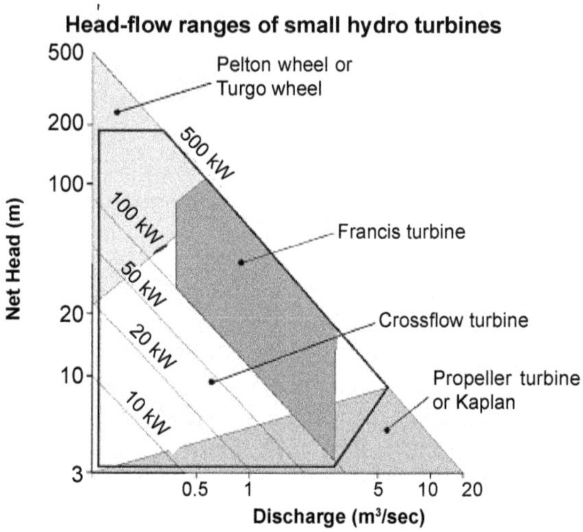

Figure 1. Turbine selection (British Hydropower Association 2005).

Management of the hydropower—irrigation system

Typically, in similar cases a usual policy is to use the reservoir water for agricultural purposes during the irrigation period and operate the hydroelectric station only during the remaining period of the year. As a result the hydropower station ceases its operation during the summer (irrigation period) or it continues its operation on a lower flow.

In the presented approach, the detailed operation of the system is calculated by the river basin management tool (MIKE BASIN) towards the maximum agricultural, financial and environmental benefits. Downstream of the mini-scale hydropower plant, the water outflows without any obstruction back to the riverbed.

A detailed classification of the types of hydropower plants is presented in Table 2.

Table 2. Classification of Hydroelectric Plants according to their capacity British Hydropower Association (2005).

Types of hydroelectric plants	Generating capacity
Large	>100 MW
Small	<10 MW
Mini	**<1000 KW**
Micro	<100 KW
Pico	<5 KW

Advantages of mini-scale hydroelectric projects

The term "small-scale" usually includes hydroelectric projects with a generating capacity lower than 10 MW. Small-scale hydros can be further subdivided into Mini, Micro and Pico hydroelectric projects. Mini hydros, which are surveyed in the present chapter, have a generating capacity smaller than 1000 KW, whereas Micro hydros usually have a generating capacity lower than 100 KW. Pico hydroelectricity is usually used off-grid, in order to cover the basic needs for electricity in isolated, developing communities. It is worth mentioning that this technology, though it has a very low capacity, according to a report prepared for the World Bank (2007), is potentially the lowest cost technology for off-grid electrification.

A Mini Hydropower Plant is not simply a reduced version of a large hydro plant. Mini-scale hydroelectricity plants produce electricity by converting the power available in flowing waters in rivers, canals and creeks into electric energy at the lower end of the scheme, where the powerhouse is located. Since the proposed hydropower scheme can be run-of-the-river, these systems can be considered an environmentally friendly energy conversion option, since they do not interfere further to the creek/river flows and fit in well with the surroundings. The advantages of small hydropower plants are numerous and include grid stability, reduced land requirements, local and regional development and good opportunities for technologies export. Because of their many advantages the number of installations has been considerably increased during the recent years. Along with their minimal environmental impact, their relatively low installation cost, has attracted the interest of those involved in the development of renewable energies in developing countries and more specifically Africa (Belward et al. 2011) sector. In particular, small hydropower comprises one of the most cost-effective and reliable energy technologies for providing clean electricity generation.

Besides, EU strategy towards hydropower energy production has turned to support small-scale hydroelectricity. According to the 2012 EEA's (European Environment Agency) report, forthcoming EU policies to promote hydropower must be compatible with the Water Framework Directive (WFD), minimizing the ecological impacts.

The minimal environmental impact of small hydroelectric projects, compared to fuel energy production, is also justified by the fact that water is not consumed. It is returned to the creek/river and its quality is not deteriorated. Moreover, compared to large-scale hydro, only a relatively small amount of water might be stored. For this reason mini hydro installations do not have the same kind of adverse effect on the local environment as large-scale hydro.

The advantages that mini hydros have over other renewable energy sources are summarized in the following bullet list, provided by the British Hydropower Association (2005):

- High efficiency.
- A high capacity factor (typically >45 %).
- A high level of predictability of the energy production.
- Gradual change of energy production.
- The output power varies only gradually from day to day.
- Hydroelectric generators respond quickly to changing conditions.
- Mini hydro systems have a good correlation with demand, i.e., the output is maximum during winter.
- It is a long-lasting technology. Systems can be engineered to last for more than 50 years, with a low maintenance cost.

Description of the Modeling Process

Introduction

The proposed modeling process includes three basic steps:

i. Geographical Information System (ArcGIS)
ii. Hydrological modeling (MIKE SHE)
iii. Economical modeling/management (MIKE BASIN)

The authors suggest the above process that has several advantages. Firstly, GIS is a common tool that is widely used. GIS databases are available in many regions, offering the required geographical data for the proposed method. More important, the GIS model offers a complete view of the studied area through detailed graphs and thematic maps. This holistic view leads to a more accurate model and finally to better management practices.

Geographical Information System (GIS)

In the first step, the geographical data of the area are inserted into a GIS (Geographical Information System) tool. The authors have used ArcGIS software which is commonly used, worldwide. Following the creation of the ground model, information such as population, crops cultivation or other activities that affect the water consumption in the area, need to be included.

Hydrological modeling

In order to predict the stream flow that flows into the reservoir, a hydrological modeling process needs to be developed. In the proposed approach the authors have used MIKE SHE, a widely used software for building and simulating surface water flow.

Geographical data are linked to those created in the GIS model. Then, the hydrological characteristics of the region are defined. The hydrological basin and the watershed are designed along with the river network. The climatological data are also inserted in the model (precipitation, snow, temperature, soil humidity, land use and others) along with data regarding the observed flow on the rivers and creeks of the studied area.

Hydrological modeling packages result in integrated hydrological modeling systems, since they can simulate the entire land phase of the hydrologic cycle and allow components to be used independently and customized to local needs. MIKE SHE can be used for the analysis, planning and management of a wide range of water resources and environmental problems related to surface water and groundwater, especially surface water impact from groundwater withdrawal, conjunctive use of groundwater and surface water, wetland management and restoration, river basin management and planning, impact studies for changes in land use and climate.

In the presented case study, the dam system is designed to operate without interruption for 20 years (design period). As a result, the created model is simulated based on recorded data for a period of 20 years, which is equal to the design period. In that way, the duration of the simulation period is equal to the forecast period. Besides, predicting the continuous operation of the system for its design period (20 years) contributes to an accurate viability study of the investment. This technique, which is typical for modelers designing water storage projects, has been extensively presented in the Technical Review by ESTIA Consulting 2012.

Then, the model is calibrated and the hydrological parameters are optimized, in order to make an accurate simulation. The aim is to train the model in order to predict as accurately as possible, the observed flow values. In that way it will be possible to predict future water discharge and calculate the surplus water. Moreover, the potential of a hydropower investment will be accurately evaluated.

Economic modeling

The technical characteristics of the dam and the topography of the area determine the generating capacity of the hydroelectric power plant, which will be installed downstream. The decisive parameters are the available Hydraulic Height (H—m) and the water flow (Q—m^3/sec). However, the water flow varies throughout the year. Thus, the turbine selection must be carefully made, in order to lead to an economically viable installation.

The use of a river basin planning and management tools can be very helpful towards the best decision-making. In the present application, MIKE BASIN model has been used, which is an effective tool towards the development and management of the rural area. Its ability to work as a Toolbar in ArcMap, secures the good cooperation between modeling steps i and iii.

As the needs of water for irrigation throughout the year play a leading role, the different types of crops and their irrigation needs should also be inserted in the river basin management tool. Along with them the results of the hydrological simulation, the characteristics of the operation of the hydropower plant and the required environmental flow are inserted.

The techno-economic analysis of the investment is made for a 20 year period, during which no need for significant maintenance works of the installed equipment is assumed.

Moreover, the detailed hydrological modeling of the river basin results in a better management of the reservoir and a deeper understanding of its characteristics. Detailed estimation of future stored volumes contributes to an optimum allocation of the water resources towards irrigation.

A Case Study in Northern Greece

Studied area

The presented approach is applied in an agricultural area of Western Macedonia, Greece, in a high elevation. The geographical, hydrological characteristics and the agricultural activities of the area, make possible the formulation and the study of the proposed method. The recent construction of the dam and the corresponding reservoir (2010) offer the required hydraulic height and volume of water. These two parameters offer an unexploited hydropower potential. Moreover, the nearby network of gauging stations that records meteorological and hydrological data, offers the required information in order to design and implement the hydrological model.

The studied area is near a small village named Ano Melas. The village is located in Northern Greece, in the district of Western Macedonia, 30 km from the city of Kastoria. The village gives its name both to the catchment and the adjacent creek. The river basin extends to both Kastoria and Florina regions.

The studied system consists of:

- "Ano Mela" creek
- the hydrological river basin
- the existing earth-fill dam
- and the corresponding reservoir

Ano Mela creek

Ano Mela creek runs mostly in Kastoria region. Its watershed starts at the top of Vitsi mountain, which is located in the west, in the region of Florina. Its main streambed has a length of 6.78 kilometers and an almost constant slope, equal to 9.18%. The total area of the river basin covers 54.35 km^2.

The flow of Ano Mela creek has been observed for a period of three years, between October 1999 and July 2002. In Fig. 2 the observed discharge values are presented in a mean monthly rate.

According to the observed values and the collected information, the flow of the stream does not stop, although during the summer months it becomes very low. Recorded evidence show that some small-scale flood events occur in the area every 6–12 years, medium-size floods occur

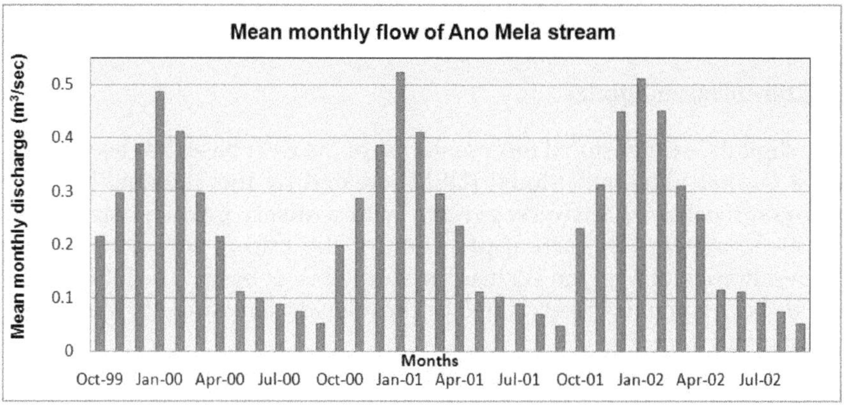

Figure 2. Mean monthly flow of Ano Mela stream (1999–2002).

periodically every 20–35 years and large-scale, dangerous floods may occur every 50–70 years. However, the dense vegetation consisting of red oak and beech trees reduces the intensity of the flood events.

The flood events occur due to the basin's geology which consists of semi-permeable and permeable geological formations, which lead to an increased surface flow. Moreover, the increased slope of both the catchment and the creek are in favor of the generation of a flood event. The flood venture in the area may become even higher considering the erosion that occurs in most of the formations and leads to a substantial amount of debris during a flood event.

Watershed characteristics

The hydrological basin of Ano Mela creek along with the adjacent basins of Makrohori and Agios Antonios creeks, compose the hydrological basin of the East course of Aliakmon river.

The watershed of Ano Mela creek is practically the basin that supplies the existing irrigation reservoir with water. The basic orographic, morphometric and hydrological data are presented in Table 3.

Table 3. Characteristics of Ano Mela basin (based on personal data).

	Symbol	Unit	
Drainage Area	F	Km^2	19.99
Watershed perimeter	Π	Km	19.40
Length of Watershed	$SL_κ$	Km	6.58
Max. elevation	H_{max}	m	1931
Min. elevation	H_{min}	m	1012
Avg. elevation	H_m	m	1372

Digital elevation model

The digital, 3-dimensional representation of the river basin has been based on a Digital Elevation Model (DEM) created by the Hellenic Military Geographical Service (www.gys.gr), with a steady grid-cell size equal to 30×30 meters. The actual precision of the grid is sufficient for the representation of a region such as the studied river basin. The DEM of the watershed region has been created by the relevant software (Arc Map) from point cloud data (Fig. 3).

The created DEM and Flow Direction Model (FDM) resulted in the precise identification of the position of the stream in the river basin, which is illustrated in Fig. 4. Moreover, the exact length of the stream has also been

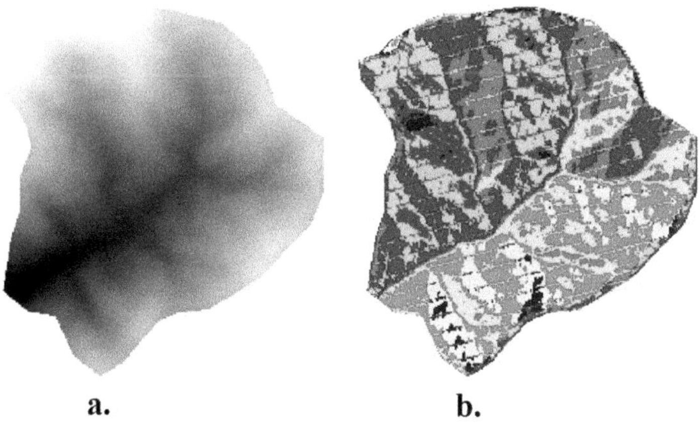

Figure 3. (a) Digital Elevation Model (DEM) and (b) Flow Direction Model (FDM) of Ano Mela river basin.

Color image of this figure appears in the color plate section at the end of the book.

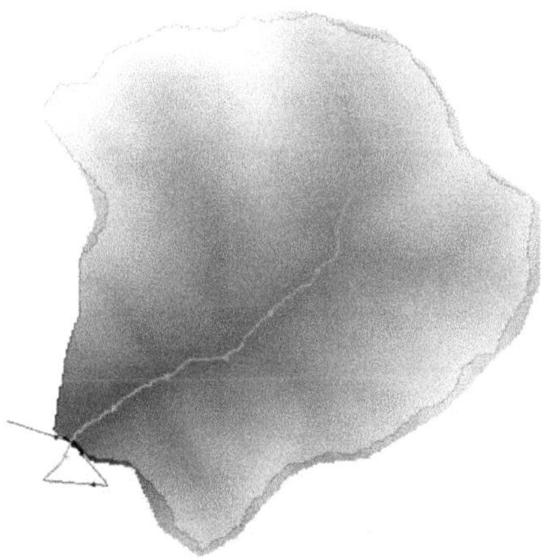

Figure 4. The position of the stream in the basin.

Color image of this figure appears in the color plate section at the end of the book.

calculated. Therefore, the source 30×30 DEM, resulted in the digitization and the representation of the basin in a topographic and hydrological manner.

The existing irrigation dam

The dam has been constructed in the south-western part of the basin, in 2010. It is an earth-fill irrigation dam that contributes, among others, to flood protection and control.

The main characteristics of the dam are:

- Elevation at the construction site: 1,036 m
- Type of dam: trapezoidal
- Maximum Height: H_{max} = 14.14 m
- Hydraulic Height: H = 12 m
- Elevation on the top of the dam: 1,050 m
- Maximum water flow: $maxQ_{100}$ = 27.00 m^3/sec
- Maximum debris flow: $maxG_{100}$ = 0 m^3/sec

The created reservoir serves for the irrigation needs of the local agricultural area. In the following Table 4 the characteristics of the reservoir's storage are presented. For various values of water level in the reservoir, the corresponding stored volume and elevation are illustrated. Important

Table 4. Characteristics of the reservoir for various water levels (based on personal data).

Water Level (m)	Surface Area (m^2)	Reservoir volume (m^3)	Elevation (m)	Comments
	370.62	133.75	1036.36	
1	1559.1	1030.49	1037.36	
2	3052.45	3296.82	**1038.36**	**Inactive Storage**
3	4642.01	7121.49	1039.36	**Active Storage**
4	5975.62	12484.69	1040.36	Active Storage
5	7412.83	19121	1041.36	Active Storage
6	10164.83	28009.66	1042.36	Active Storage
7	12670.47	39369.83	1043.36	Active Storage
8	15949.73	53685.79	1044.36	Active Storage
9	18833.48	71096.43	1045.36	Active Storage
10	21544.98	91352.08	1046.36	Active Storage
11	23927.54	113816.93	1047.36	Active Storage
12	26197.72	138857.88	**1048.36**	**Spillover**
13.06	28941.64	168057.37	**1049.42**	**Max. Water Level**
13.64	30539.42	185305.15	1050	
14.64	33503.22	217359.38	1051	

water levels correspond to the spillover (1048.36 m), the maximum allowed water level (1049.42 m) and the water level that corresponds to the inactive storage (1038.36).

Having an active storage of 140,000 m^3, the existing reservoir offers an attractive hydroelectric potential. Due to the high elevation (more than 1000 m) and the corresponding climatological conditions, the reservoir has minimal losses from evaporation. Moreover, these conditions secure its repletion even during drought periods.

Using the data included in Table 4 it is possible to create the Water level–Surface and Water level–Volume diagrams of the reservoir, which are illustrated in Fig. 5.

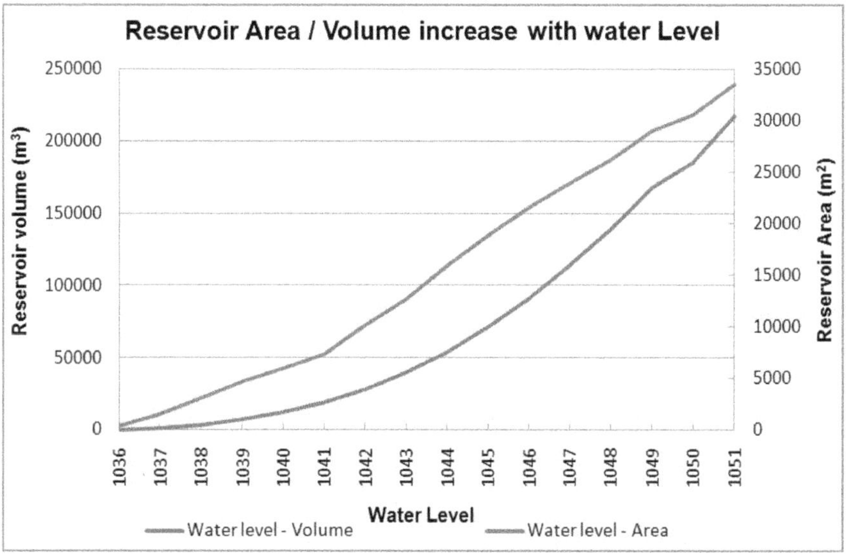

Figure 5. Water level–Surface and Water level–Volume diagrams.

Color image of this figure appears in the color plate section at the end of the book.

Hydrological modeling

In the studied area, MIKE SHE has been used towards the integrated hydrological modeling. The parameters that affect the basin in a hydrological manner have been included in the created model. The structure of the model includes a simulation period between 1/10/1999 and 30/9/2002 (3 years) with an initial time step of 6 hours.

The catchment has been inserted in the model as a shape file. Then the topography of the studied area has been inserted. The Digital Elevation

Model (DEM) as presented in paragraph 4.3 has been included in the hydrological model (Fig. 6).

The next step has been the insertion of the precipitation data in the hydrological model. The created model has been adjusted to process daily data. There are several rain gauge stations in the area, some of which provide data on a monthly basis. Using relevant software, the monthly data generated and provided daily precipitation data. The result of this simulation is considered to be reliable, since the generated information has been compared and cross-checked with observed, known values.

Evapotranspiration was given a uniform and constant value for all the extent of the stream basin. This value was based in the literature concerning similar mountainous areas and is equal to 2 mm/day.

In the studied agricultural area, the terms Land Use and Land Cover are identical, since both refer to the cultivated crops. Corine 2000 European Union project has derived Land Use data for the specific area, which have been used and are illustrated in Fig. 8.

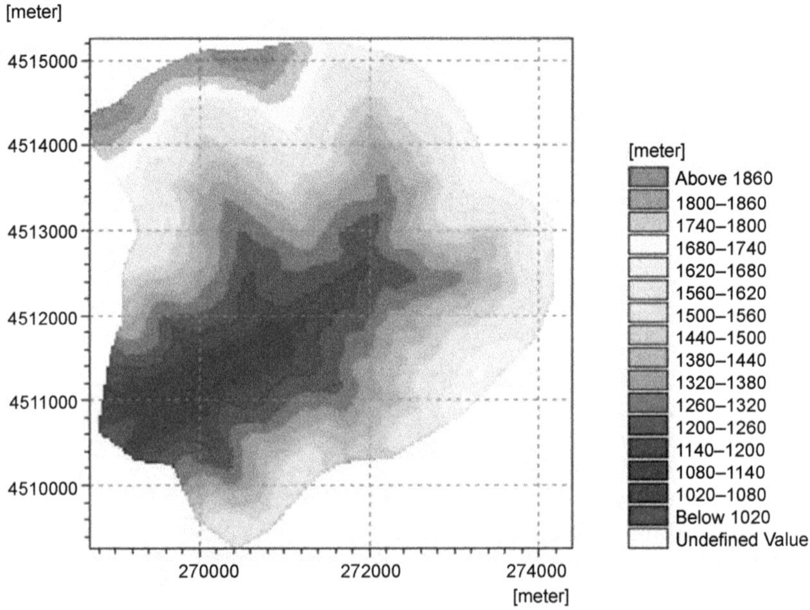

Figure 6. The Ano Mela DEM in MIKE SHE.

Color image of this figure appears in the color plate section at the end of the book.

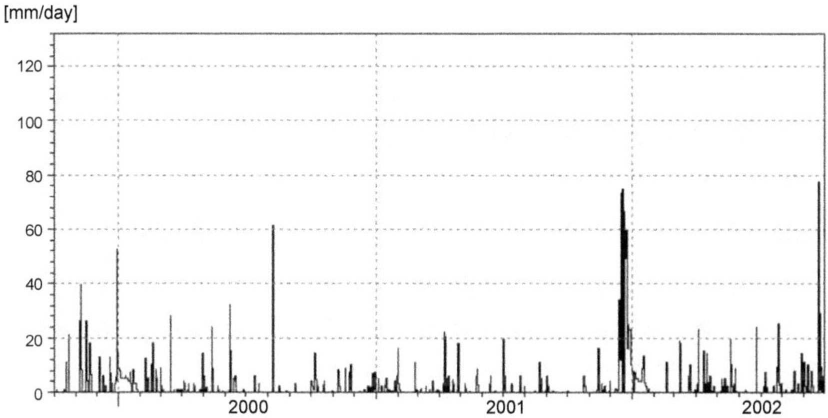

Figure 7. Precipitation data.

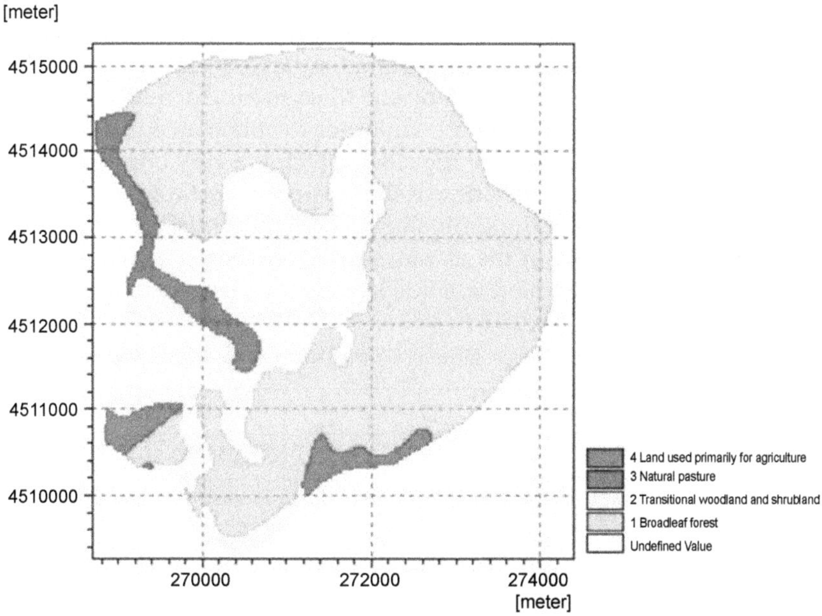

Figure 8. Land Use.

Color image of this figure appears in the color plate section at the end of the book.

Results Obtained by the Hydrological Model

After the initial setting and calibration of the model, a PP (post process) simulation has been executed. Since no error occurred, the final run of the model has been executed. The total computation time of the simulation was 15 minutes.

Water balance

The model, after completing the calculations, has provided the water balance of the Ano Mela basin for the period 1999 to 2001. These results are illustrated in Table 5.

The model has also estimated the water flow of Ano Mela stream. This estimation is illustrated in Fig. 9 along with the observed values of the flow. It is obvious that the hydrological model has managed to simulate the characteristics of the flow. Regarding the peak flow which typically occurs in January, the model has made a very accurate estimation, especially for the years 2000 and 2002.

The results of the 1999–2002 simulation have been used in order to expand the calculation of the flow of Ano Mela stream for a 20-year period. This longer period has been chosen in order to obtain an equal long-term estimation of the future flow of the creek which is a very important factor towards the optimum installation of the hydropower station.

The time interval of the simulation is daily and MIKE SHE provided over 7300 daily values of the stream flow (Q). These values have been processed and are illustrated in a flow-frequency diagram in Fig. 10.

Table 5. Water Balance (as Resulted from MIKE SHE Modeling process).

Precipitation	2455 (mm)
Flow to river	950 (mm)
Evapo–transpiration	893 (mm)
Infiltration	613 (mm)
Base flow to River	508 (mm)
Groundwater Storage change	42 (mm)

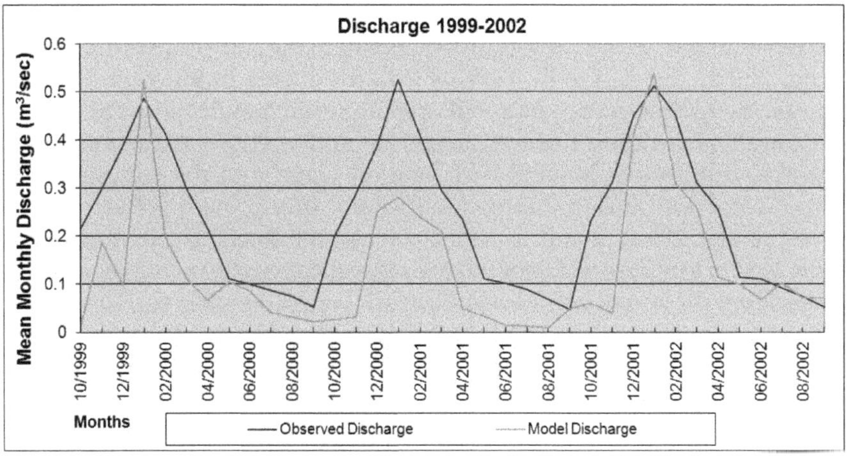

Figure 9. Discharge of Ano Mela stream, as calculated by MIKE SHE.

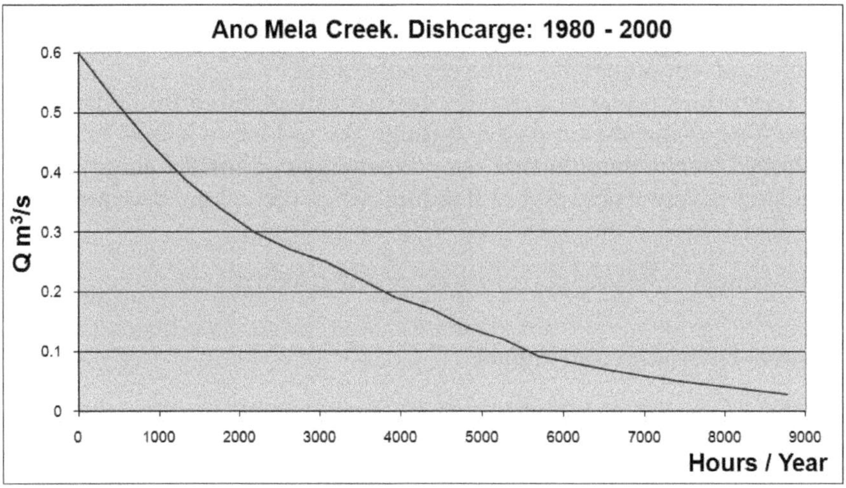

Figure 10. Flow Duration Curve (FDC) of Ano Mela creek.

Installation of mini-scale hydroelectric plant

The presented simulation results in a reliable estimation of the future stream flow and its characteristics. Thus, it is possible to design the hydroelectric project and choose its basic technical characteristics.

According to the proposed scheme, the stored water will be directed to irrigation channels/pipes in the period between May and the end of September (five months). This is the period that the crops are irrigated and thus the reservoir will be exclusively used for irrigation purposes. During the remaining seven months, from October until the end of April, the stream water will be directed to the turbine towards energy production. In other words, the hydroelectric plant will operate seven months per year.

Parallel operation towards irrigation and energy production is not possible due to the demand of irrigation water and the simultaneous lower water flow during summer. At the same time providing the required Environmental Flow quantity further increases the demands. Environmental Flow is a very important factor towards the sustainability of the ecosystem. In the letter the potential for a combinatorial operation only for September, when the irrigation demand is minimized, is examined.

The analysis of the hydrological data and the available hydraulic height led to a selection of a turbine with a Power Capacity of P = 120 kW. The type of the turbine is Cross-Flow and has a potential of a 680 MWh of energy production, annually. The aforementioned production can only be achieved in case of an exclusive reservoir use towards energy production and results in a capacity factor (CF) of the Mini Hydroelectric project equal to 65%. Obviously, ceasing the operation of the hydropower plant during the irrigation period, results in a substantial reduction of the produced energy and -consequently- of the capacity factor.

The turbine is placed 20 meters downstream of the earth-fill dam. The entire flow of the stream passes through the turbine, while its release is controlled by the dam. In that way, the energy potential of the stream is exploited, except from cases of flooding, when the volume of water can't be utilized. Briefly, the creek water follows the route:

Water course → Reservoir → Turbine → Water course

Minimal environmental impact

The designed mini-scale hydro project causes no additional diversion of the creek. After the removal of any debris and gravel in the grave deposition channel, the water is directed through a 50 meter long penstock to the turbine. There, the dynamic energy is converted to kinetic and the moving generator produces the electricity, which can be either sold to the National Grid or directed to the local community or industries for local consumption. The water return back to the creek flow, in perfect quality.

Environmental Flow regime

The calculation of the Environmental Flow (EF) in the studied creek needs to take into account both the two different stages of operation of the reservoir (Ganoulis and Skoulikaris 2011). This means that the required EF must be provided in both winter, when the reservoir will be used towards energy production and in summer, when the reservoir covers the demand for irrigation water.

During the design of the specific dam, a detailed calculation of the EF flow has been made, according to the Greek/EU legislation and practices. The current Greek law defines the minimum EF as the 30% of the average summer creek flow. This definition results in a minimum value for the Ano Mela creek in the range of 50–70 l/sec.

The operation of the hydroelectric station is scheduled for the period between October and April. In that period the water flow of the river has increased values that can cover sufficiently the energy production and the environmental flow constraints. According to Fig. 9, the average flow in that period ranges between 200 and 500 l/sec. As a result, it can be safely said that the construction of the hydroelectric plant does not prevent the provision of the required EF regimes.

The critical period to secure the provision of EF regimes is summer, when the inflows to the creek are reduced. Ensuring that sufficient EF is released has been an important issue of the dam design and construction process. Extending the operation of the reservoir towards energy production doesn't affect the summer Environmental Flow regime, as defined in the dam's construction and feasibility study.

Covering the irrigation—agricultural needs

The cultivation area, irrigated from the water that is stored in the reservoir, is 250 hectare. The main cultivations are beans (202 ha) and secondarily, maize (18 ha). In the remaining 30 ha medic and potato are the main cultivations.

In Table 6 the irrigation needs of the area are illustrated. The Table includes the water demand per 1,000 m^2, the number of irrigation periods for each product and finally, the total amount of the required irrigation water.

The total required water for the cultivation of the crops is 931,600 m^3. This demand is partially covered by groundwater sources with the use of existing wells and springs. The extent that these sources contribute to the irrigation needs is 337,600 m^3 annually.

The remained annual needs for irrigation are equal to 594,000 m^3. Since the available volume in the reservoir is 595,398 m^3, it can be estimated that the reservoir can sufficiently cover the remaining annual irrigation needs, offering a surplus quantity equal to 1398 m^3. Detailed information regarding the volume of stored water in the reservoir is illustrated in Table 4.

Table 6. Irrigation water needs in the Ano Melas basin (based on personal data).

		Demand for irrigation water						Water demand		Total surplus water
Cultivated Product	Cultivated Area	Required Water (per 1000m^2)	Irrigations per season	Total required water per season	Interval between succeeding irrigations (days)	Irrigation season	Soil Moisture contribution (m^3)	m^3	Water Vol. pumped from the Reservoir	m^3
Beans:	3100	40	5.6	694400	18–25	5/5–31/8	241000	594000	595398	+1398
Maize:	400	40	4.2	67200	18–25	1/6–31/8	36300			
Rest:	500	50	6.8	170000	16–28	1/5–30/9	58700			
Total:	4000			931600			135000			

Combinatorial operation of the reservoir

Interesting information derived from Table 6 is that the irrigation needs during September are quite low. From the total demand of 931,600 m^3, only 3–5% is needed for September. The cultivation of beans and maize is completed in August. As a result, the water demand in September is low. Considering that a part of this demand is to be covered from groundwater sources, it can be deduced that the reservoir has a small demand to cover during September.

Thus, it is possible to start the operation of the hydroelectric station earlier, before the irrigation period has ended. The operation of the hydro station will be under a smaller water supply (Q), leading to a lower rate of energy production, compared to the one in winter. Still, the produced energy can offer important financial benefits. This energy is estimated between 15–30 MWh, depending on the summer precipitation.

Conclusions

The proposed scheme suggests the seasonal operation of a mini-scale hydropower station in an existing earth fill irrigation dam. The studied reservoir has been constructed in 2010, in order to cover the local agricultural needs in a rural area in North-Western Greece.

Although the proposed scheme results in the closure of the hydro stations for five months annually, it still offers an important economic potential. MIKE BASIN package provided some management options that have been described in paragraph 7. These scenarios aim for the best possible management of the systems. They foresee that the total expected cost of the hydro station will range between 200,000 and 220,000€, which is economically viable, since the largest and most expensive part of the SHP (dam) is already built.

Moreover, the 7-month operation of the hydroelectric station is expected to result in an annual energy production of almost 400 MWh. As a result, a depreciation of the investment is expected after 6–8 years, depending on the price per KWh. As a result, existing small irrigation dams produce energy during the non-irrigation months in a sustainable manner, adding a green use to the existing water-storage infrastructure. Since hydroelectricity is a long-lasting technology, it is obvious that the financial terms of the investment are very good and economic benefits from the proposed application are substantial (Kaldellis et al. 2005).

Since in such mountainous-remote catchments land uses are limited (e.g., livestock, small-scale agriculture) the GIS modelling can be facilitated with open, available data (DTM, Corine 2000, Google Earth, etc.) or public-service datasets (ministry, HMGS, IGME, etc.). However, calculating the water discharge requires delicate calculations and on-site observation and measurements.

This approach is important since it offers many alternatives for utilizing the produced energy. As already said, the usual decision is to sell the produced electricity through a connection to the National Grid. Another funding strategy requires an investment by local cattlemen, in order to provide energy to their cattle farms. In that way, the produced energy is directed to the nearby cattle productions, providing renewable, low-cost energy. A third option is to provide energy to the irrigation systems which are connected to the reservoir (pumping stations, irrigation networks), reducing the cost of the agricultural production.

Besides the above, the proposed scheme, through the integrated management of mountainous agricultural areas, contributes to the management, protection and flood control of the greater basin. At the same time the production of green, renewable energy with minimal environmental impacts follows the recent EU practices towards sustainable development.

Acknowledgements

The authors are indebted to DHI Software Company for kindly providing MIKE SHE and MIKE BASIN products, towards research purposes. Moreover, the authors would like to thank Mr. Elias Moussoulis, DHI Greece Country Manager, for his guidelines and well-aimed comments regarding the modeling procedure. His valuable assistance is highly appreciated.

References

Belward, A. et al. 2011. Renewable energies in Africa, Joint Research Centre Scientific and policy Reports. 67752, EUR 25108 EN. Luxembourg: Publications Office of the European Union.

British Hydropower Association. 2005. A Guide to UK Mini-hydro Developments. Available online: Britishhydro.org.

Efstratiadis, A. and Hadjibiros, K. 2011. Can an environment-friendly management policy improve the overall performance of an artificial lake? Analysis of a multipurpose dam in Greece. Environmental Science and Policy 14(8): 1151–1162.

ESTIA Consulting, S.A. 2012. Technical Review/Energy Study on the Design of Small-scale Hydroelectric Projects.

European Environment Agency—EEA. 2012. European Waters—Current Status and Future Challenges—A Synthesis EEA Report, No. 9.

Ganoulis, J. and Skoulikaris, C. 2011. Impact of Climate Change on Hydropower Generation and Irrigation: A Case Study from Greece, Climate Change and its Effects on Water Resources. NATO Science for Peace and Security Series C: Environmental Security, Springer–Netherlands 3: 87–95.

Kaldellis, J.K., Vlachou, D.S. and Korbakis, G. 2005. Techno-economic evaluation of small hydro power plants in Greece: a complete sensitivity analysis. Energy Policy 33(15): 1969–1985.

Kollias, P. and Kollias, V. 2009. Artificial Lakes, Reservoirs, Dams, Hydroelectric Installation: Novel Energy Production Technologies (in Greek language), Private ed., Athens.

Kougias, I.P. and Theodossiou, N.P. 2012. Application of the harmony search optimization algorithm for the solution of the multiple dam system scheduling. Optimization and Engineering 1(14): 1389–4420.

Kougias, I.P. and Theodossiou, N.P. 2013. Application of the Harmony Search optimization algorithm for the solution of the multiple dam system scheduling. Optimization and Engineering 14(2): 331–344.

Papantonis, D. 2001. Small-Scale Hydropower Projects (in Greek language). Simeon ed. Athens.

Stephens, Tim. 2010. Manual on small earth dams: a guide to sitting, design and construction. Food and Agriculture Organization of the United Nations, Rome.

Texas Commission on Environmental Quality. 2009. Rule 299, Dams and Reservoirs, Subchapter B: Design and Evaluation of Dams.

World Bank. 2007. Report: Technical and Economic Assessment of Off-Grid, Mini-Grid and Grid Electrification Technologies, Vol. 1/2.

SECTION 4

Water Demand

CHAPTER

13

Estimating an Average Dairy Farm's Demand for Water in New Zealand

Alexey Kravchenko

Introduction

Freshwater is fast approaching over-allocation in many catchments in New Zealand [NZ] and regional councils are struggling to cope with the outdated, first-come first served principle of allotment enacted by the 1991 Resource Management Act (Land and Water Forum 2011). In all likelihood, some sort of demand management is going to be required to encourage efficiency of use among competing users, through instruments such as tariffs or regulated water markets. Either system will effectively raise the cost of water to users. Whichever system wins governmental support, it will require understanding of water users' responses to such increases. While this paper does not attempt to champion any particular method of solving the problem of water allocation, it does attempt to answer the question of response to changes in water cost to NZ's largest consumptive freshwater users—dairy farmers.

Department of Economics, Waikato Management School, The University of Waikato, Private Bag 3105, Hamilton 3240, New Zealand.
Email: oak1@waikato.ac.nz

NZ water issues: background

Water is an essential building block of life. Not only is it a pivotal element in a vast number of economic activities such as agriculture, horticulture, industry, electricity, and tourism, but it is also a spiritual substance and is a taonga[1] for Māori. NZ is seemingly relatively abundant in freshwater. It has the fourth highest per capita total renewable freshwater resources among OECD countries of over 80 thousand cubic metres per capita (Fuentes 2011). By comparison, Australia and the US have approximately 15 thousand and 10 thousand, respectively. Even when coupled with its highest water abstraction per capita level in OECD, NZ is still third lowest in terms of abstraction vis-à-vis its relative freshwater endowment (Ministry for the Environment 2010a). However, "much of it needs to be retained in the rivers, lakes and aquifers to maintain the ecological, recreational, or cultural values", with only a relatively small portion allocated for consumptive use (Ministry for the Environment 2010a). For example, in the Waikato Region the default allocations for freshwater are only 5% for upland and 10% for lowland catchments of Q5—the low flow statistic derived from analyzing the frequency of seven consecutive day annual low flow in a catchment that has a 20% chance of occurring in a particular year (Waikato Regional Council, n.d.).[2]

The current system of water use allocation in New Zealand has been described as "first-come-first-served": whoever applies first for a resource consent obtains it first. There is a nominal fee for the application, and a limit is set on the maximum allowable intake, but otherwise the water from freshwater bodies and aquifers is virtually free. This system has been established by the Resource Management Act [RMA] of 1991 (Scrimgeour 1997). According to this legislation, regional authorities are entrusted with managing their territories' natural resources, including water. The Act stipulates a number of provisions specifically addressing the issues pertaining to freshwater management, including the settlement of limits of freshwater intakes, allocation of rights of freshwater intakes and other functions to maintain quality of water (NZ Parliament 2011).

When this system was established, there was little need for an alternative solution as freshwater was deemed to be an inexhaustible resource in NZ. However, with the proliferation of irrigated farming, as well as a general

---

[1] A **taonga** in Māori culture is a treasured thing, whether tangible or intangible.

[2] The council determines each catchment's environmental flow—the level deemed necessary for a particular catchment to maintain its environmental and ecological health—through setting it proportional to Q5. For example, the Waikato River at Hamilton is deemed to need an environmental flow of 140 cubic metres per second [cms]. Its Q5 is 156 cms, which means that 16 cms is available for allocation. The likelihood of flow falling below the environmental level is 20%, during which time water intake restrictions will apply.

growth from other competing needs of water such as hydropower generation (which accounts for approximately 60% of all electricity generated in NZ), ecosystem management, and recreation, among others, this system is fast becoming unable to keep up with its objective.

Table 1 shows the areas under the irrigation system in 2002 and 2007. Approximately 4.2% of all agricultural land in New Zealand is reported to be under an irrigation system.[3] South Island, Canterbury and Otago in particular, account for most of the irrigated land in NZ. Still, most regions

Table 1. Irrigable land by region (000's ha).

	2002	2007	2002	2007	2002	2007
	Total Agricultural Land[4]		Total Area Equipped for Irrigation		Share of Total Ag. Land Equipped for Irrigation	
Northland	810	765	7.0	8.7	0.9%	1.1%
Auckland	302	245	6.2	6.3	2.1%	2.6%
Waikato	1,730	1,600	12.7	16.6	0.7%	1.0%
Bay of Plenty	600	531	8.8	10.0	1.5%	1.9%
Gisborne	643	615	1.3	2.3	0.2%	0.4%
Hawkes Bay	962	952	18.2	25.2	1.9%	2.6%
Taranaki	497	470	2.9	3.4	0.6%	0.7%
Manawatu-Wanganui	1,545	1,417	8.0	11.7	0.5%	0.8%
Wellington	504	491	9.5	12.9	1.9%	2.6%
TOTAL North Island	*7,593*	*7,086*	*74.7*	*97.1*	*1.0%*	*1.4%*
Tasman	277	253	10.0	10.7	3.6%	4.2%
Nelson	21	18	N/A	0.3	N/A	2.0%
Marlborough	696	507	20.1	26.7	2.9%	5.3%
West Coast	225	200	2.5	0.6	1.1%	0.3%
Canterbury	3,151	3,080	287.2	385.3	9.1%	12.5%
Otago	2,379	2,331	68.9	91.1	2.9%	3.9%
Southland	1,198	1,178	4.1	7.5	0.3%	0.6%
Chatham Islands	49	47	N/A	N/A	N/A	N/A
TOTAL South Island	*7,997*	*7,615*	*393.0*	*522.2*	*4.9%*	*6.9%*
TOTAL NEW ZEALAND	**15,590**	**14,701**	**467.6**	**619.3**	**3.0%**	**4.2%**

Source: Stats NZ 2002, 2007a.

[3] According to Aqualinc 2010, the area of land consented to be irrigated differs slightly to this figure as not all area has been actually equipped to be irrigated, as defined by Statistics NZ.

[4] Farms using land for: tussock and danthonia for grazing; grassland; arable crop, fodder crop and fallow; horticulture; planted production forest; mature native bush; native scrub and regenerating native bush; and other (Stats NZ 2007b; Stats NZ 2004).

experienced double digit percentage growth of irrigated land area within the five year period.

Aqualinc Research reported in 2004 that in Waikato, surface water is close to full allocation and the current surface water allocation processes do not account for variations in seasonal demand (Aqualinc Research 2004, p. 15). So much so, that in 2006 Environment Waikato [EW] declined two applications to take "significant volumes of water from the Waikato River for the purposes of dairy farm irrigation" (EW 2008, p. 26). The applications were particularly opposed by hydroelectricity generators and municipal water suppliers (EW 2008). Similarly, other regions experience a growing number of declined resource consents due to increasing scarcity of allocative water.

In 2010, there were 20,500 active freshwater consents in NZ, 75% of which was for the purposes of irrigation (Aqualinc Research 2010). In terms of annual consumptive allocation,[5] irrigation constituted just over half of the amount. When considering top weekly consumption[6]—irrigation constitutes 78% of allocation (Fig. 1):

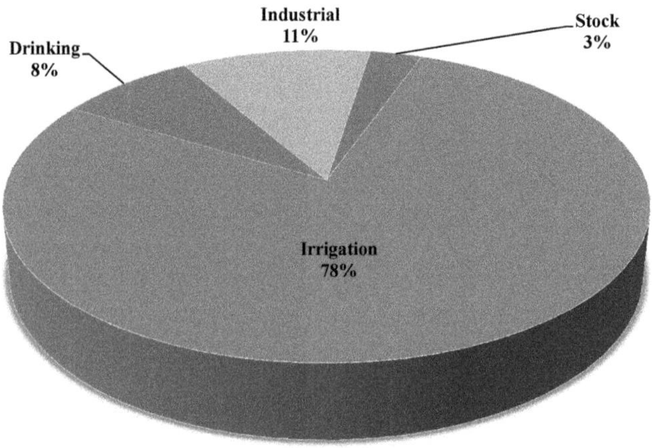

Figure 1. Top weekly consumptive allocation shares by sector.

[5] For drinking, stock water and industrial users, annual rate is calculated as weekly times 52. For irrigation, annual rate depends on the number of irrigating weeks typically 12 to 22 weeks.

[6] Consumptive use means the usage of water after which no other users can use the same water, like irrigation or domestic and industrial water use. Hydropower is generally not considered to be consumptive since water is later made available to users downstream, except in the case of Manapouri Hydropower Plant, which outlets the water to the Doubtful Sound. For the purposes of this analysis, the consumptive use of water by this power plant is omitted.

This higher relative and absolute demand is due to the seasonal nature of farming—planting and dairying seasons are predominantly in warmer months from October to March. Of all volumetric annual irrigation allocation, 81% of it is for pasture irrigation (or 76% of consented irrigated area in 2010). On top of the extra demand from farming, hydropower stations, due to the increased demand for air-conditioning in the summers, require more water to generate electricity. Unfortunately, this also coincides with the periods of lowest rainfall levels (Fig. 2):

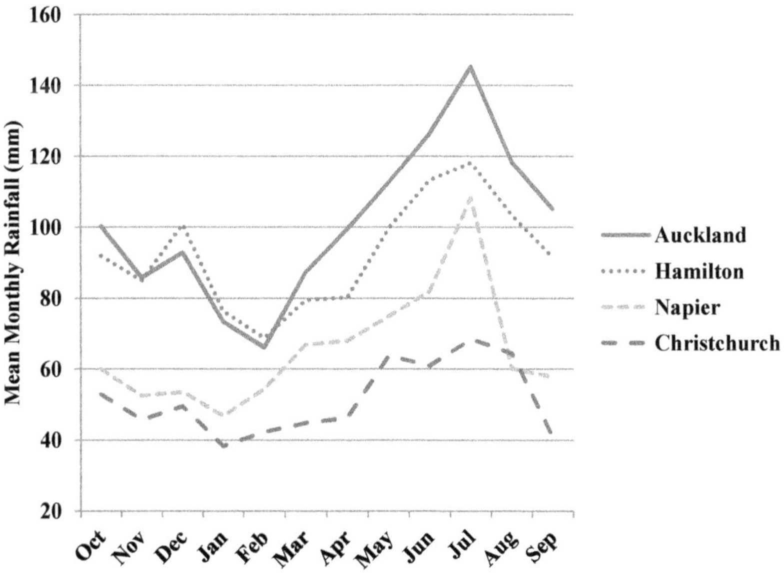

Figure 2. NZ monthly rainfall for selected cities.

Color image of this figure appears in the color plate section at the end of the book.

Fresh start for fresh water

Recognizing the growing need to better manage all the more scarce water, the NZ government commissioned a task force, Land and Water Forum [LAWF], to study the current freshwater situation in NZ and advise as to how it should be managed better (Land and Water Forum 2010). LAWF also concluded that many water catchments are fully allocated, or close to full allocation (Land and Water Forum 2011). They recognized freshwater's growing scarcity and recommended establishing more effective allocative processes than the "first-come-first-served" basis system currently in place.

The most recent report from the Forum specifically mentioned water pricing as a desirable mechanism to allocative efficiency (Land and Water Forum 2012).

Since there are competing users of water (such as domestic and industrial users, farms and hydroelectricity generators), who have a rivalrous water demand, it is imperative to establish a system that distributes water to users who hold it in the highest value. This requires actually knowing what that value is to various users. Since intakes of water so far have not been measured (in relationship to how much a user was consented to take or otherwise) and water being virtually free, for most users the value of water has not yet been adequately measured.

In order to establish a functioning water market, it is first necessary to measure the value of water, and, subsequently, estimate users' sensitivity to its pricing. At this stage, however, little research has been conducted with regards to the actual of value of water in NZ, much less with regards to users' sensitivity to pricing.

White et al. (2004) reported that estimates of the value of water in New Zealand, let alone elasticities, are few. This is partly due to the perception that water quantity is not an issue in New Zealand and partly due to lack of suitable data (as discussed below). They put the figure at $\$0.2/m^3$ for field and stock watering. It is derived from observing the land values without stock watering as well as per animal consumption of water, and contrasting it against property values with stock water (hedonic pricing method).

McDonald and Patterson (1998, as cited in Ford et al. (2001)) presented results of using a value added technique to determine the value each cubic meter of water generated through various industries. These estimates ranged from an average[7] of \$2,783 per m^3 in wood and wood product category to \$12.3 per m^3 in horticulture. The authors, among others (see Niewoundt et al. 2004; Young 2005; Schiffler 1998), cautioned against reaching any conclusion based on the figures obtained through this method since production in different sectors also requires other inputs.

In one of the few NZ academic studies, Grimes and Aitken (2008) addressed the subject and used a hedonic pricing approach to value irrigation water in a drought-prone area in McKenzie District, Canterbury. This method valued irrigation through estimating the difference between irrigated and non-irrigated farms' sales price and valuation, while controlling for spatial differences, such as distance from towns, rainfall, soil and slope characteristics. They found that flatter areas with poorly draining soils received the most benefit from irrigation, suggesting that it may be due to water being able to stay longer periods in these lands. Drier

[7] Averages based on Northland, Auckland and Waikato Regions (Ford et al. 2001).

areas benefited more than wetter areas. The authors joined the criticism of the RMA allocation mechanism by suggesting that some farms that may benefit from irrigation cannot get access to water rights because of existing regulation and lack of mechanisms of transferring water rights. The study found that net returns of irrigation were negative to farms due to high investment costs.

Ministry of Primary Industries [MPI] conducted an extensive study attempting to quantify the value of irrigation to New Zealand as a whole (Doak et al. 2004). They put the economic value of irrigation at $820 million[8] (in 2002/2003 dollars) by estimating a counter-factual scenario where irrigated land was hypothetically used as dry land instead. Their method was as follows: they classified all agricultural land into 14 agricultural sectors in each region, subdividing each sector into irrigated and non-irrigated portions. Next, the authors acquired the difference in yields between irrigated and dryland production for each sector in each region based on specialist opinions. Finally, they decreased the yield on the irrigated farms to match dryland yields and thereby estimated the effect of irrigation. In their subsequent analysis they used yields to estimate the impacts of new irrigation systems, and considered the effect of varying output on sector output prices.

Since the recent emphasis of freshwater management restructuring, MPI commissioned the New Zealand Institute of Economic Research to conduct a study using their proprietary Dynamic CGE model to measure the impact of increased irrigation in New Zealand (Kaye-Blake et al. 2010). While this study did not consider pricing of water per se, it did consider the changes in productivity of various sectors' post-irrigation schemes installations, as well as the costs of installing the schemes.

As Doak (2005) noted, "the value of water per cubic metre cannot be calculated as water use data is not yet available" (p. 2). Indeed, it was only in November 2010 that regulations requiring recording of volumetric intake of water came into effect for new consents (Ministry for the Environment 2010b). Still, this study targets to provide a starting point estimate of the farms' short-run (annual) responses to at-site (irrigation cost inclusive) changes of water costs based on panel data analysis of dairy monitory survey data.

Analytical Framework

Scheierling et al. (2004), in their meta-analysis study, summarized the price elasticities of the derived demand for irrigation water using various

[8] This figure includes their analysis of price changes resulting from sectoral output changes.

techniques since the 1960s. Their conclusion was that results obtained through different methods vary to a great extent partly because different methods were used: mathematical programming studies over-estimated elasticities, and field experiments produced the least elastic estimates (econometric studies were in between). Schiffler (1998) noted that the most common way of determining the value of water as an intermediary good is through residual imputation method.[9] In this method "the value of all non-water factor inputs is subtracted from the total value of products generated by an agricultural activity" (Schiffler 1998, p. 42). However, due the lack of accurate water usage data such approaches seemed unfeasible, and hence a unique approach has been developed specifically for the case at hand.

The main premise of this study is that farmers are rational economic agents and respond to changes in incentives by altering their production—the higher the expected profit the more they produce. Indeed, a qualitative study by Watters et al. (2004) partially confirmed this as the authors concluded that there seems to be a "wide-spread inclination for [dairy] farmers to respond to increasing prices through increasing input and production outputs" (p. 22). As one of their respondents suggested, "if payout allows" s/he maintains or increases the use of fertilizer and brought-in feed to increase the milk solid [MS] production.

Perhaps a more economically rational observation is that higher profitability (measured as an output-input price ratio) induces higher levels of production, and vice versa. Hence, an increase in the cost of water would essentially be the equivalent of a reduction in profitability, thus lowering the incentives for extra production. One possible way to visualize this relationship is considering what would happen if a hypothetical water tax for each milk-solid sold was introduced on the portion of the farm's supply relying on irrigation (Fig. 3). If the output price remained unchanged, quantity supplied would fall from Q_0 (quantity of MS produced due to irrigation prior to water tax) to Q_1 (post introduction of water tax). An important feature to note is that production due to irrigation would cease altogether if output/input price ratio falls below unity since the cost of paying for one unit of production would exceed the revenue received (i.e., average variable cost would become higher than marginal revenue). Note too, that the quantity of MS produced in the rain-fed production process would remain unchanged.

Finding a relationship between the volumetric cost of water and farmers' responses one needs to know:

1. the relationship between the quantity of water required for production of each milk-solid, and;

[9] Also known as farm budget residual method when estimating value in the agricultural sector (Schiffler 1998).

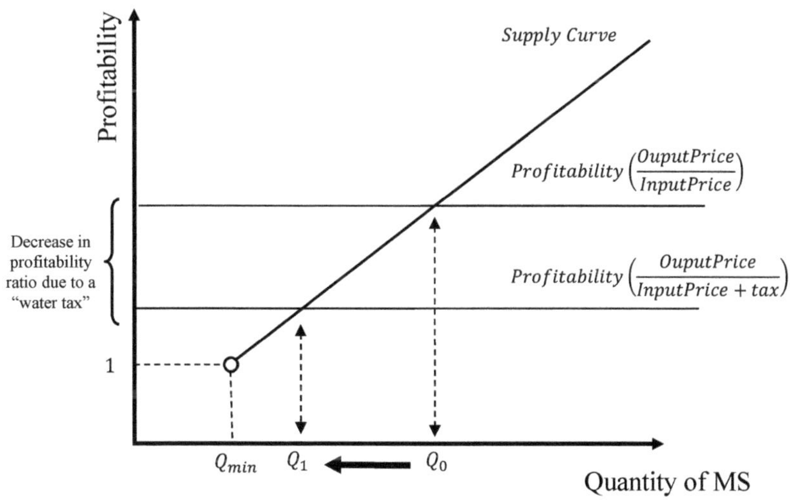

Figure 3. Effect of a hypothetical water tax on production.

2. the relationship between the output variations due to changes in the expected output-input price ratio.

The relationship between a volumetric unit of water and corresponding yields of kgMS production can be derived from the literature on pasture response to irrigation. It is conditional on the slope of the land, soil type, irrigation rates, grass type, fertilizer regime, climatic conditions as well as time of the year (Brown and Haigh 2005; Thomson 1996). Average responses will be used in the following explanation. In the study of predicting future demand for irrigation in Waikato, Brown and Haigh (2005) find that, on average, an extra millimeter of irrigation yields an additional 9.3 kg Dry Matter per hectare (DM/ha). In Canterbury, using an average of 7 irrigations of 100 mm per season yielded an increase from of 6.7 t DM/ha to 11.9 t DM/ha on average, or 5,200 kg/700 mm = 7.4 kg DM/ha per 1 mm (McBride 1994). In Taranaki, the average yield response to 1 mm of irrigation is similarly 7.56 kg DM/ha/year, ranging from 3.9 kg to 10.1 kg DM/ha/year on average across zones (Rout 2003).

In terms of relating DM to milk-solids, numerous factors affect cow productivity, such as cow weight, breed, distance needed to walk, topography of pasture, etc. (as well as the quality of DM itself). DairyNZ (2010) suggests that annual dry matter requirements for 350 kgMS/year producing Jersey weighting 400 kg that walks 4 km/day on flat land and is in milk for 270 days requires 4.6t DM + 6% of wastage = 4.9 t DM. Hence, each kg of DM would yield 350/4,876 = 0.072 kgMS. It follows that, *on average and conditional on a range of factors*, if 1mm of irrigation yields 7.4 kg DM/ha annually (in Canterbury), it is transferred into 7.4 * 0.072 = 0.52 MS/ha/year. Since 1 mm on a hectare is equivalent to 10 m^3, then it follows that it takes approximately 10/0.52≈20 m^3 of irrigated water to produce 1 kgMS.

The relationship between the change in the expected output-input price ratio and corresponding change in output is the subject of subsequent data analysis. It seeks to establish a correspondence between expected profitability (as measured by the output-input price ratio) and its effect on a farm's output in terms of kgMS, while controlling for other factors. Once such relationship is established, it would mean that the coefficient on the output-input price ratio could be interpreted as the expected change of an average farm to a change in profitability, due to an introduced "water tax wedge". Since only a portion of production on farms is due to irrigation, the effect would only apply to that portion (rain-fed production would remain unchanged).

Data

The data has been provided by the Ministry of Primary Industries [MPI] for the purposes of this research. It is an unbalanced panel data of a sample of dairy farms throughout New Zealand's main dairying regions over 11 financial years (from 2000 to 2011), with a total of 1,508 observations. Farm-level data available and used includes the total kgMS produced, effective farming area (in hectares), number of cows and total expenditure (see Table 2 for summary statistics). Additional series, namely precipitation, price indices and payout data were merged as described below.

Table 2. Summary statistics.

	No. of cows	kgMS	area	Total expenses
Mean	384	137,683	153	438,597
Median	330	110,116	127	330,518
SD	236	101,130	95	367,731
Min.	79	15,000	30	56,723
Max.	2,200	800,000	884	3,339,402
Normality Test Statistic	5,225*	6,070*	3,968*	5,991*

*indicates p-value <0.0001.

Output-weighted expected payout and profitability ratio

New Zealand dairy farmers' largest source of income is through the sale of MS to their cooperatives, the biggest being Fonterra (accounting for approximately 90% of all milk production in NZ). The majority of famers do not have the scale to exercise market power, and hence are bound by the payouts. The payout per milk-solid consists of a farmgate milk-solid price as well as a profit share (Distributable Profit—formally known as "value added components") from the profit of value-added activities of the cooperative.

Although farmers receive advance payments to aid their yearly cash flow, the final payout is usually announced well into the next production season (usually around September, whereas the milking season coincides with fiscal calendar and ends at the end of June), hence it has no effect on farm production in the corresponding milking season. What motivates short-run variability in the production is the forecasted payout—or how much the cooperative predicts the final payout to be. After the opening forecast at the start of each season, the cooperative updates its forecast, which is driven by such factors as currency fluctuations, international dairy auction prices as well as the expected profit from the value-added activities.

As per Fig. 4, initial forecasts sometimes substantially differ from the final payout. For instance, in the 2009/2010 season, the opening forecast was only $4.55 whereas the final payout was actually $6.55, making the actual payout an inadequate measure of farmer short-run incentive.

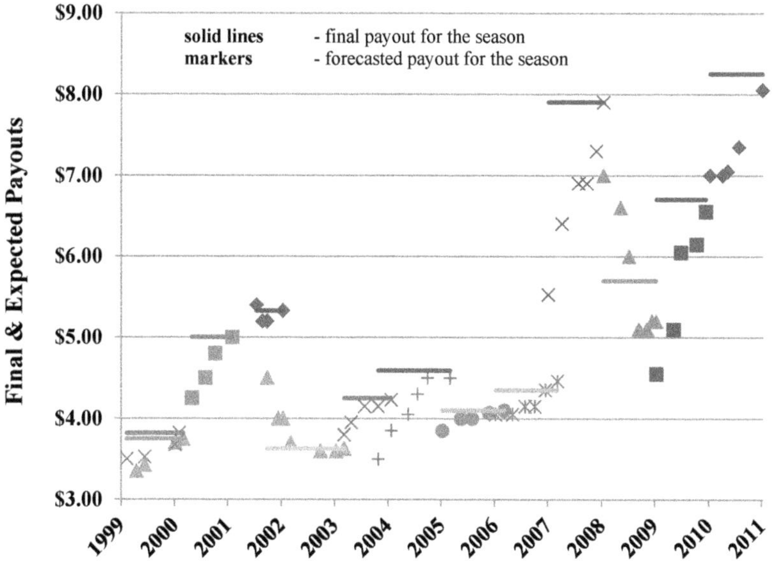

Figure 4. Forecasted vs. Actual Payout.
Source: Fonterra 2000–2012.

Color image of this figure appears in the color plate section at the end of the book.

To obtain a more reliable incentive indicator, an output-weighted forecast (O-W forecast) measure was developed, where the forecasts were weighted by the quantity of MS produced NZ-wide when each forecast was in effect. For example, the opening forecast for the 2009/2010 season of $4.55 was updated on 22nd September, 2009 to $5.1. Between the start of the season and 22nd September, approximately 205,340 thousands of kgMS was produced NZ-wide (DCANZ 2012).[10] This corresponds to 14% of the 1,438,496 thousands kgMS produced in the 2009/2010 season. Hence, for 14% of the total production, the expected payout was $4.55 (see Fig. 5). Similarly, for 23% of the total 2009/2010 production, the forecasted payout was $5.1, for 43% it was $6.05, for 17% it was $6.15, and for 4% it was $6.55. Weighting each forecast by the proportion of milk produced NZ-wide in the period the forecast was effective results in average weighted expected payout for the season (output-weighted forecast) of 0.14*$4.55 + 0.23*$5.1 + 0.43*$6.05 + 0.17*$6.15 + 0.04*$6.55 = $5.65.[11] Detailed data on NZ-wide total MS production was available only for seasons 2008 through 2011, hence for other years, the average of four years of available total production record was used. Table 3 summarizes the disparity between the final payout and the OW forecast.

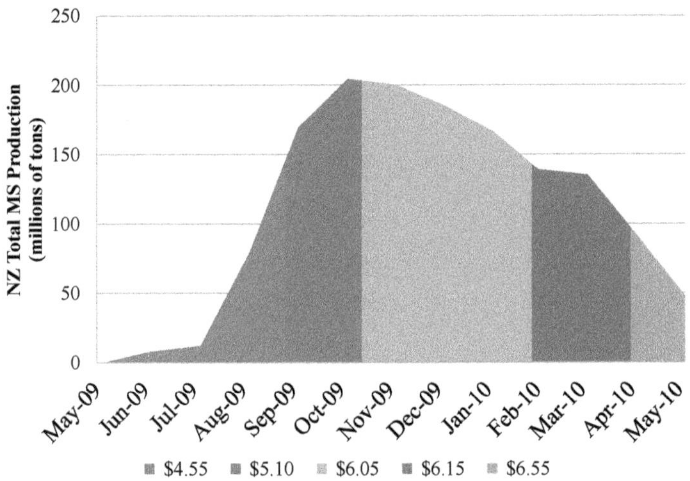

Figure 5. Output-weighted forecast estimation.
Data Sources: DCANZ (2012); Fonterra (2009, 2010).

Color image of this figure appears in the color plate section at the end of the book.

[10] DCANZ (2012) only published monthly MS production data, hence the cumulative production up to 22nd September was linearly extrapolated using the end of August and end of September cumulative production figures.

[11] Figures may not add up due to rounding.

Table 3. Output-weighted Forecast vs. Final Payouts.

Season	O-W Forecast	Final	Difference
2010/2011	7.19	8.25	1.06
2009/2010	5.65	6.70	1.05
2008/2009	6.05	5.70	–0.35
2007/2008	6.60	7.90	1.30
2006/2007	4.10	4.35	0.25
2005/2006	3.98	4.10	0.12
2004/2005	4.18	4.59	0.41
2003/2004	4.03	4.25	0.22
2002/2003	3.68	3.63	–0.05
2001/2002	5.30	5.33	0.03
2000/2001	4.43	5.00	0.57
1999/2000[12]	3.40	3.75	0.35

While the nominal payout more than doubled between 1999/2000 and 2010/2011 seasons, the costs of production and costs of living have likewise risen (Fig. 6). The cost of producing (Producer Price Index (PPI))

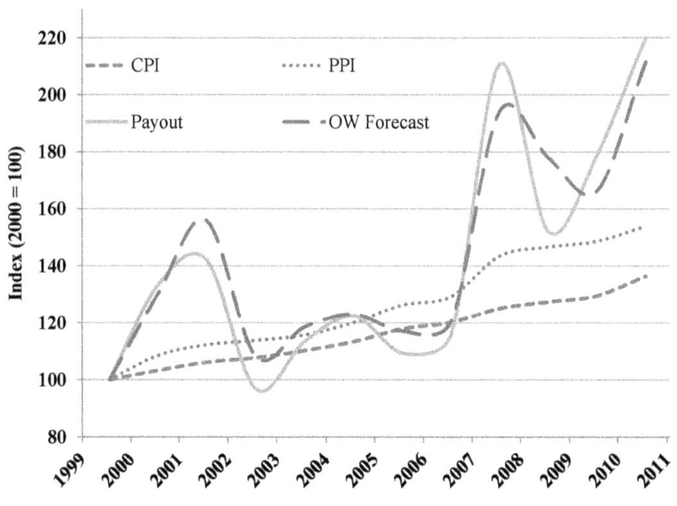

Figure 6. CPI, PPI and Payout Indices (1999 = 100).

Color image of this figure appears in the color plate section at the end of the book.

[12] For the 1999/2000 season the NZ Dairy Group forecasts and final payout were used for the purposes of the analysis since this was prior to the establishment of Fonterra.

has risen at a substantially faster pace than cost of living (Consumer Price Index (CPI)).

To adjust for the changing rates of price increases, as well as to mitigate for multicollinearity which would arise since the year and region dummy variables would be perfectly collinear with the same payout experienced by each farm, an output price/input price ratio (O/I ratio) was calculated for each farm. This ratio could be interpreted as profitability ratio, and hence changes in profitability due to either changes in payout or costs per each MS could be interpreted as having the same effect. In lieu of higher payout (output price) or lower expense (input price), the ratio would increase and hence motivate higher levels of production—a supply curve. Moreover, logistic transformation of the ratio could be interpreted as price elasticity of supply (Tauer 1998). As per Fig. 7, the expectation adjusted O/I ratio is centered just above 1.5 and is relatively steady over time except for the low payout year of 2003 and high payout year of 2008.

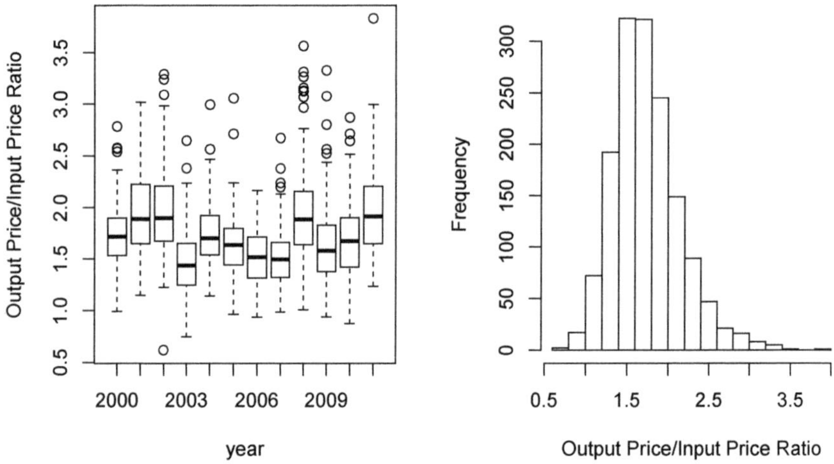

Figure 7. Output price/input price ratio across years and its distribution.
Data Sources: Statistics NZ (2012); Fonterra 2000–2012.

Precipitation

For each dairy region in the sample, a representative weather station was selected from the National Institute of Water and Atmospheric Research [NIWA] weather database and corresponding monthly total rainfall (in mm) was obtained. For each region and each production season, months November through April were selected, deemed to most impact the variation in production. Since both seasons, extremely wet (as in 2003) and extremely dry (in 2008), can potentially negatively affect DM growth, the rainfall variable for each region was first centered on the mean in each corresponding region, then split into negative and positive deviations from it.

Results

To derive the relationship between the O/I ratio and output, total production of kgMS in a year from individual farms was regressed on the available explanatory variables. Table 4 summarizes the results of the model. The following describes each variable and their significance.

The number of cows (*cows*) was included to control for the scale of farms. Having the most explanatory power, the coefficient suggests that an additional cow can add extra 413 kgMS. This is somewhat larger than the average MS production per cow in the sample (344), but in line with the averages from recent years. Interestingly, variables attempting to control for the intensity of dairying—*stocking_rate* and *area* were not found to be statistically significant in most regressions. While farming area is highly

Table 4. Regression output.

Regressor	Dependent Variable: **Total kgMS**
cows	413*** (15.6)
dry_year_rain (mm)^{CR}	52.8*** (10.5)
wet_year_rain (mm)^{CR}	–26.0** (8.0)
OI_ratio	10,865*** (1,704)
OI_ratio × *stocking_ rate*^{C}	–4,183** (1,368)
OI_ratio × *area* (ha)^{C}	–48.7** (18.9)
Standard Error of Regression	23,390
R^2	0.947

Robust standard errors are given in parentheses under coefficients. Individual coefficients are statistically significant at the **1% or ***0.1% significance level. ^{C} denotes that the variable was centered by subtracting the mean of all observations in the sample, while ^{CR} indicates that the variable was centered with the mean of the corresponding region. 1999/2000 season and CANDY region dummy variables were omitted from estimation to avoid perfect multicollinearity with the intercept. Heteroskedasticity adjusted *F*-statistic testing whether all year dummy variables are zero is 5, (*p*-values < 0.0001); and 44.0 testing that all region dummy variables are jointly insignificant.

correlated with the number of cows (r=0.82), suggesting low efficiency due to multicollinearity, lack of explanatory power of the stocking rate is harder to explain.

Next, positive and negative deviations from the region's average rainfall for months November through April were added to control for weather. Note that the coefficient on extra mm of rain in a dry year above the average (*dry_year_rain* (mm)) is 52.8, which, when divided by the average farm size (153 ha) yields a marginal effect of 0.35. This is smaller than the effect of 1mm/ha of *irrigation* on MS production derived earlier—0.52 MS/ha/year (page 308)—and is in line with expectations since watering from *rainfall* (unlike from *irrigation*) does not follow an optimal schedule intended for maximum pasture growth.

The effect of rain in a wet year has expected signs (exceedingly wet seasons slow down grass growth and bog down cows) and is less than half the size of the effect in a dry year.

The model also includes dummy variables for time and region specific effects. The rationale behind this fixed effects specifications is that in each year there are explanatory factors omitted that are shared among all farms (such as economic outlook and confidence); whereas some effects are likely to remain constant across time, but shared among neighboring farms (e.g., regional climatic attributes). Inclusion of the dummy variables ensured that these time and region specific effects (although unobserved) are controlled for.

The coefficient on *OI_ratio* has an expected sign, but a comparatively low magnitude, suggesting a low responsiveness of farms to changes in output and input prices. Logistic transformation of both sides of the regression yielded a coefficient of 0.16, which can be interpreted as the price elasticity of supply—a 1% change in price ratio triggers only a 0.16% change in quantity supplied. This inelastic response suggests that farms have low flexibility in the short-run, due to constrained fixed resources (number of cows and land) and diminishing marginal returns to variable inputs (irrigation, fertilizer and feed).

Farm size (*area*) and farming intensity (*stocking_rate*) were included as interaction terms with the *OI_ratio*, and their significance suggests that the effect of expected profit varies with farm sizes and farming intensity. Each interaction was centered by subtracting their respective means, so that interpretation of *OI_ratio* can be taken as that of a farm with an average stocking rate and farm size. Smaller farms and those with lower farming intensity tended to be more flexible when output/input prices changed.

The main advantages of centering of *area* and *stocking_rate* variables at their respective mean values are that the estimates on the variable of interest (*OI_ratio*) remain comparable with estimates from the models that do not include such interactions; and that the coefficient on the variable of interest itself remains meaningful—as opposed to if the interactive terms were unscaled, and the coefficient on the *OI_ratio* would be evaluated for a farm with zero area and zero stocking rate (Woolridge 2003). Although interactions under linear transformation lack scale in variance, Aitken and West (1991) show that post-hoc analysis of interaction is not affected by such scaling. Furthermore, the authors demonstrate that centering has an additional advantage of reducing the chance of multicollinearity of with the original multiplicative terms, thereby increasing the efficiency of the estimates.

The conditional effect of the *OI_ratio* is estimated to be 10,865 kgMS for a unitary change in the *OI_ratio* for an average farm. Since interaction terms were included, it must be qualified by stating that coefficient holds for a farm of 153 hectares and a stocking rate of 2.64. This reduces to approximately 10,865/153 ha = 71 kgMS/ha. The marginal effects of a unitary increase in the *OI_ratio* for larger/smaller farms as well as those with higher/lower stocking rate can be calculating by adding the average *OI_ratio* effect with a product of required values for area and stocking rate and their respective coefficients—see Table 5.[13]

As per Cohen and Cohen (2003), +/– 1 Standard Deviations (StDev) from the mean values were used to estimate the simple slope coefficients in Table 5. Standard errors (in parenthesis below the coefficient) were derived using a technique from Aitken and West (1991, pp. 24–25):

$$\sigma_b^2 = \mathbf{w}'\,\Sigma_\mathrm{b}\mathbf{w}$$

Where **w** is the vector matrix of the coefficients included in the interaction and zeros for others ($\mathbf{w}' = [0 \quad 0 \quad 0 \quad 1 \quad stocking\_rate \quad area \quad 0 \cdots 0]$); and Σ_b is the heteroskedasticity consistent variance covariance matrix (White's). For each simple slope coefficient corresponding values of *area* and *stocking_rate* were substituted in **w**.

Table 5. Marginal effects of O/I ratio with interaction terms.

		stocking rate		
		1.92 (–1 StDev)	**2.64** (mean)	**3.36** (+1 StDev)
***area* (ha)**	**58** (–1 StDev)	18,504*** (2,525)	15,488*** (1,870)	12,472*** (1,600)
	153 (mean)	13,881*** (1,663)	10,865*** (1,704)	7,849*** (2,233)
	248 (+1 StDev)	9,258*** (2,359)	6,242 (3,717)	3,226 (3,717)

[13] Note that coefficients are for centered variables, hence, the –1 StDev multiplier for stocking rate interaction term, for example, is 1.92–2.64 = –0.72.

Coefficients on the interaction terms suggest that smaller and less intense farms are more responsive to changes in profitability. This makes economic sense as there is inevitably "excess capacity" within farms with lower stocking rates, and they are more likely to be flexible if there is a short-run change in either input or output prices.

Application to Water Demand

To predict the response of a farm to an increase in a volumetric pricing of pasture irrigation water, it is first necessary to include a number of parameters and assumptions, some of which may be changed in accordance to application requirements. As an example, suppose there is a farm with the following attributes:[14]

Number of cows	425
Farm Size	144 ha
- *Stocking Rate*	*2.95*
Payout	$7.23
Cost / MS	$3.87
- *O/I Ratio*	*1.87*
MS Production	150,000 kgMS
Pasture Irrigation Response	7.4 kg DM/ha/mm
Proportion of DM grown due to irrigation	10%
DM requirements/cow	5 tons

Each cow requires 5 tons of DM to produce 150,000/425 = 353 kg of MS, so each kg of DM yields 353/5,000 = 0.0706 kgMS. Since 1 mm/ha of irrigation produces 7.4 kg DM/ha/mm, it results in 7.4*0.0706 = 0.522 kgMS/ha. 1 mm/ha of irrigation is equivalent to 10 m^3, then it takes 10/0.522 = 19.14 m^3 to produce 1 kgMS. In absence of water tariffs, the farm would consume 19.14*0.1*150,000 = 287,162 m^3 of water.

Now suppose a 5 cent/m^3 tariff is introduced. Assuming no input substitution (e.g., for brought-in feed), the farm now faces a 287,162*0.05 = $14,358 bill for irrigation water. The overall farm working expense/MS rises from $3.87 to $3.87 + $14,358/150,000 = $3.97. The O/I ratio falls from 1.87 to $7.23/$3.97 = 1.82, hence the O/I ratio changes by 1.87–1.82 = 0.05.

Since the stocking rate and area are not at their mean values, the simple slope coefficient on the O/I ratio needs to be derived by including the multiplicative terms, i.e., 10,865 + (–4,183*(2.95–2.64)) + (–48.7*(144–153)) = 10,015. Using this simple slope coefficient on the O/I ratio, the consequent

[14] Based on averages from the 2010/2011 production season.

predicted fall in the production is 0.05*10,015 = 452 kgMS. As the increase in cost is only for the irrigated production, for the farm to produce 901 fewer kgMS, it would require 452*19.14 = 8,645 fewer m^3 of water (or 8,645/287,162 = 3% less water).

This formula can be extended to include any number of assumptions about the parameters of the farm in question, or the tariff levels. For instance, Figure 8 traces the quantities of water consumed by the farm with average parameters used in this illustrative example under various hypothesized water tariffs. Table 6 presents relative changes in water consumption.

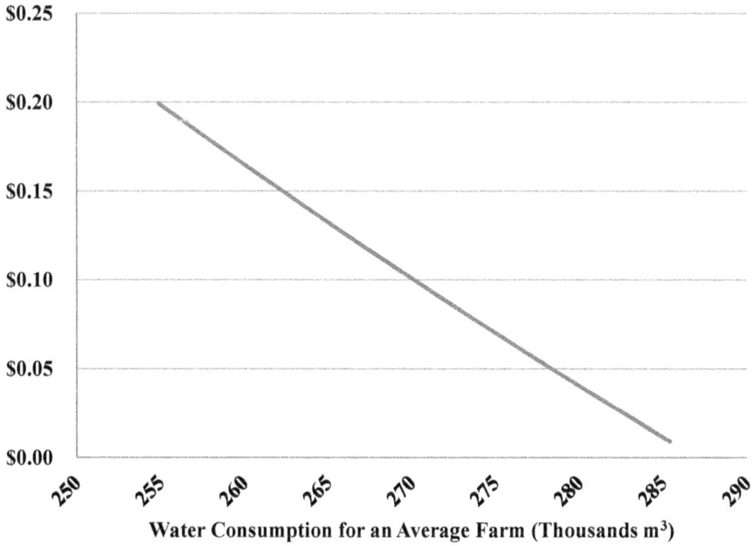

Figure 8. Water consumption for an average farm vs. price (per 1 m^3).

Table 6. Changes in water consumption for an average farm.

Price	Change in Water Consumption
$0.05	–3.0%
$0.10	–5.9%
$0.15	–8.6%
$0.20	–11.2%

Conclusion & Limitations

This study's aim was to produce a "starting point" estimate of the response curve to water price tariffs of dairy farmers and should be treated as such. It requires prior knowledge of a number of sensitive parameters and is based on restrictive assumptions including that all farms employ the same production function, there is linearity in DM yield in response to irrigation, and there is no substitution among factors of production.

In reality, faced with increasing water costs, farmers are likely to substitute to brought-in feed and water usage efficiency technologies. Allowing for substitution would theoretically yield much sharper responses (i.e., more production would be shifted towards using brought-in feed, less irrigated water). Indeed, in Australia farmers have to decide every year before the start of production season whether to invest in "temporary water" and make a loss if the year ends up to be wet, or risk it and face the prospect of having to purchase expensive feed (O'Connor, n.d.). Further study should be carried out to examine the trade-off between brought-in feed and irrigation.

Nevertheless, notwithstanding the limitations of the data and restrictive assumptions, useful conclusions can be drawn: smaller and less intense farms are likely to be more flexible with production should they be faced with a freshwater tax (equivalent of a lower expected payout). Larger farms and those that operate closer to full capacity, on the other hand, are likely to internalize the costs in the short run, and hence their demand for freshwater is likely to be less susceptible to influence in the face of levies.

In conclusion, rather than relying on tools such as water intake restrictions and arbitrary distribution of water resource consents, theoretical rationale suggests that a pricing mechanism can be a viable alternative for water demand management in the face of scarcity. It is expected that this study adds perspective to discussion on the topic.

Acknowledgment

Special thanks to Prof. John Gibson (Waikato University dept of Economics) for his many helpful suggestions and help with this manuscript.

References

Aitken, L.S. and West, S.G. 1991. Multiple Regression: Testing and Interpreting Interactions. Sage, London.

Aqualinc Research. 2004. Assessment of Improvements to Environment Waikato's Water Allocation Process and Procedures. Retrieved April 6, 2011, from Environment Waikato: http://www.ew.govt.nz/publications/Technical-Reports/Assessment-of-improvements-to-Environment-Waikatos-water-allocation-processes-and-procedures/.

Aqualinc Research. 2010. Update of Water Allocation Data and Estimate of Actual Water Use of Consented Takes 2009–10. Retrieved May 5, 2011, from Ministry for the Environment: http://www.mfe.govt.nz/publications/water/water-allocation-2009-10/update-of-water-allocation-data-and-estimate-of-actual-water-use-of-consented-takes.pdf.

Brown, E. and Haigh, A. 2005. Future Water Demand from Pasture Irrigation in the Waikato Region. Environment Waikato Technical Report.

Cohen, J. and Cohen, P.W. 2003. Applied Multiple Regression—Correlation Analysis for the Behavioral Sciences (3 ed.). Lawrence Erlbaum, Mahwah, NJ, USA.

DairyNZ. 2010. Facts and Figures for Dairy Farmers. Retrieved May 1, 2011, from DairyNZ: http://www.dairynz.co.nz/file/fileid/33397.

DCANZ. 2012. New Zealand Milk Production 2008–2012. Retrieved December 10, 2012, from Dairy Companies Association of New Zealand: http://www.dcanz.com/files/New%20Zealand%20Milk%20Production%2019Nov2012.pdf.

Doak, M., Parminter, I., Horgan, G., Monk, R. and Elliot, G. 2004. The economic value of irrigation in New Zealand. MAF Technical Paper 04(01).

Doak, M. 2005. Value of irrigation in New Zealand. In: OECD Workshop on Agriculture and Water Sustainability, Markets and Policies. 14–18 November, 2005, Adelaide, Australia.

EW. 2008. Proposed Waikato Regional Plan: Proposed Variation No. 6—Water Allocation. Retrieved December 10, 2010, from Environment Waikato: http://www.waikatoregion.govt.nz/PageFiles/7062/1382630-RPV6_Water_Allocation_Hearing_Committee_s32_Analysis.pdf.

Fonterra. 2000–2012. Reports and Publications. Retrieved April 20, 2011, from Fonterra: http://www.fonterra.com/cn/en/Financial/Reports+and+Publications.

Ford, S., Butcher, G., Edomnds, K. and Braggins, A. 2001. Economic efficiency of water allocation. MAF Technical Paper No.: 2001/7. Retrieved from MAF Policy Paper: http://www.maf.govt.nz/mafnet/rural-nz/sustainable-resource-use/water-efficiency/economic-efficiency-of-water-allocation/water-allocation-technical-paper-7.pdf.

Fuentes, A. 2011. Policies towards a sustainable use of water in Spain. OECD Economics Department Working Papers No. 840.

Grimes, A. and Aitken, A. 2008. Water, water somewhere: the value of water in a drought-prone farming region Motu. Working Paper 08–10. Motu Economic and Public Policy Research, Wellington, New Zealand.

Kaye-Blake, B., Schilling, C. and Zucollo, J. 2010. The Economic Impact of Increased Irrigation: A Dynamic Computable General Equilibrium Analysis of Increased Irrgation in New Zealand. Retrieved May 2, 2011, from Ministry of Agriculture and Forestry: http://www.maf.govt.nz/portals/0/documents/environment/water/irrigation/nzier-economic-impact-increased-irrigation.pdf.

Land and Water Forum. 2010. Report of the Land and Water Forum: A Fresh Start for Fresh Water. Retrieved October 29, 2010, from Land and Water Forum: www.landandwater.org.nz/land_and_water_forum_report.pdf.

Land and Water Forum. 2011. Report of the Land and Water Forum: A Fresh Start for Fresh Water. Retrieved from Land and Water Forum: http://www.landandwater.org.nz/summary_report.pdf.

Land and Water Forum. 2012. Retrieved October 30, 2012, from Third Report of the Land and Water Forum: Managing Water Quality and Allocating Water: http://www.landandwater.org.nz/includes/download.aspx?ID=124767.

McBride, S.D. 1994. Pasture yield responses to irrigation in Canterbury. Proceedings of the New Zealand Grassland Association (56): 165–168.

Ministry for the Environment. 2010a. Ministry for the Environment. Retrieved July 12, 2011, from Freshwater demand (allocation): http://www.mfe.govt.nz/environmental-reporting/freshwater/demand/.

Ministry for the Environment. 2010b. Ministry for the Environment. Retrieved November 25, 2010, from Measuring and reporting water takes: An introduction to the Resource Management (Measurement and Reporting of Water Takes) Regulations 2010: http://

www.mfe.govt.nz/publications/water/measuring-and-reporting-water-takes/index.html.

Niewoundt, W.L., Backeberg, G.R. and Du Plessis, H.M. 2004. The value of water in the South African economy: some implications. Agrekon 43(2): 162–183.

NZ Parliament. 2011. Resource Management Act 1991 No. 69. Retrieved April 2, 2011, from New Zealand Legislation: Acts: http://www.legislation.govt.nz/act/public/1991/0069/latest/DLM232560.html.

O'Connor, R. n.d. What's Water Worth? Echca, Department of Primary Industries, Australia.

Rout, R. 2003. Optimisation of Farm Irrigation. Prepared for Taranaki Regional Council.

Scheierling, S., Loomis, J.B. and Young, R.A. 2004. Irrigation water demand: a meta analysis of price elasticities. American Agricultural Economics Association Annual Meeting, Denver CO, August 1–4, 2004.

Schiffler, M. 1998. The Economics of Groundwater Management in Arid Countries: Theory, International Experience and a Case Study of Jordan. Frank Class Publishers, Portland.

Scrimgeour, F. 1997, October. Water pricing experiences: an international perspective: New Zealand. Retrieved November 3, 2010, from World Bank Technical paper no. 386: Water Pricing Experiences: An International Perspective: New Zealand.

Stats NZ. 2002. NZ Statistics. Retrieved May 01, 2011, from 2002 Agricultural Production Census (Final Results): June 2002: http://www2.stats.govt.nz/domino/external/web/prod_serv.nsf/929f646420956813cc256b16006b9ec1/898a7ddd91d3065dcc256d58000dd215/$FILE/Irrigated%20Land%20by%20Region.xls2F%24FILE%2FIrrigated%2520Land%2520by%2520Re.

Stats NZ. 2007a. Irrigable Land by Region and Type. Retrieved May 1, 2010, from NZ Statistics: http://www.stats.govt.nz/methods_and_services/access-data/tables/2007-agricultural-census-tables/~/media/Statistics/Methods%20and%20Services/Tables/2007-agriculture-census/land%20treatments/7agprod-irrigable-land-by-region07.ashx.

Stats NZ. 2004. Hectares Used and Farms by Land Use by Region. Retrieved May 15, 2013, from Statistics New Zealand: http://www.stats.govt.nz/browse_for_stats/industry_sectors/agriculture-horticulture-forestry/2002-agricultural-census-tables/~/media/Statistics/browse-categories/industry-sectors/agriculture-horticulture-forestry/census-tables-2002/1-land-farm.xls.

Stats NZ. 2007b. Agricultural Areas in Hectares by Farm Type. Retrieved May 12, 2013, from Statistics New Zealand: http://www.stats.govt.nz/browse_for_stats/industry_sectors/agriculture-horticulture-forestry/2007-agricultural-census-tables/~/media/Statistics/browse-categories/industry-sectors/agriculture-horticulture-forestry/ag-census-2007/ag-1areas-hect-farm07.xls.

Tauer, L.W. 1998. Estimates of individual dairy farm supply elasticities. Working Paper, Department of Agriculture, Resource, and Managerial Economics, Cornell University.

Thomson, N.A. 1996. Irrigation and Pasture Quality. Retrieved July 27, 2011, from DairyNZ: http://www.dairynz.co.nz/file/fileid/27188.

Waikato Regional Council. n.d. Appendix 1a: Water Allocation in Relation to Q5. Retrieved August 15, 2011, from http://www.waikatoregion.govt.nz/PageFiles/7062/appendix1.pdf.

Watters, A., Rowan, G. and Williams, L. 2004. Incentives for intensification: a report based on farmer case studies. Parliamentary Commissioner for the Environment, Wellington.

White, P.A., Sharp, B.M. and Reeves, R.R. 2004. New Zealand Water Bodies of National Importance for Domestic Use and Industrial Use. Retrieved November 3, 2010, from Ministry of Economic Development: http://www.med.govt.nz/upload/24450/gns.pdf.

Woolridge, J.M. 2003. Introductory Econometrics: A Modern Approach. South Western, Cengage Learning, Mason, Ohio.

Young, R.A. 2005. Determining the Economic Value of Water: Concepts and Methods. Resources for the Future, Washington, DC.

CHAPTER

14

Price Volatility and Water Demand in Agriculture

A Case Study of the Guadalquivir River Basin (Spain)

Carlos Gutiérrez-Martín,[1,a,*] *Carlos Dionisio Pérez Blanco,*[2,c] *Carlos Mario Gómez Gómez*[2,d] and *Julio Berbel*[1,b]

Introduction

Fresh water is a scarce good that is essential for sustaining life, development and the environment and it should be managed accordingly (ICWE 1992). However, water policy has failed to consider water as an economic good and has focused instead on guaranteeing the provision of this resource

[1] Universidad de Córdoba. Campus Rabanales. Ctra N-IV, km 396, Edificio C5. 14014 Córdoba.

[a] Email: carlos.gutierrez@uco.es

[b] Email: berbel@uco.es

[2] Madrid Institute for Advanced Studies in Water Technologies (IMDEA). C/Punto Net 4, 2nd Floor, ZYE Building, Parque Científico Tecnológico de la Universidad de Alcalá. 28805, Alcalá de Henares.

[c] Email: dionisio.perez@imdea.org

[d] Email: mario.gomez@uah.es

* Corresponding author

at subsidized prices. In Southern Europe, this paradigm has resulted in a significant irrigation expansion that in many basins has led to a hypertrophic irrigated agriculture as compared to the scant water resources available (Ward and Pulido-Velazquez 2008). In some of these basins, agricultural water use has grown to such an extent that it is now larger than renewable water resources, meaning that groundwater stocks are being depleted (EEA 2009). This ingrained overexploitation of water resources has reportedly been aggravated by more recurrent and intense droughts as a result of climate change. The resultant water crisis is causing EU institutions to make new laws and policies to foster water conservation in the agricultural sector (EC 2000; EC 2008). In this context, there is an increasing concern over the effects that agricultural policies (e.g., the Common Agricultural Policy) or exogenous agricultural shocks (e.g., world food prices fluctuations) may have over water use. While the former issue is frequently addressed in the literature (see for example Kampas et al. 2012; Dono et al. 2010), to the best of our knowledge there are no studies available on the effect that fluctuations in world food prices may have over water use in the EU.

Agriculture is a sector especially sensitive to price variability. Nonetheless, in the EU this variability has traditionally been absorbed by the Common Agricultural Policy (CAP), limiting the impact of price shocks over income and uncertainty and thus over farmers' decisions. This situation started to change after the year 1999, when a series of CAP reforms (and particularly the decoupling that began in 2003) made agriculture progressively more exposed to price variability. This liberalization of the EU agricultural market was followed by some relevant shocks over international agricultural markets, namely, the growing demand of agricultural products from countries such as Brazil, India or China (Gilbert 2010; Timmer 2012), speculation in agricultural markets (Cooke and Robles 2009; Gilbert 2010), the transformation of traditional agricultural land into land for biofuel production (Mitchell 2008), the depreciation of the US dollar (Abbott et al. 2008; Roache 2009), export restrictions and high oil prices (FAO 2011) and climate change and weather-related crop losses (Quiroga and Iglesias 2009).

Even though the specific role played by each shock is still being discussed (Mitchell 2008; Gilbert and Morgan 2010), the result was evident: since 2007 international prices experienced an unprecedented increase that made wheat and maize prices rise by 150%, rice by 200%, sunflower by 220% and cotton by 140% as compared to the pre-crisis prices (World Bank 2012).

The instability of agricultural prices worldwide was not fully endured by EU farmers, who still enjoyed a moderate degree of protection. As a result, in spite of the previous CAP reforms, EU farmers faced relatively more stable prices and income than those observed worldwide. This is not to say that EU farmers were completely isolated from these shocks: domestic agricultural prices did experience fluctuations as large as 64% for rice, 51.1% for wheat, 47% for sunflower, 41% for maize and 25% for cotton (MAGRAMA 2009). The effects of this sharp increase of agricultural prices were twofold: i) in the short run farmers benefited from higher agricultural prices and income; ii) however, in the medium run, price instability resulted in income variability and uncertainty, which in turn moved farmers' decisions and agricultural outcomes away from the social optimum (Sumpsi 2011). This kind of sub-optimum allocation is reported to have negative effects over employment, consumers (who may have a reduced accessibility to articles of first necessity), macroeconomic stability and also over environment (Zezza et al. 2009). The exposure of EU farmers to price variability is expected to increase in the next years, as the liberalization of agricultural markets in the EU and worldwide continues. Accordingly, there is a growing concern on the impact that price volatility may have over economy and also over the environment: apart from the effects over agricultural income and employment, price shocks may have an impact on water demand, a matter of special relevance in the context of scarcity and droughts characteristic of many Southern EU areas.

The objective of this research is to assess the effects that agricultural price shocks may have over relevant agricultural variables (such as employment and gross margin), with a special emphasis on water use. To achieve this objective, we have developed a Revealed Preferences Model (RPM) that represents farmers' behavior, which we use to simulate farmers' response to different price shocks that are defined according to the price fluctuations observed since 2007. This model aims to be general and applicable to different contexts. In this chapter, the model is calibrated for the particular case of the Guadalquivir River Basin in Southern Spain. Our results show that price variability has a significant effect over farmers' decisions, resulting in a relevant change in the crop portfolio that increases the expected income, reduces labor use and makes water demand more inelastic.

This chapter is structured as follows: Case Study Description section introduces the area where the case study is applied (the Guadalquivir River Basin in Southern Spain) and key concepts related to the model proposed. Methodology section presents the RPM. Price Volatility Scenarios section specifies the price volatility scenarios that we use in the simulations. Results section shows and discusses the results obtained under the alternative scenarios considered and Conclusion section concludes the chapter.

Case Study Description

The Guadalquivir River Basin (GRB), located in the south of Spain, is one of the largest river basins in the country (57,071 km^2). Agriculture is a relevant activity in the area in terms of GDP (4.2%, as compared to the Spanish average of 2.5%) and employment generation (7%, as compared to the Spanish average of 4%) (GRBA 2010). Agriculture is also the main water consumer and demands, on average, 3, 485 million cubic meters (hm^3) every year, 86.8% of the total water use in the basin of 4,016 hm^3/year. As renewable resources in the GRB amount to 3,028 hm^3/year, there is a severe water overexploitation that in an average hydrological year equals 987.7 hm^3 (GRBA 2010). In order to reduce water use and prevent further water overexploitation, the Guadalquivir River Basin Authority has developed a new River Basin Management Plan (RBMP) (GRBA 2010), which includes a program of measures that imposes more restrictive conditions to access to water for irrigation. As a result, although according to the Spanish law RBMPs should guarantee irrigators a water access reliability of 90%, in the case of the GRB it is not possible to guarantee a failure rate below the target of 10% (Berbel et al. 2012).[1] Consequently, agriculture in the GRB is highly exposed to price shocks both from an economic (employment, income) and from an environmental (water use) perspective.

There are more than 840,000 ha of irrigated lands in the GRB. The main irrigated crop in the GRB is by large olive trees (55.4% of the total irrigated land). Other relevant irrigated crops include cereals (9.1%), citrus and fruit trees (6.9%), cotton (6.3%), horticulture (5.2%), rice (4%), industrial crops (3.5%), maize (2.2%) and others (7.3%). These irrigated lands are managed by over 235,000 farmers (see Table 1). In this chapter we follow the work by Gómez-Limón et al. (2012) and we group farmers into local aggregation units that share a common water source and water infrastructure (hence, showing a similar management technique). Thus, we obtain 1,603 aggregation units. Each aggregation unit is different and may include a large group of individual farmers, a group of Water Users Associations including several farmers each, or a combination of both. Finally, we group these aggregation units at a basin level into five clusters, again according to the water source and water infrastructure observed (Gómez-Limón et al. 2012): C1—Traditional herbaceous crops of the fertile lowland; C2—Modernized fertile lowland; C3—Olive grove; C4—Traditional horticulture; and C5—Rice. These clusters will be the agents of our RPM. Figure 2 shows the location of the clusters and Table 1 characterizes them.

Cluster C1 includes traditional herbaceous crops in the fertile lowland mostly located in the lower GRB and in the upper Genil River, largely

[1] With the new RBMP, irrigators have to stand a "guarantee gap" of 534 hm^3 before the special clauses of the RBMP to guarantee that a minimum agricultural water supply is implemented; after the implementation of the program of measures, this "guarantee gap" should be reduced to 276 hm^3, implying that with the present level of demand and the "stored" resources (reservoirs, non-regulated rivers and groundwater) it is not possible to guarantee a failure rate below the target of 10% (Berbel et al. 2012).

Table 1. Main crops in the GRB and distribution among clusters.

Variable/Cluster	C1	C2	C3	C4	C5	Total
Number of aggregation units	137	252	892	288	34	1,603
Number of irrigators	17,916	56,348	117,119	42,741	1,499	235,623
Total irrigated surface (ha)	39,002	310,849	406,354	51,069	34,782	842,056
Irrigated surface: olive tree (ha)	4,586	86,219	361,833	14,027	12	466,677
Irrigated surface: cereals (ha)	6,008	48,950	12,975	8,322	145	76,401
Irrigated surface: cotton (ha)	8,899	40,253	3,105	68	347	52,672
Irrigated surface: horticulture (ha)	3,468	19,741	8,444	11,784	127	43,564
Irrigated surface: citrus and fruit trees (ha)	1,825	46,284	5,831	4,574	8	58,521
Irrigated surface: rice (ha)	380	1,112	15	0,0	32,542	34,049
Irrigated surface: industrial crops (ha)	3,085	24,695	1,477	294	279	29,830
Irrigated surface: maize (ha)	4,866	11,038	1,571	989	18	18,482
Irrigated surface: other crops (ha)	5,884	32,556	11,102	11,013	1,305	61,860
Gravity irrigation (% of the total surface)	61.3	7.2	5	75	96.1	48.92
Sprinkler irrigation (% of the total surface)	15.9	48.1	3.1	3.8	3.5	14.88
Drip irrigation (% of the total surface)	22.8	44.7	91.9	21.2	0.4	36.2
Irrigation efficiency (% estimated)	69.2	80.6	88	66.9	60.5	73.5

Source: Authors' elaboration from Gómez-Limón et al. (2012).

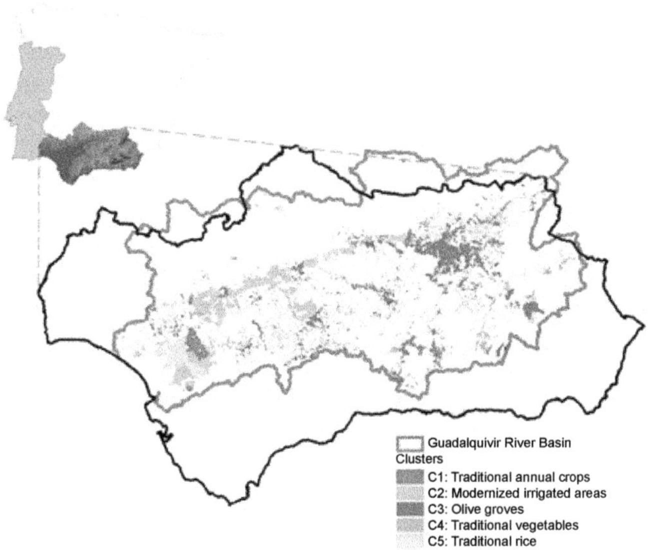

Figure 1. Clusters identified in the GRB.
Source: Gómez-Limón et al. (2012).

Color image of this figure appears in the color plate section at the end of the book.

occupied by herbaceous crops and olive trees. Modern large irrigated areas (C2) are mainly located on the banks of the Guadalquivir and Genil rivers and include mainly herbaceous crops, citrus and fruit trees and olives. Drip (44.7%) and sprinkler (48.1%) irrigation devices are widespread. (C3) includes a monoculture of irrigated olive grove and it is scattered all around the GRB, although the most important concentrations of olive trees are located by the main branch of the river upstream. The traditional horticulture (C4) is limited to the areas located upstream of the Genil and the Guadalquivir rivers. Finally, rice (C5) is limited to the humid marshes in the estuary of the Guadalquivir.

Methodology

The economics literature has devoted much effort in the last decade to develop behavioral models with the ability to anticipate famers' response to different shocks, such as agricultural prices fluctuations. The most commonly used methods so far have been Linear Programming (LP), Positive Mathematical Programming (PMP) and Multi-Criteria Decision Methods (MCDM). All these methods have succeeded because of their ability to provide numerical results, although many authors have argued that they do not comply with some basic economic principles. In what follows we present the advantages and disadvantages of these methods and our proposed alternative: the RPM.

The need to represent complex decision problems with limited information has extended the use of Linear Programming (LP) and Positive Mathematical Programming (PMP) to simulate farmers' response against different shocks and to obtain water demand functions. Although the origin of LP dates back to the '50s, it has been widely used recently to assess the impact of agricultural policies due to its low data requirements and flexibility (see for example the work of Kampas et al. 2012; Dono et al. 2010). However, this method has been strongly criticized as a result of its inability to approximate, even roughly, realized farm production plans and, therefore, to become a useful methodology for policy analysis (Paris 2010). This criticism is grounded on the linear nature of this method, which often results in overspecialization and corner solutions. In addition, LP might be criticized by the way it deals with the parameter specification problem: there is an infinite set of parameters and functions able to lead the model to a perfect calibration, and each set of parameters and functions leads to a different behavior in response to changing agricultural prices and policy constraints.

PMP came as a response to the critiques above. PMP offers many advantages over LP, including full calibration, a significant reduction in the number of resource, technical, economic and policy constraints, and the use of nonlinear cost functions that guarantee smooth simulation results. The use of these models to simulate farmers' behavior and to obtain water demand functions can be found, for example, in Heckelei and Britz (2005) and in De Frahan et al. (2007). The general idea of these models consists, first, in using information contained in dual variables of the calibration constraints to bind the solution of the linear profit maximizing problem to the observed activity levels. Once these dual variables are identified, they are used to specify a nonlinear objective function, such as the production cost, provided that the marginal cost of the activities is equal to its price in the observed activity levels. This action guarantees that both the profit maximization and the cost minimization problems lead simultaneously to an optimal solution which exactly matches the baseline activity levels (Howitt 1995; Paris and Howitt 1998; Heckelei and Britz 2005). Although effective, this calibration mechanism is not rooted in explicit economic principles and this constitutes the main critique against PMP.

Finally, Multi-Criteria Decision Methods (MCDM) also played a major role in the study of farmers' behavior (see for example Berbel and Rodríguez-Ocaña 1998; Berbel and Gómez-Limón 2000; Gómez-Limón and Riesgo 2004). Contrary to PMP methods, in MCDM farmers do not act simply as profit maximizing agents; instead of that, agents consider other relevant attributes in their decision. Therefore, MCDM assume that farmers' preferences can be represented by a weighted sum of different criteria, such as expected profits, risk, management issues and/or others, which provides a better explanation of current decisions. Although this method has succeeded in reproducing the baseline decision, the assumption that farmers respond with linear preferences to changes in the policy is again an issue prone to discussion.

Therefore, the construction of models to simulate farmers' behavior is confronted with a tradeoff between the model's capability to provide numerical results for policy evaluation and its coherence with basic economic principles. However, it is still possible to develop a methodology that is consistent with these principles and offers useful results for policy analysis through the use of methods that reveal farmers' preferences (or RPM). These applied models try to provide a clearer intuition of the logic behind farmers' decisions by using standard economic analysis and by implementing a multi-attribute utility function. Moreover, RPM do not need to assume linear preferences (as in LP and MCDM) or implicit costs functions that are not observable (as in PMP). Although the complex programming and optimization procedure and the high data requirements of these models have made their use difficult as a policy assessment and project analysis tool (to find studies using RPMs we need to go back three decades, to Rausser and Yassour 1981; Delforce and Hardaker 1985), the advances in computational methods of the last two decades and the recent proliferation of high quality agricultural microeconomic databases in several developed countries now make their implementation feasible.

This paper develops an RPM to explore the effects of price volatility in agricultural markets over water use and agricultural income and employment. In this section we present the steps that we follow in the calibration process of the RPM, as well as the calibration errors. This model is calibrated for each one of the five clusters in which we have divided the irrigated lands of the GRB. Therefore, clusters are the agents in our model and they act as a representative farmer of the homogeneous set of farmers that constitute each cluster.

The Revealed Preferences Model (RPM)

In this chapter we present an RPM able to calibrate observed decisions with a procedure rooted in basic microeconomic theory. This method not only allows us to obtain simulation results but also offers a clear interpretation of farmers' responses to changing incentives and resource and policy environments (Gutierrez-Martin and Gomez Gomez 2011). In our model, agents decide on crop land surfaces trying to maximize their utility, which is a function of a set of relevant attributes that may contain expected profit, risk avoidance, complexities management and/or others. We assume that the explanation of any decision, consisting in a distribution of the available land among the different crop options, relies on an underlying utility function formed by the many attributes that agents use to assess all the alternatives they have, given crop prices and costs, resource availability and the other relevant economic, agronomic and policy constraints. According to that, we may assume that observed decisions respond to a decision problem of the following kind:

$$\underset{x}{Max}\ U(x) = U(z_1(x);\ z_2(x);\ z_3(x)\ \ldots\ z_m(x)) \quad [1]$$

$$s.t.:\quad 0 \leq x_i \leq 1 \quad [2]$$

$$\sum_{i=1}^{n} x_i = 1 \quad [3]$$

$$X \in F(x) \quad [4]$$

$$z = z(x) \in R^m \quad [5]$$

Where $x \in R^n$ is the decision profile or the crop portfolio (a vector), showing one way to distribute the land among crops, and each x_i measures the share of land devoted to the crop *i*, including a reservation option (x_n) consisting of rainfed agriculture. From the agent's perspective any particular crop may be considered as an asset with a known present cost and an uncertain value in the future (as crop yields are not known in advance). As the available land is taken as given, this investment may be represented as a percentage (x_i) of the available land. F(x) represents the space of feasible decision profiles, given the different constraints:[2] policy, economic, agronomic and environmental.[3] Finally z_i, or alternatively the vector z, are the attributes that farmers value. For example, farmers might prefer decisions with high expected profits, highly predictable yields and prices and not too many managing actions apart from planting and harvesting. To accept taking high risk options, risk averse farmers will ask for a compensation, for example, higher expected profits, and the same can be said about the willingness to accept crop decisions that demand additional management skills.

Let us assume that we have an observed decision profile and we know the whole set of constraints defining the feasible decision set. Assume also that we can measure a set of potentially relevant decisions attributes such as, for example, the expected profit, the variance of the expected profit, the hired labor demanded, the cost of inputs over the total cost and all the variables that might be relevant from the farmers' point of view. Then, the first problem we need to deal with to reveal farmers' preferences is to know which among the potentially relevant attributes are relevant to explain the observed decision. Our method, in answer to this question, is to say that the relevant set of attributes is the one to which the observed decision is closest to the attributes possibility frontier. In real situations, this efficiency frontier cannot be defined analytically with a closed mathematical function and the only way to represent it is by numerical methods. One practical solution consists in extending a ray from the origin, passing through the observed decision attributes and extending them as far as possible in the space of feasible attributes. This way we can measure the distance from the observed attributes to the efficiency frontier attributes. We can repeat this procedure for any set of potentially relevant attributes and the best candidate to reveal farmers' preferences will be the one whose observed values were closest to its associated efficiency frontier. Formally, this problem must be solved for every member of the Power set (P(z)) (which comprises all the possible

[2] These constraints vary for each cluster. In our model we consider the following: land availability, available water resources, agricultural vocation (crops that have not been planted in an area before cannot appear in that area in the short run), crop rotation, CAP restrictions and ligneous crops restrictions (the surface of ligneous crops cannot change significantly in the short run).

[3] We only consider one environmental constraint: water availability. Although the clusters comprise different farmers scattered along the GRB, farmers within each cluster share similar water sources and water infrastructures, and thus environmental constraints work properly even though clusters are not in the same area.

combinations of potentially relevant attributes for the farmer) and for its associated observed attributes in the Power set ($P(z_o)$).[4]

The solution for this problem is an application assigning a distance φ_l ($l=1,\dots,2^m$) to each member of the power set P(z). Each member of the power set (i.e., each possible combination of potentially relevant attributes) is denoted by $\tau(x)$, and its associated observed attributes by $\tau_o(x)$. The relevant set of attributes (τ^*) will be the one with the lower distance to the efficiency frontier measured by the parameter ($\varphi - 1$). Summing up, the preference eliciting problem can be presented as:

$$\underset{\tau}{Min}\ \varphi_l - 1 \qquad [6]$$

Where:

$$\varphi_l = ArgMax\ [(\varphi) s.t. \tau(x) = \varphi(\tau_o(x)); 0 \le x_i \le 1; \sum_{k=1}^{n} x_k = 1; X \in F(x); for\ all\ \tau \in P(z)] \qquad [7]$$

$$l = (1 \dots 2^m) \qquad [8]$$

By solving this problem we obtain the set of attributes that better explains current farmers' decisions (τ^*). Among the many factors that might be of relevance in farmers preferences, this set of attributes is the one which takes the observed decision closer to the attributes efficiency frontier.

Once a farmer's decision is shown as close as possible to the efficiency frontier, the second problem consists in obtaining the farmers' preferences that explain the observed decision as a utility maximizing choice. Taking into account the relevant decision attributes obtained in the calibration stage, the multi-attribute utility function is the one that is able to represent farmers' preferences in such a way that the observed decision becomes the optimal choice. Using basic economic principles and knowing the efficiency frontier in the surroundings of the observed decision allows one to integrate such a utility function. Rational decisions imply that in equilibrium farmers' marginal willingness to pay in order to improve one attribute with respect to any other is equal to the marginal opportunity cost of this attribute with respect to the other. In other words, the marginal transformation relationship between any pair of attributes over the efficiency frontier (MTR_{kp}) is equal in equilibrium to the marginal substitution relationship between the same pair of attributes over the indifference curve tangent to the observed decision (MSR_{kp}).

[4] A power set P(Z) is the set of all the 2^m subsets of the set Z and the power set $P(Z_o)$ is the set formed by the 2^m subsets of the numerical set of observed attributes.

The calibration model allows us to obtain the relative opportunity cost of each of the relevant attributes with respect to the others. This opportunity cost is measured by the marginal transformation relationship between any pair of attributes (β_{kp} = MTR_{kp} = MSR_{kp}). This value can be obtained numerically by solving partial optimization problems in the proximity of the observed decision (as for example, searching by how much expected profits would need to be reduced in order to have a 1% less uncertainty or, equivalently, what is the maximum expected profit attainable with a slightly lower risk level). The numerical results of the marginal relationship of transformation of any pair of attributes in a reference point over the efficiency frontier (β_{kp}) are the basic information to integrate the farmers' utility function. Provided that farmers act rationally, in equilibrium, the value (β_{kp}) representing the relative opportunity cost of any attribute in terms of any other, is equal to the marginal substitution relationship between the same pair of attributes (which represents the farmers' willingness to pay for marginal improvement of a given attribute in terms of any other). In other words, in equilibrium, decisions over crop surfaces are such that:

$$\beta_{kp} = \mathrm{MTR}_{kp} = \mathrm{MSR}_{kp} = -\frac{\partial U / \partial z_p}{\partial U / \partial z_k}; \; p, k \in (1,\ldots 1); p \neq k \qquad [9]$$

This information for the reference point over the efficiency frontier is enough to integrate a utility function leading to the observed decision as the optimal decision given the existing resource, economic, balance and policy constraints. For example, if we assume constant returns of scale such as the Cobb Douglas utility function below:

$$U(\tau) = \prod_{r=1}^{l} z_r^{\alpha_r}; \qquad \sum_{r=1}^{l} \alpha_r = 1 \qquad [10]$$

Then the marginal substitution relationship among any pair of attributes is:

$$-\frac{\partial U / \partial z_p}{\partial U / \partial z_k} = -\frac{\alpha_p \, z_k}{\alpha_k \, z_p} \qquad [11]$$

And the parameters of the Cobb-Douglas utility function are obtained from the following system:

$$-\frac{\alpha_p \, z_k}{\alpha_k \, z_p} = \beta_{kp} \qquad [12]$$

$$\sum_{r=1}^{l} \alpha_r = 1 \qquad [13]$$

In the results section we use this type of function, which offers the advantage of having a unique solution according to the Walras' Law (a condition which is guaranteed by the constant returns of the utility function represented above). Then the model is calibrated for each cluster using the high quality microeconomic data offered by MAGRAMA (2009). This database contains data on land use, water demand, productive factors, agrarian income, prices, costs, employment and all the variables needed in the calibration stage for the period 2004–2009 and covers 92.5% of the irrigated surface in the GRB.

Calibration errors

Our model represents farmers' decisions. These decisions are simulated in accordance to the observed crop portfolio, which is the crop portfolio that maximizes the representative farmer's utility function (in accordance to a set of relevant attributes). Therefore, deviations of the model's crop portfolio (x_i^*) from the observed crop portfolio (x_i^o) during the calibration stage may result in prediction errors in our model, and this is our first calibration error (e_x). The second source of error is the distance between the observed attributes and the attributes' efficiency frontier (e_f). A large distance would mean that our representative farmer is actually taking a sub-optimal decision, and this goes against our main assumption that farmers are individuals that seek to maximize their utility. Finally, the third calibration error (e_τ) is the distance between the observed attributes (z_r^o) and the calibrated ones (τ_r^*). If this distance is large, it would mean that we are not capturing the real source of utility for the representative farmer, and therefore our model would be simulating someone else's utility function.

Summing up, our RPM provides three types of calibration errors that give an idea of the accuracy of the model's adjustment:

- The relative distance between the observed crop pattern and the model's one:

$$e_x = \frac{1}{n}\sum_{i=1}^{n}\left(\frac{(x_i^{o2} - x_i^{*2})^{1/2}}{x_i^o}\right) \quad [14]$$

- The distance between the observed attributes and the attributes' efficiency frontier:

$$e_f = (\varphi - 1) \quad [15]$$

- The distance between the observed attributes and the calibrated ones:

$$e_\tau = \frac{1}{l}\sum_{r=1}^{l}\left(\frac{(z_r^{o2} - \tau_r^{*2})^{1/2}}{z_r^o}\right) \quad [16]$$

Finally, the mean calibration error is defined as a combination of these three calibration errors:

$$e = \frac{\sqrt{e_x^2 + e_r^2 + e_f^2}}{3} \quad [17]$$

Price Volatility Scenarios

After the calibration process, we carry out a simulation in which we progressively increase the volatility of agricultural prices in order to assess farmers' response. Prices scenarios are defined according to the international and domestic prices observed since 2007. Since 2007, agricultural prices worldwide experienced a significant increase, especially those of wheat (150%), rice (200%), maize (150%), cotton (140%) and sunflower (220%) (World Bank 2012). The price of these commodities in the agricultural markets within the GRB[5] also increased, but less sharply: wheat prices rose in average by 51.1%, rice by 64%, maize by 41%, cotton by 25% and sunflower by 47% (MAGRAMA 2009). Table 2 shows the maximum and minimum world prices of these crops for the period 2002–2012 (World Bank 2012) as

Table 2. Maximum and minimum international agricultural prices and average domestic prices in the GRB.

	Wheat	Maize	Rice	Sunflower	Cotton
Maximum world price	288.21	271.26	648.83	650.78	738.00
Minimum world price	108.89	71.30	160.38	200.81	162.00
Average price, GRB	182.04	162.31	246.05	288.69	502.58
Max. World price/average price in the GRB	1.58	1.67	2.64	2.25	1.47

Source: World Bank (2012); MAGRAMA (2009).

[5] Domestic prices refer to those prices observed in the agricultural markets of the GRB, as reported in MAGRAMA (2009). These prices are homogeneous along the entire basin.

well as the average domestic prices in the GRB for the period 2004–2009 (MAGRAMA 2009).

Based on this data, this section defines a set of price volatility scenarios that simulate higher prices and thus a higher income and price volatility within the EU. In these scenarios domestic prices get progressively closer to and finally match international prices. We define five scenarios, with two extreme situations: the baseline scenario, where agricultural prices in the GRB remain constant; and the S4, in which total domestic prices match international prices. The remaining three scenarios are between these two extreme cases.

For every scenario we define a *lambda* coefficient which represents the price increase for every commodity in each scenario. The coefficient value ranges from 1 (Baseline Scenario, BL) to the quotient of the maximum world price and the average price in the GRB (Scenario 4, S4). Table 3 shows the coefficient values in the five scenarios considered.

In the next section, we calibrate the model and we use these coefficients to modify the prices vector and its corresponding expected income (profit

Table 3. Price volatility scenarios and *lambda* coefficients.

	Wheat	Maize	Rice	Sunflower	Cotton
Scenario 4 (S4)	1.58	1.67	2.64	2.25	1.47
Scenario 3 (S3)	1.44	1.50	2.23	1.94	1.35
Scenario 2 (S2)	1.29	1.34	1.82	1.63	1.24
Scenario 1 (S1)	1.15	1.17	1.41	1.31	1.12
Baseline Scenario (BL)	1.00	1.00	1.00	1.00	1.00

Source: Authors' elaboration from World Bank (2012); MAGRAMA (2009).

attribute) vector and the income variance and covariance matrix (risk attribute). Then we study how this shock affects the farmers' decision.

Results

The methodology above is now applied to the particular case of the GRB. First, we calibrate the RPM for every cluster. Second, we conduct an exploratory assessment of the price elasticity of the crop portfolio (effects of a price increase in a single crop over the crop portfolio). Finally, we simulate the effect of the price volatility scenarios defined in the previous section over the crop portfolio, water demand, income and employment.

Model calibration and validation

Farmers have to decide over the combination of crops to plant, subject to a set of feasible options. It is reasonable to think that farmers will choose the crop portfolio that maximizes their income and minimizes their risk and management complexities. Accordingly, we consider the following variables in our model:

i) *Expected profit per hectare*, measured by the gross variable margin.

$$z_1(x) = \Sigma_i x_i \pi_i \qquad [18]$$

Where π_i is the gross variable margin per hectare of the crop i.

ii) *Avoided risk*, measured by the standard deviation of the expected profit per hectare.

$$z_2(x) = \bar{\sigma} - \sigma(\pi(x)) \qquad [19]$$

Where $\bar{\sigma}$ is the risk associated to the crop decision $\bar{x}$ leading to the maximum expected profit and $\sigma(\pi(x))$ is the risk associated to the alternative crop decision x, which can be defined as:

$$\sigma(\pi(x)) = x^{\tau} VCV(\pi(x))x.$$

Being $VCV(\pi(x))$ the variance and covariance matrix of the per hectare crop profits $(\pi(x))$ of the crop decision x.

iii) *Total labor avoidance*, the first way to measure management complexities avoidance through the reluctance to use too much labor.

$$z_3(x) = \bar{N} - N(x) \qquad [20]$$

Where $N(x) = \Sigma_i x_i N_i$ is the total labor used per hectare, being N_i the total labor required per hectare for a crop i, and $\bar{N}$ is the labor required to implement the crop decision leading to the maximum expected profit.

iv) *Direct cost avoided*, the third way to measure management complexities, which includes all the seeds, fertilizers, hired equipment and all other intermediate expenditures required to implement a particular crop decision.

$$z_4(x) = D(x) - \bar{D} \qquad [21]$$

Where $D(x) = \Sigma_i x_i D_i$ is the direct cost of a crop decision x, being D_i the direct cost per hectare for a crop i, and $\bar{D}$ is the direct cost corresponding to the maximum expected profit decision.

As a result, our Cobb-Douglas Utility Function adapts the following form:

$$U(z_1, z_2, z_3, z_4) = z_1^{\alpha_1} z_2^{\alpha_2} z_3^{\alpha_3} z_4^{\alpha_4}; \qquad \sum_{r=1}^{4} \alpha_r = 1$$

Where there are five unknown variables (α_r; $r = 1, ...4$). Following the methodology above, we assess the relevance of these attributes and we estimate the values of the *alpha* coefficients for every cluster, which are used to calibrate the Cobb-Douglas Utility Function. Finally, we obtain the calibration errors. The results are displayed in Table 4.

Table 4. *Alpha* coefficients and calibration errors.

Cluster/Variable	α_1	α_2	α_3	α_4	e_f	e_τ	e_x	e
Cluster 1	0.61	0.36	-	0.03	19.75%	0.76%	13.61%	6.8%
Cluster 2	0.71	0.26	-	0.03	9.54%	1.17%	9.05%	4.6%
Cluster 3	0.99	0.01	-	-	1.12%	0.28%	0.93%	0.5%
Cluster 4	0.87	0.1	0.03	-	2.51%	1.13%	2.13%	1.2%
Cluster 5	1	-		-	1.00%	27.93%	1.00%	14.0%

Source: Authors' elaboration from MAGRAMA (2009).

The most relevant attributes explaining the agents' decision are expected profit (z_1) and avoided risk (z_2). Total labor avoidance (z_3) and direct cost avoided (z_4) are less relevant, but still significant. This means that the farmers in the GRB aim to achieve the maximum possible income with a minimum risk, giving little consideration to management complexities.

Price elasticity of the crop portfolio

This section describes an exploratory assessment of the price elasticity of the crop portfolio, which shows the effect of a price increase in a single crop over the aggregate crop portfolio. This is an intermediate step that allows us to isolate the effect that the price variation of a single crop has over the crop portfolio, as opposed to the overall effect of the price volatility scenarios in which several agricultural prices are modified simultaneously, which will be the subject of the next section.

Therefore, in this section, we implement a simulation in which we increase individually the prices of crops. Higher agricultural prices imply a greater expected income but also greater income variability. Thus, we increase the gross variable margin and we modify the variance and

Table 5. Price elasticity of the crop portfolio, GRB.[a]

1%	**W**	**M**	**C**	**S**	**V**	**F**	**O**	**RF**
W	1.88	0.00	–0.08	0.23	–1.04	–0.06	0.00	–0.94
M	0.02	0.00	–0.01	0.00	0.00	–0.01	0.00	0.00
C	0.12	0.00	0.46	–0.03	–0.25	–0.04	0.00	–0.26
S	0.21	0.00	0.06	0.08	–0.14	–0.02	0.00	–0.19
V	–0.81	0.00	–0.26	–0.23	0.73	0.03	0.00	0.53
F	0.00	0.00	0.00	0.00	0.00	0.00	0.00	0.00
O	0.00	0.00	0.00	0.00	0.00	0.00	0.00	0.00
RF	–1.48	0.00	–0.44	–0.45	0.90	0.06	0.00	1.41
10%	**W**	**M**	**C**	**S**	**V**	**F**	**O**	**RF**
W	0.49	0.00	–0.13	0.02	–0.24	–0.01	0.00	–0.13
M	0.00	0.00	0.00	0.00	0.00	–0.01	0.00	0.00
C	–0.14	0.00	0.19	–0.15	0.00	–0.03	0.00	0.12
S	0.10	0.00	0.02	0.04	–0.06	–0.01	0.00	–0.08
V	–0.26	0.00	–0.23	–0.12	0.45	0.01	0.00	0.14
F	–0.10	0.00	–0.13	–0.13	0.11	0.07	0.00	0.18
O	0.09	0.00	–0.04	–0.01	0.07	–0.06	0.00	–0.05
RF	–0.14	0.00	0.02	–0.10	0.16	–0.01	0.00	–0.07

30%	**W**	**M**	**C**	**S**	**V**	**F**	**O**	**RF**
W	0.17	0.00	–0.05	0.01	–0.07	–0.01	0.00	–0.04
M	0.00	0.00	0.00	0.00	0.00	0.00	0.00	0.00
C	–0.16	0.00	0.21	–0.11	–0.05	–0.03	0.00	0.01
S	0.04	0.00	0.01	0.01	–0.02	–0.01	0.00	–0.03
V	–0.10	0.00	–0.12	–0.06	0.25	0.00	0.00	0.02
F	0.07	0.00	–0.10	–0.08	–0.01	0.03	0.00	0.08
O	0.09	0.00	–0.08	–0.02	0.08	–0.05	0.00	–0.02
RF	–0.05	0.00	0.00	–0.06	0.10	–0.01	0.00	–0.09
90%	**W**	**M**	**C**	**S**	**V**	**F**	**O**	**RF**
W	0.06	0.00	–0.03	0.00	–0.01	–0.01	0.00	–0.01
M	0.00	0.00	0.00	0.00	0.00	0.00	0.00	0.00
C	–0.04	0.00	0.11	–0.04	–0.05	–0.03	0.00	–0.02
S	0.02	0.00	0.01	0.00	–0.01	–0.01	0.00	–0.01
V	–0.01	0.00	–0.05	–0.03	0.10	0.00	0.00	–0.01
F	0.02	0.00	–0.03	–0.03	0.01	0.00	0.00	0.03
O	0.03	0.00	–0.03	–0.02	0.05	–0.04	0.00	–0.01
RF	0.00	0.00	–0.02	–0.03	0.05	–0.01	0.00	–0.03

Source: Authors' elaboration from MAGRAMA (2009).

[a] W = Wheat, R = Rice, M = Maize, C = Cotton, S = Sunflower, V = Vegetables, F = Fruit trees, O = Olive trees, RF = Rainfed crops.

covariance matrix accordingly. We build tables to present the effect of higher agricultural prices over the crop portfolio in the GRB (see Table 5). We consider a price increase of 1%, 10%, 30% and 90% for each crop. Main diagonal shows direct effect of the price increase of a specific crop over its own surface. We would expect that a price increase has a positive effect over the number of hectares devoted to this crop, thus showing a positive value. The remaining cells show the effects of the price increase of a specific crop over the remaining crops. These values have a positive value when a price increase augments the surface devoted to other crops (complementary crops) and a negative value when a price increase reduces the surface devoted to other crops (substitutive crops).

Complementarities arise due to the agronomic restrictions in the model. These rotational constraints are based on the observed agricultural practices in the GRB (MAGRAMA 2009). For example, we would expect that wheat is a complementary good to cotton, and that when the surface of cotton increases, so does the surface of wheat. However, we cannot be sure of the opposite, as the complementarity of two goods is not necessarily reciprocal.

Overall, the price elasticities of the crop portfolio are low, and this is largely explained by the role played by olive trees. Olive trees are a ligneous crop and thus decisions over land use cannot be solely taken in the short run, so we have fixed a maximum land use variability of 10% based on the historical maximum variability observed in two consecutive years (MAGRAMA 2009). We do the same with the other ligneous crops. As olive trees represent more than 50% of the irrigated surface, price volatility has a limited effect over crop portfolio. We would expect different results in clusters where ligneous crops are less relevant. For example, in Table 6 we calculate the price elasticity of the crop portfolio for the modernized fertile lowland Cluster (C2).

We can draw many conclusions from the tables above: (1) Elasticities in the diagonal in C2 are significantly greater than those of the whole GRB, as olive tree in the C2 only represents 28% of the surface. (2) As the price increases, the effect over the crop portfolio is reduced. This is a consequence of the higher income variability, which makes the price elasticity of the crop portfolio decrease, but also of the existing agronomic constraints (each crop has a maximum surface in the model to allow rotations and avoid land exhaustion). (3) There are cross elasticities of both positive and negative sign. Most of the crops are substitutive crops, but some are complementary, such as the wheat and cotton (and cotton and wheat) and the fruit trees and wheat (but not the opposite due to the land use maximum variability described above).

Table 6. Price elasticity of the crop portfolio, Modernized fertile lowland Cluster (C2).[a]

1%	**W**	**M**	**C**	**S**	**V**	**F**	**O**	**RF**
W	3.84	–0.04	0.61	–2.09	–0.17	0.00	–2.15	0.00
M	0.82	1.11	0.02	–1.13	–0.09	0.00	–0.73	0.00
C	0.56	0.17	0.21	–0.38	–0.05	0.00	–0.51	0.00
S	–1.59	–0.85	–0.60	1.46	0.10	0.00	1.49	0.00
V	–1.70	–0.49	–0.61	1.10	0.31	0.00	1.39	0.00
F	0.31	–0.13	–0.04	0.17	–0.15	0.00	–0.17	0.00
O	–4.28	–1.25	–1.26	2.51	0.20	0.00	4.07	0.00
RF	0.00	0.00	0.00	0.00	0.00	0.00	0.00	0.00
10%	**W**	**M**	**C**	**S**	**V**	**F**	**O**	**RF**
W	1.03	–0.31	0.05	–0.53	–0.03	0.00	–0.22	0.00
M	–0.29	0.44	–0.34	–0.10	–0.04	0.00	0.33	0.00
C	0.25	0.05	0.10	–0.14	–0.03	0.00	–0.22	0.00
S	–0.44	–0.66	–0.32	0.98	0.04	0.00	0.41	0.00
V	–0.18	–0.38	–0.33	0.29	0.19	0.00	0.41	0.00
F	0.30	–0.15	–0.04	0.19	–0.14	0.00	–0.16	0.00
O	–0.40	–0.20	–0.21	0.41	–0.01	0.00	0.41	0.00
RF	0.00	0.00	0.00	0.00	0.00	0.00	0.00	0.00

30%	**W**	**M**	**C**	**S**	**V**	**F**	**O**	**RF**
W	0.34	–0.13	0.02	–0.14	–0.02	0.00	–0.07	0.00
M	–0.14	0.35	–0.23	–0.12	–0.06	0.00	0.12	0.07
C	0.09	0.02	0.04	–0.06	–0.02	0.00	–0.07	0.00
S	–0.07	–0.27	–0.14	0.41	0.00	0.00	0.07	0.00
V	0.26	–0.27	–0.21	–0.01	0.09	0.00	0.14	0.00
F	0.29	–0.25	–0.06	0.21	–0.12	0.00	–0.07	0.00
O	–0.12	–0.13	–0.14	0.26	–0.02	0.00	0.14	0.00
RF	0.00	0.00	0.00	0.00	0.00	0.00	0.00	0.00
90%	**W**	**M**	**C**	**S**	**V**	**F**	**O**	**RF**
W	0.11	–0.07	0.01	–0.01	–0.02	0.00	–0.02	0.00
M	–0.02	0.21	–0.08	–0.13	–0.04	0.00	–0.02	0.09
C	0.04	0.02	0.00	–0.02	–0.01	0.00	–0.02	0.00
S	0.03	–0.09	–0.07	0.16	–0.01	0.00	–0.02	0.00
V	0.09	–0.09	–0.08	0.03	0.01	0.00	0.05	0.00
F	0.11	–0.09	–0.05	0.14	–0.08	0.00	–0.02	0.00
O	0.00	–0.09	–0.08	0.12	–0.01	0.00	0.05	0.00
RF	–0.18	–0.09	–0.02	0.03	–0.01	0.00	–0.02	0.28

Source: Authors' elaboration from MAGRAMA (2009).

[a] W = Wheat, R = Rice, M = Maize, C = Cotton, S = Sunflower, V = Vegetables, F = Fruit trees, O = Olive trees, RF = Rainfed crops.

Price volatility scenarios

Finally, calibrated model is used to obtain the outcomes in terms of crop portfolio, water use, income and employment in every price volatility scenario considered. In the following sections, we assess the results obtained for each cluster as well as for the basin as a whole.

Crop portfolio

The observed crop portfolio is mostly driven by the expected profits and risk avoidance (see Table 4), with the exception of the farmers in the Clusters 3 (olive grove) and 5 (rice), whose decisions are based on profit maximization. Figure 2 shows the optimum crop portfolio for each cluster and for the entire GRB under the different scenarios considered, namely, Baseline (BL), Price Volatility Scenario 1 (S1), Price Volatility Scenario 2 (S2), Price Volatility Scenario 3 (S3) and Price Volatility Scenario 4 (S4).

In Cluster 1, the combined price increase of maize, wheat, cotton and sunflower significantly increases the expected income of these crops and the surface of maize and wheat, though the result is the opposite for cotton and sunflower. This is explained by the fact that, while wheat and maize are reliable crops with low income variability, cotton and sunflower are comparatively more risky, and thus the representative farmer reduces their surface to reduce income variability. It is also remarkable that with the first price increase (S1), the surface of sunflower rises while that of wheat decreases, and then the trend is reverted. This is a result of the tradeoff between risk and income: in the first increase, the income effect is higher than the income variability effect and thus the surface of sunflower increases and that of wheat decreases; with the subsequent price increases (S2, S3 and S4), the result is the opposite. The scenarios considered also show a decrease in the surface of vegetables and olive trees as prices increase.

In Cluster 2 results are very similar to those of Cluster 1, though the tradeoff between risk and income in the case of wheat and maize, on the one hand, and sunflower and cotton, on the other, is smoother. Also, the surface of vegetables and olive trees is reduced.

Cluster 3 (olive grove) has interesting results. Agents' decisions depend almost exclusively on the expected profit, but there is still space

[6] Our RPM works in the short and medium run. However, decisions over the surface devoted to ligneous crops cannot be solely taken in the short run, as they have large fixed costs associated that can only be recovered in the long run. This is why in our model we have fixed a maximum land use variability of 10% for ligneous crops. This figure is based on the historical maximum land use variability observed in two consecutive years (MAGRAMA 2009).

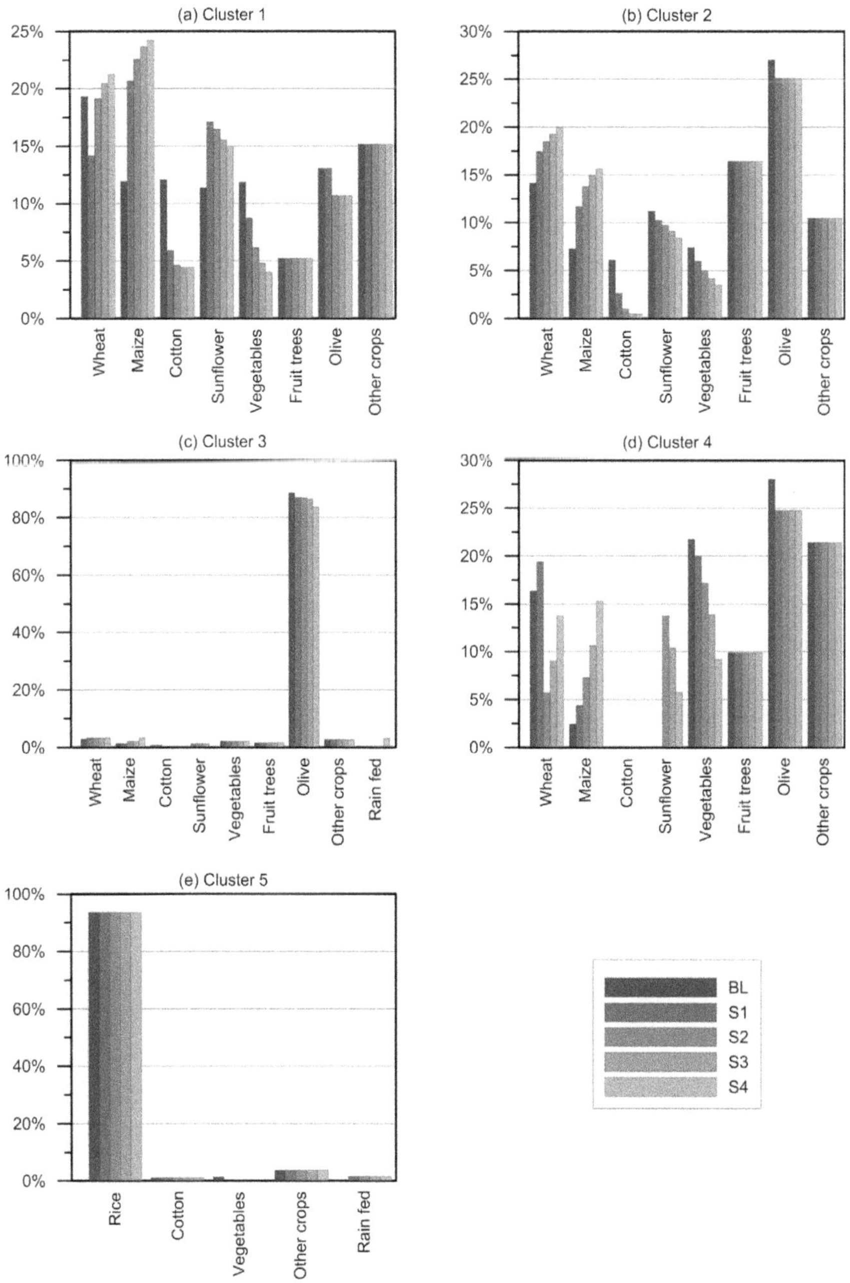

Figure 2. contd....

Figure 2. contd.

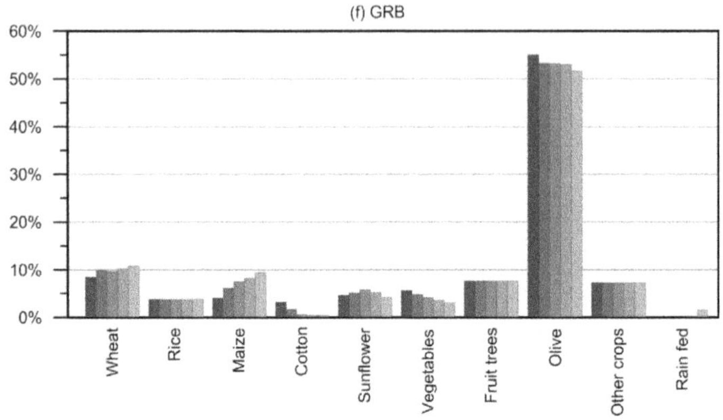

Figure 2. Crop portfolio under alternative price volatility scenarios: Cluster 1 (a), Cluster 2 (b), Cluster 3 (c), Cluster 4 (d), Cluster 5 (e) and GRB (f).
Source: Authors' elaboration from MAGRAMA (2009).

for risk aversion, which explains why the farmers still maintain a share of herbaceous crops. Therefore, as price increases so does income variability and the demand for herbaceous crops. However, given the scarce surface devoted to herbaceous crops and the agronomic constraints of the model, according to which the maximum land use variability of trees is 10%,[6] the solution to the model is not feasible for the S1, S2, S3 and S4. At this point, there are two alternatives: either we remove the agronomic constraint and allow the farmers to cut down a portion of the olive grove to increase the share of low risk crops, or the farmers adapt to the new situation. We choose the second alternative. This result is common in the literature and it is related with utility dynamics: as our methodology reveals static preferences and does not allow a change in these preferences, we get a sub-optimum solution (Collins et al. 1991).

In Cluster 4, the effects of the combined price increase of maize, wheat, cotton and sunflower over the crop portfolio are of particular interest. In S1, the surface of wheat and maize increases, while that of vegetables and olive trees is reduced and the small share of cotton that existed in the BL disappears. However in S2, while the surface of maize continues increasing and that of vegetables continues decreasing, a large share of wheat is substituted by sunflower (which had almost no relevance in the previous scenarios). Thereafter, in S3 and S4, sunflower is progressively substituted by wheat, although it ends up maintaining a relevant share over 5%. On the other hand, maize continues increasing and vegetables decreasing.

In Cluster 5, rice surface is enclosed and cannot be further expanded due to water and soil restrictions. As a result, price shocks only affect the remaining crops (less than 10% of the surface) and produce changes of little significance.

Overall, results show that a higher exposition to price variability in the GRB will increase the surface devoted to wheat and maize and reduce that of cotton, vegetables and olive trees as compared to the BL. The surface of sunflower increases in the S1 and S2, but in the S3 it starts decreasing and under the S4 its share is smaller than in the BL. Also, a small share of formerly irrigated areas is transformed into rainfed areas. The reduction in the area of olive trees is especially remarkable, as it is opposed to the observed evolution of the surface of irrigated olive trees in the basin during the last two decades (GRBA 2010) and it is also opposed to the prediction of López-Baldovin et al. (2005), whose model forecasts an increase in the surface of irrigated olive trees in the GRB; the explanation may be the market crisis for olive oil since 2008, with lower prices that are forcing some farmers to stop irrigation and return to rainfed cultivation. Moreover, it should be noted that this dynamics is the result of price volatility and that during the last two decades and in the model of López-Baldovin et al. (2005) agricultural prices were stable and income variability was low.

This new crop portfolio is largely the result of the two attributes that mostly define the clusters' utility functions: profit and risk. The larger share of wheat and maize can be partially explained by their increasing prices, but it is also the result of their reduced risk as compared to other competing crops that also show prices on the rise, such as rice, cotton and sunflower. Indeed, if we focus exclusively on the prices, rice (164%) and sunflower (125%) experience larger price increases than wheat (58%) and maize (67%) (see Table 3), but their joint surface share shrinks as a consequence of greater income variability. The sunflower is a paradigmatic example of this: in the S1 and S2 the price effect prevails over the volatility effect and its surface expands, but in the S3 and S4, the volatility effect prevails over a price increase as large as 125% and its surface is reduced. Also, the reduction in the number of hectares of cotton and vegetables and the increase in the surface of rainfed crops are a strategic response to reduce risk.

Water demand, income and employment

The resultant crop portfolios have direct effects over water demand, agricultural income and employment. Overall, the higher price volatility results in a higher expected gross variable margin, which grows as much as 31% in the S4 as compared to the BL. However, as price volatility increases the amount of total labor used decreases by 9%. This means a displacement from labor intensive crops to alternative crops, with a higher expected gross

Table 7. Water demand, agricultural income and employment in the GRB.[a,b]

	Irrigated area(ha)	Water use (hm^3)	Average dose(m^3/ha)	Gross Variable Margin (€/ha)	Hired labor (day's wages)	Household labor (day's wages)	Water apparent productivity (€/m^3)
BL	778,660	2,571	3,301	786	11,289	5,840	0.17
S1	778,090	2,578	3,313	843	10,987	5,557	0.18
S2	778,090	2,578	3,313	914	10,885	5,351	0.2
S3	776,731	2,578	3,319	991	10,810	5,200	0.23
S4	765,290	2,578	3,368	1,071	10,620	5,048	0.26

[a]Data covers 92.5% of the irrigated surface in the GRB.
[b]Water demand refers to water metered in the entrance of the farm.
Source: Authors' elaboration.

variable margin and a moderate risk. According to Sumpsi (2011), this can be explained by the higher exposure to price volatility of labor intensive crops as compared to alternative crops. Table 7 shows the results.

On other hand, water use remains almost constant in the five scenarios considered. This is a result of the significant restrictions in place over water use in the GRB. The GRB is a severely overexploited area (EEA 2009) and the river basin authority does not consider the assignment of new water use rights for agriculture. In addition, existing water allocations for irrigation are being nearly fully used (GRBA 2010). Consequently, in our scenarios, water demand barely increases and the growing net gross variable margin results in a water productivity increase of 43%.

Although water use in our scenarios remains constant, this is not to say that price volatility has no effect over water demand. To assess the effect of price volatility over the water demand curve, we simulate successive increases of water costs and we assess the farmers' response in terms of water consumption, thus building a demand curve for every scenario.

Results show that higher price volatility significantly affects the water demand curve, which becomes more inelastic: as expected prices and income variability rise, farmers' profits are increasingly sensitive to water supply restrictions and thus the farmers' willingness to pay for water is higher. This has at least two general implications for water policy: first, agriculture will be more vulnerable to drought events, creating incentives towards illegal abstractions (Gómez Gómez and Pérez Blanco 2012); second, water demand policies will need to be readapted to the higher willingness to pay for water, which in areas where water rights are not fully utilized may significantly increase water use.

We also assess the impact of the increase of water costs over labor, gross variable margin and productivity for every scenario. As expected, the gross variable margin becomes more elastic and thus more sensitive

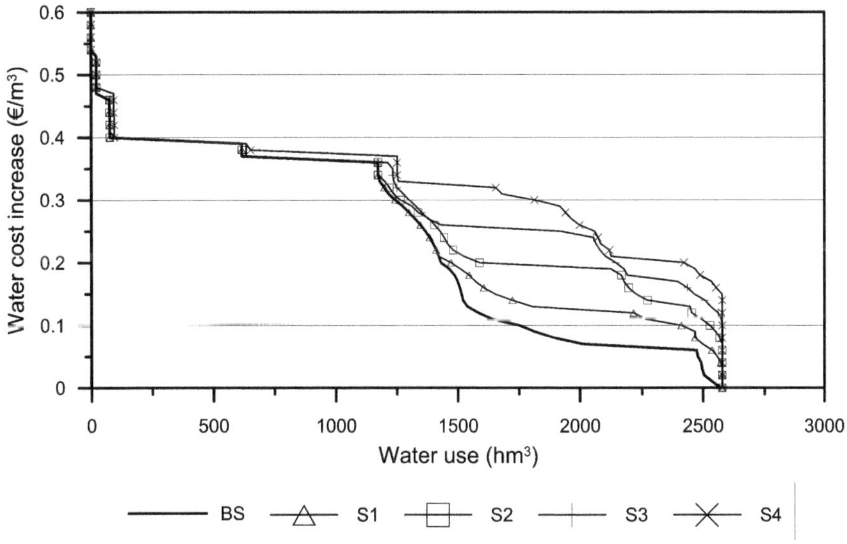

Figure 3. Water demand curve in the GRB.[a]
Source: Authors' elaboration from MAGRAMA (2009).
[a] Available microeconomic data covers 92.5% of the GRB irrigated surface (MAGRAMA 2009). Water demand in this figure is referred to the water that effectively reaches the farm, excluding transportation and distribution losses.

to increases of the water cost. Labor demand, although at first is reduced, becomes more inelastic and changes very little with increases of the water cost under 0.35 €/m³. Finally, water productivity is greater in the S4 with cost increases under 0.10 €/m³; after that, the highest productivity corresponds to the BL. With a cost increase of 0.10 €/m³, water intensive rice fields have a very low gross margin, and with subsequent water cost increases they are abruptly replaced by rainfed crops with a higher gross margin. Therefore, water productivity increases sharply. In the price volatility scenarios, this dynamics is not observed.

Conclusion

This chapter has shown the relevant impact that price volatility has over farmer decisions. Price volatility has significant effects over the crop

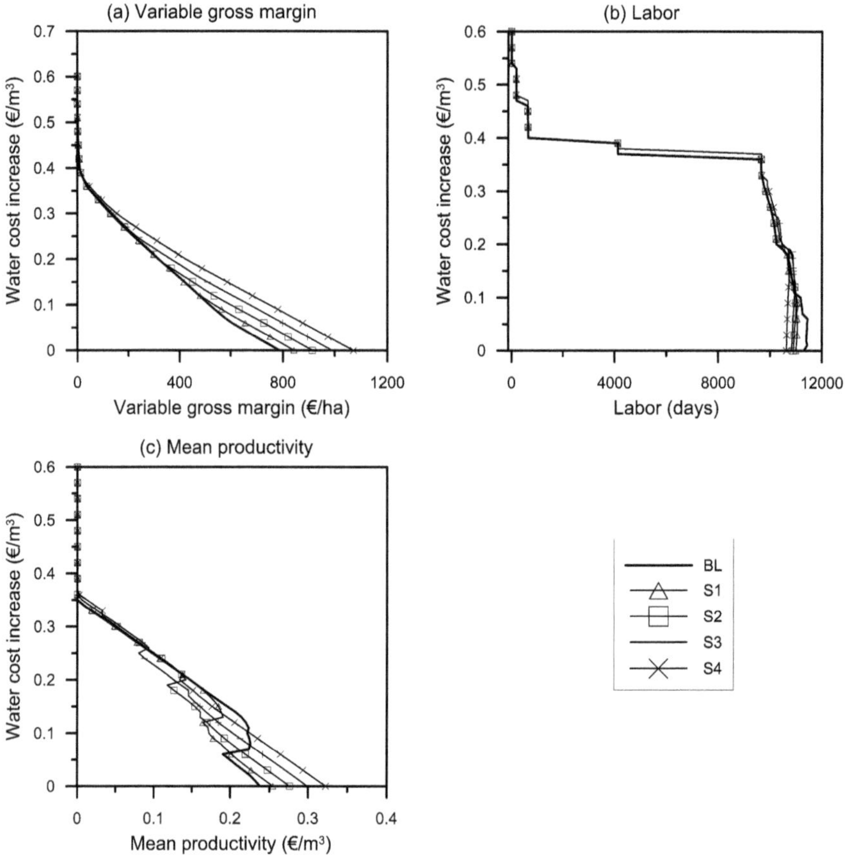

Figure 4. Gross variable margin, labor and average water productivity in the GRB.
Source: Authors' elaboration from MAGRAMA (2009).

portfolio, and thus over agricultural variables such as income, employment and also over water demand (even if water use remains almost constant, as in our case study area). Price volatility favors the expansion of the irrigated area of crops with higher prices but also with lower income variability, such as wheat and maize, and reduces that of crops with higher income variability, such as cotton and sunflower (though the latter increases its surface with a limited degree of variability, i.e., in the BL, S1 and S2). The resultant new crop portfolio significantly increases the expected income, by 36% in the S4 as compared to the BL. This seems a remarkable increase in the agricultural income, which contradicts the findings by Gilbert and Morgan (2010), according to which the effects of higher prices over agricultural income in developed countries like Spain would be small. Although a higher income implies a higher income variability, farmers may significantly reduce this risk exposure through a range of different mechanisms such as forward and future markets and crop insurance (Gilbert and Morgan 2010; Pérez Blanco and Gómez Gómez 2013). In addition, better market information and analysis could reduce uncertainties and assist producers to make better decisions (FAO 2011). However, it should be noted that this scenario would also provoke the substitution of labor intensive crops by alternative crops, resulting in a decrease of labor demand of 9% in the S4 as compared to the BL. This is expected to have a significant impact over employment and income distribution for an area where 7% of the workers are employed in the agricultural sector (GRBA 2010). Nonetheless, the public sector could partially redistribute this agricultural income increase through taxes and subsidies in order to alleviate this negative effect. Finally, although water use in the GRB barely changes with the new portfolio, the water demand curve is significantly altered and becomes more inelastic with the price shocks.

Our findings have many policy implications. For example, subsidies to the agricultural sector in OECD countries still represent 22% of the agricultural income, and over 50% of these subsidies are considered to directly distort trade and competition (OECD 2010). However, our results show that a more liberalized agricultural market would have positive effects over the agricultural income and could therefore benefit farmers. This finding may serve as an argument to reassess the justification of some subsidies. In any case, this should not be interpreted as an argument to indiscriminately remove agricultural subsidies, not even the most distorting ones, which in some cases may be justified (for example to prevent the negative impacts of sudden cost increases in the agricultural inputs, such as fertilizers (FAO 2011)). This is rather a suggestion to reconsider the need for those subsidies that are exclusively used to isolate farmers from international prices and competition.

This chapter also has some relevant implications for water policy. First, higher prices in international markets make water demand more inelastic and thus reduce the effectiveness of water pricing, which is one of the key instruments to reduce water use in the EU (EC 2012). For example in our case study area, water use in the S4 is reduced only slightly (by 0.8%) after an increase in the cost of water of 0.15 €/m^3, while the same cost increase allowed to reduce water use by 41.7% in the BL. Therefore, although water pricing would become a more useful tool for cost recovery or even for tax revenue, it would lose most of its ability to reduce water use in agriculture. This is a key finding that implies that water saving/conservation policies should be redefined in order to consider alternative instruments.

In addition, higher agricultural prices will result in higher incentives towards water use. Although water use cannot formally increase because of the current legal framework, which avoids the allocation of further water rights due to the exhaustion of local water sources (GRBA 2010), the higher expected agricultural income could be an important incentive to increase illegal water abstractions, so better governance and a strict application of the law should be enforced.

The methodology that we develop in this paper is flexible and can be used to assess the impact of different agricultural policies over farmers' decisions. However, it should be noted that this method has some drawbacks that need to be explored in future studies. First of all, the calibration of the model is not exempt of calibration errors, which in some cases may demand a relaxation of some of the assumptions or the imposition of additional restrictions in order to obtain a solution. However, this situation is not very common and in our case study area only appears in the calibration of the Cluster 3 (olive grove). The second critique, maybe more important to the present study, refers to the design of the scenarios. This design is based on the observed evolution of international and domestic agricultural prices, and they are not integrated in the model. As a result, they do not consider the collateral effects that a price increase of agricultural products worldwide may have over other relevant agronomic variables apart from domestic prices, such as input prices. Depending on the relevance of this secondary effect, the results of our model could need a reassessment. Therefore, the next steps in this research should try to fully integrate the stages in which scenarios are designed with the stage in which the model is calibrated.

References

Abbott, P.C., Hurt, C. and Tyner, W.E. 2008. What's Driving Food Prices? Issue Report. Farm Foundation. Available at http://www.farmfoundation.org/.

Berbel, J. and Rodríguez-Ocaña, A. 1998. An MCDM approach to production analysis: an application to irrigated farms in southern Spain. European Journal of Operational Research 107(1): 108–118.

Berbel, J. and Gómez-Limón, J.A. 2000. The impact of water-pricing policy in Spain: an analysis of three irrigated areas. Agricultural Water Management 43(2): 219–238.

Berbel, J., Kolberg, S. and Martin-Ortega, J. 2012. Assessment of the draft hydrological basin plan of the Guadalquivir river basin (Spain). International Journal of Water Resources Development 28(1): 43–56.

Collins, A., Musser, W.N. and Mason, R. 1991. Prospect theory and risk preferences of Oregon seed producers. American Journal of Agricultural Economics 73(2): 429–435.

Cooke, B. and Robles, M. 2009. Recent food price movements: a time series analysis. IFPRI Discussion Paper, 00942, IFPRI, Washington DC.

De Frahan, H.B., Buysse, J., Polomé, P., Fernagut, B., Harmignie, O., Lauwers, L., van Huylenbroeck, G. and van Meensel, J. 2007. Positive mathematical programming for agricultural and environmental policy analysis: review and practice. pp. 129–154. *In*: Weintraub, A., Bjorndal, T., Epstein, R. and Romero, C. (eds.). Handbook of Operations Research in Natural Resources. Springer, New York.

Delforce, R.J. and Hardaker, J.B. 1985. An experiment in multiattribute utility theory. Australian Journal of Agricultural and Resource Economics 29(3): 179–198.

Dono, G., Giraldo, L. and Severini, S. 2010. Pricing of irrigation water under alternative charging methods: possible shortcomings of a volumetric approach. Agricultural Water Management 97: 1795–1805.

EC—European Commission. 2000. Water Framework Directive. Directive 2000/60/EC, 23 October 2000.

EC—European Commission. 2008. Drought Management Plan Report. Including Agricultural, Drought Indicators and Climate Change Aspects. Water Scarcity and Droughts Expert Network. http://ec.europa.eu/environment/water/quantity/pdf/dmp_report.pdf. Last accessed 07/05/2013.

EC—European Commission. 2012. A Blueprint to Safeguard Europe's Water Resources. COM (2012) 673 final. Brussels.

EEA—European Environment Agency. 2009. Water Resources Across Europe—Confronting Water Scarcity and Drought. EEA Report 2/2009.

FAO. 2011. Price volatility in food and agricultural markets: policy responses. Available at http://www.fao.org. Last accessed 07/05/2013.

Gilbert, C.L. 2010. How to understand high food prices. Journal of Agricultural Economics 61: 398–425.

Gilbert, C.L. and Morgan, C.W. 2010. Food price volatility. Philosophical Transactions of the Royal Society B: Biological Sciences 365(1554): 3023–3034.

Gómez Gómez, C.M. and Pérez Blanco, C.D. 2012. Do drought management plans really reduce drought risk? A risk assessment model for a Mediterranean river basin. Ecological Economics 76: 42–48.

Gómez-Limón, J.A. and Riesgo, L. 2004. Water pricing: analysis of differential impacts on heterogeneous farmers. Water Resources Research 40: W07S05.

Gómez-Limón, J.A., Arriaza, M. and Villanueva, A. 2012. Tipología de las unidades de agregación de la Demarcación Hidrográfica del Guadalquivir. XXX Congreso Nacional de Riegos. Albacete, Spain.

GRBA—Guadalquivir River Basin Authority. 2010. Propuesta de Proyecto de Plan Hidrológico de la Demarcación Hidrográfica del Guadalquivir. Available from: http://www.chguadalquivir.es/opencms/portalchg/servicios/prensaComunicaciones/boletinNoticias/noticias/noticia0084.html. Last accessed 07/05/2013.

Gutierrez-Martin, C. and Gomez Gomez, C.M. 2011. Assessing irrigation efficiency improvements by using a preference revelation model. Spanish Journal of Agricultural Research 9(4): 1009–1020.

Heckelei, T. and Britz, W. 2005. Models based on positive mathematical programming: state of the art and further extensions. 89th EAAE Seminar. Parma, Italy, February 3–5, pp. 48–73.

Howitt, R.E. 1995. Positive mathematical programming. American Journal of Agricultural Economics 77(2): 329–342.

ICWE—International Conference on Water and Environment. 1992. The Dublin statement on water and sustainable development. Available from: http://www.inpim.org/files/Documents/DublinStatmt.pdf. Last accessed 07/05/2013.

Kampas, A., Petsakos, A. and Rozakis, S. 2012. Price induced irrigation water saving: unraveling conflicts and synergies between European agricultural and water policies for a Greek Water District. Agricultural Systems 113: 28–38.

López-Baldovin, M.J., Gutiérrez-Martin, C. and Berbel, J. 2005. Multicriteria and multiperiod programming for scenario analysis in Guadalquivir river irrigated farming. Journal of the Operational Research Society 57: 499–509.

Ministerio de Agricultura, Alimentación y Medio Ambiente (MAGRAMA). 2009. Resultados Técnicos-económicos de Sistemas de Explotación Agraria. Madrid.

Mitchell, D. 2008. A note on rising food prices. Policy Research Working Paper 4682. World Bank, Development Prospects Group, Washington DC.

OECD—Organization for Economic Co-operation and Development. 2010. Agricultural Policies in OECD Countries: At a Glance, OECD, Paris.

Paris, Q. 2010. Economic Foundations of Symmetric Programming. Cambridge University Press, Cambridge.

Paris, Q. and Howitt, R.E. 1998. An analysis of Ill-posed production problems using maximum entropy. American Journal of Agricultural Economics 80(1): 124–138.

Pérez Blanco, C.D. and Gómez Gómez, C.M. 2013. Designing optimum insurance schemes to reduce water overexploitation during drought events: a case study of La Campiña, Guadalquivir River Basin, Spain. Journal of Environmental Economics and Policy 2(1): 1–15.

Quiroga, S. and Iglesias, A. 2009. A comparison of the climate risks of cereal, citrus, grapevine and olive production in Spain. Agricultural Systems 101: 91–100.

Rausser, G.C. and Yassour, J. 1981. Multi attribute utility analysis: the case of Filipino rice policy. American Journal of Agricultural Economics 63(4): 484–494.

Roache, S.K. 2009. What explains the rise in food price volatility? IMF Working Paper. Research Department.

Sumpsi, J.M. 2011. Volatilidad de los mercados agrarios y crisis alimentaria. Revista española de estudios agrosociales y pesqueros (229): 11–35.

Timmer, C.P. 2012. Rice price formation in the short run and in the long run: the role of market structure in explaining volatility. pp. 151–186. *In*: Munier, B.R. (ed.). Global Uncertainty and the Volatility of Agricultural Commodities Prices. IOS Press. Amsterdam, Netherlands.

Ward, F.A. and Pulido-Velazquez, M. 2008. Water conservation in irrigation can increase water use. Proceedings of the National Academy of Sciences of the United States of America 25; 105(47): 18215–18220.

World Bank. 2012. Agricultural prices database. http://databank.worldbank.org. Last accessed 07/05/2013.

Zezza, A., Davis, B., Azzarri, C., Covarrubias, K., Tasciotti, L. and Anriquez, G. 2009. The Impact of Rising Food Prices on the Poor. International Association of Agricultural Economists Conference, Beijing, China, August 16–22.

Index

Color Plate Section

Chapter 3

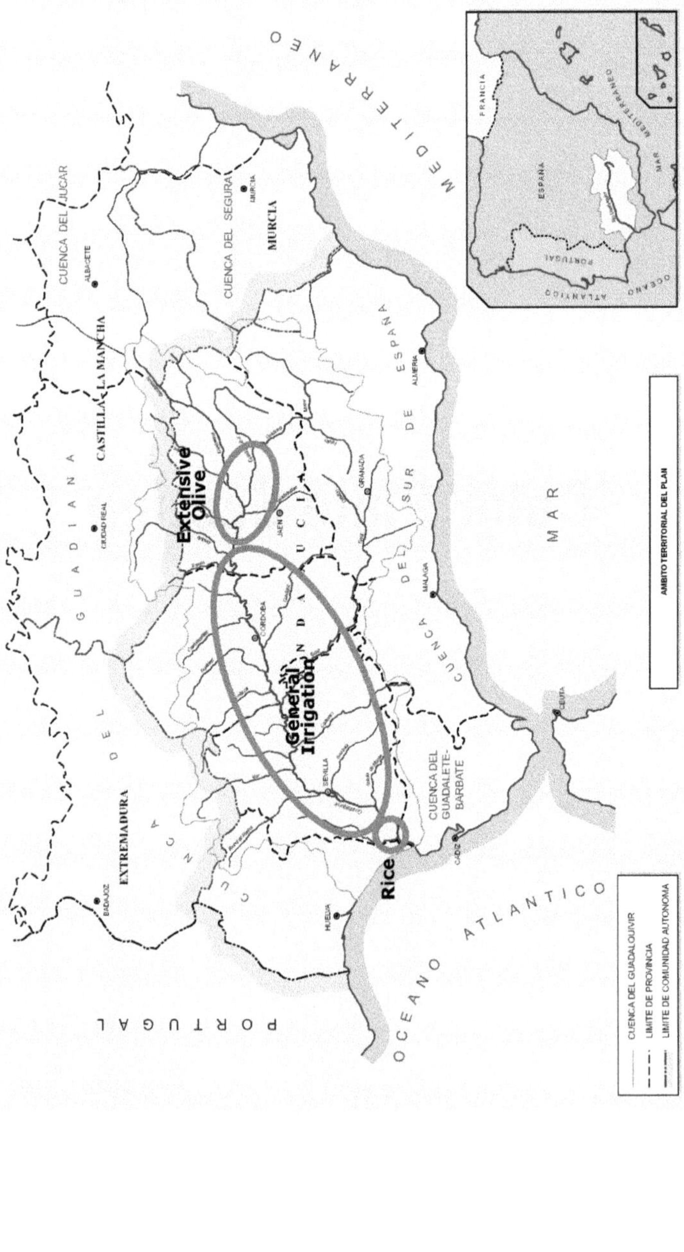

Figure 2. Map of the Guadalquivir River Basin in southern Spain.
Source: CHG (2009).

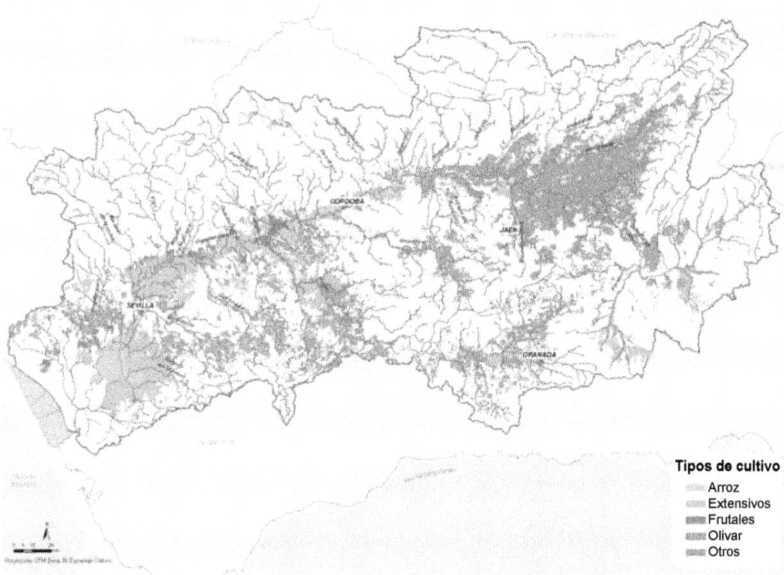

Figure 3. Spatial distribution of crop types in the Guadalquivir RB in 2008.
Source: (CHG 2010b).

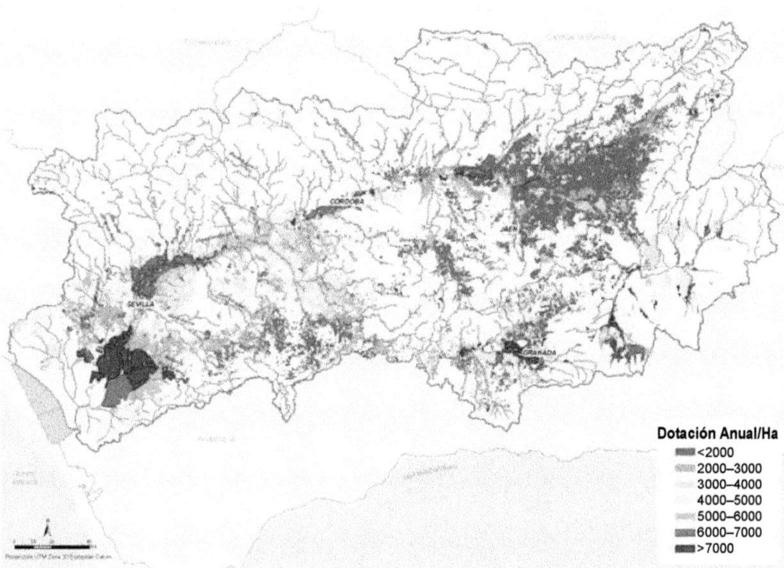

Figure 4. Spatial distribution of annual water allocation (m^3 ha^{-1}).
Source: CHG (2010b).

Chapter 4

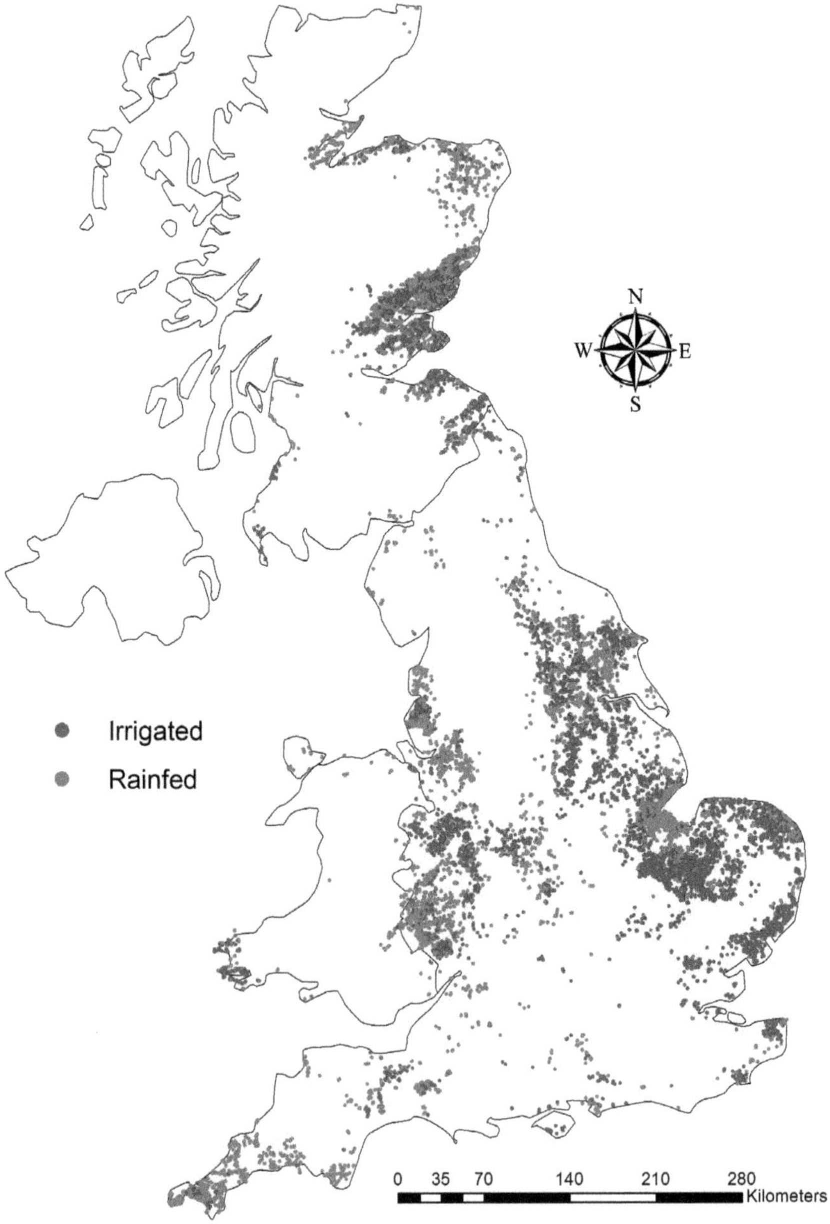

Figure 2. contd....

Figure 2. contd.

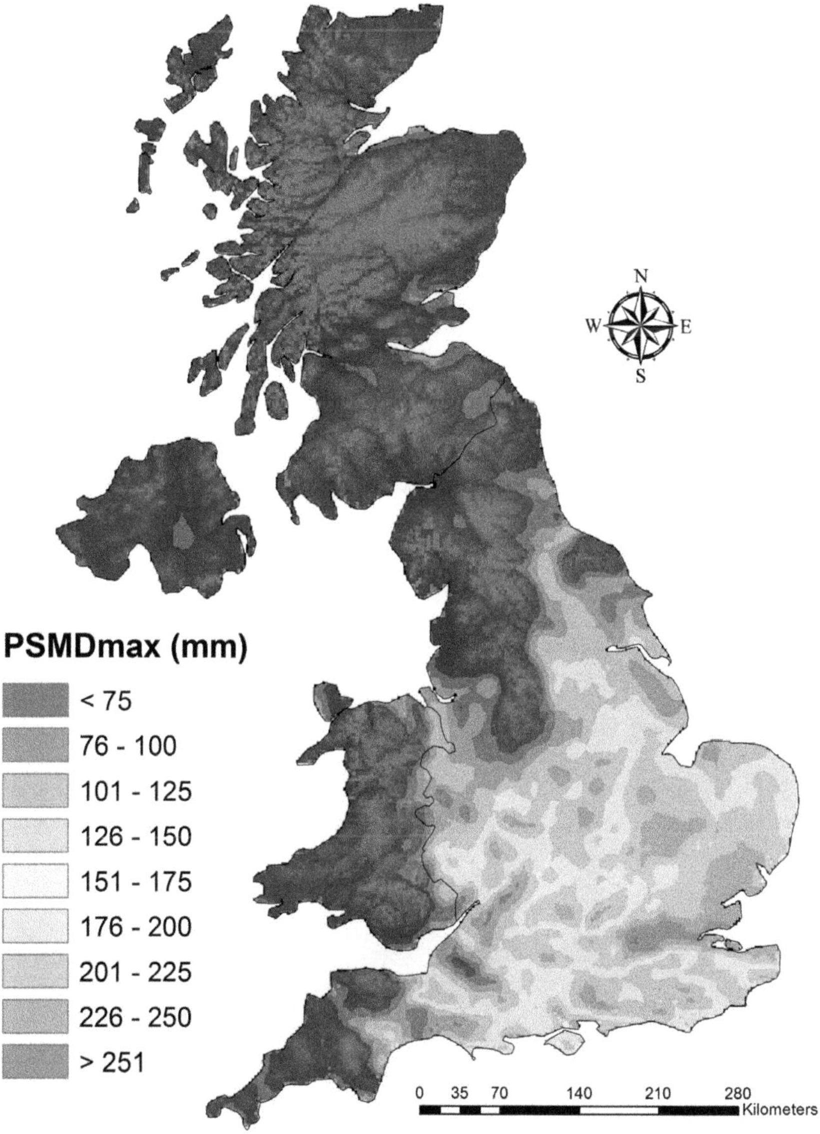

Figure 2. Spatial distribution of irrigated and rainfed potato farms across the UK in 2009 (left panel) and long-term average (1961–1990) spatial variation in agroclimate, using potential soil moisture deficit (PSMD) as an aridity index (right panel).

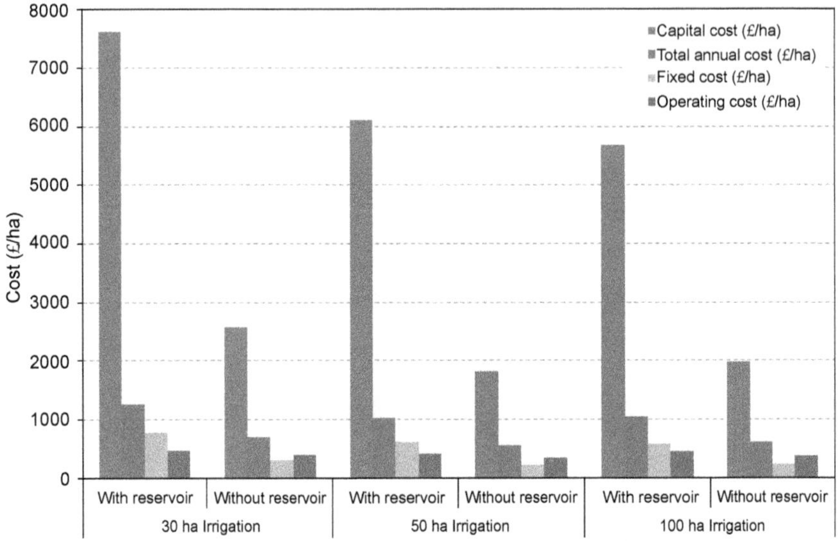

Figure 6. Irrigation costs by area of irrigation and use of unlined on-farm reservoirs providing 100% of water requirements.

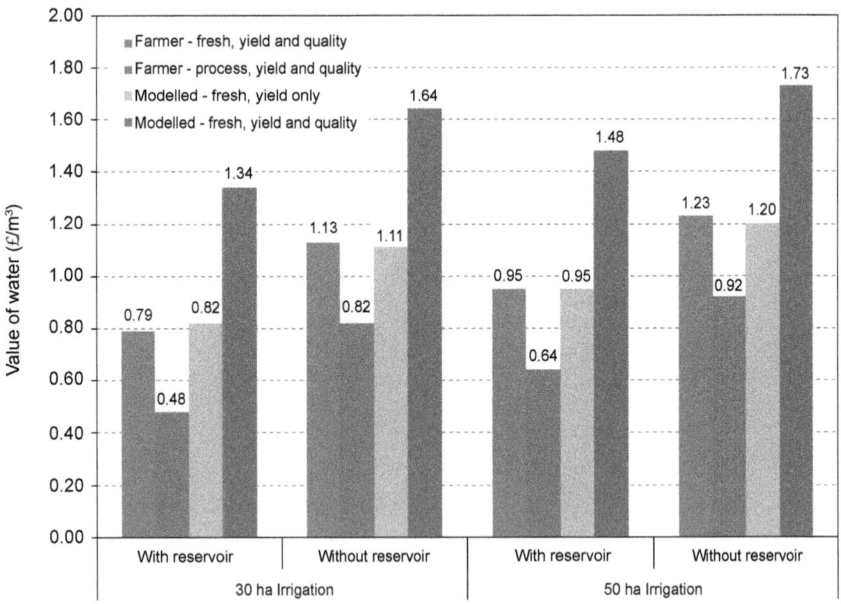

Figure 8. Average value added by irrigation water on potatoes (£/m^3). Note: Farmer estimates based on Shropshire. Modeled estimates based on Cambridgeshire.

Chapter 12

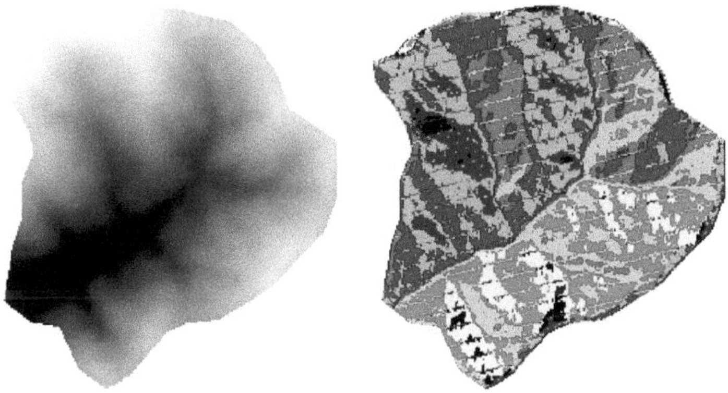

a. b.

Figure 3. (a) Digital Elevation Model (DEM) and (b) Flow Direction Model (FDM) of Ano Mela river basin.

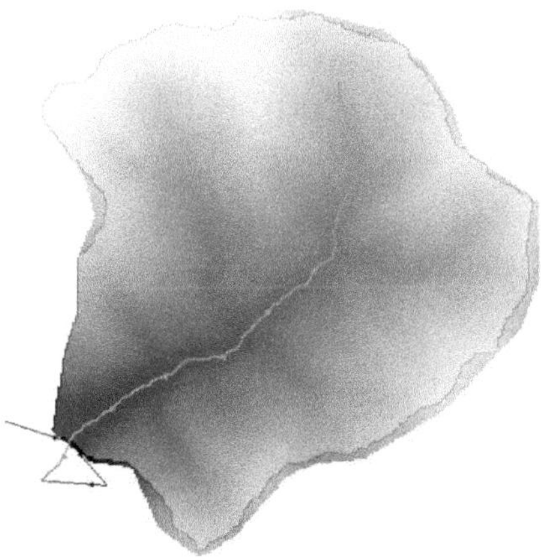

Figure 4. The position of the stream in the basin.

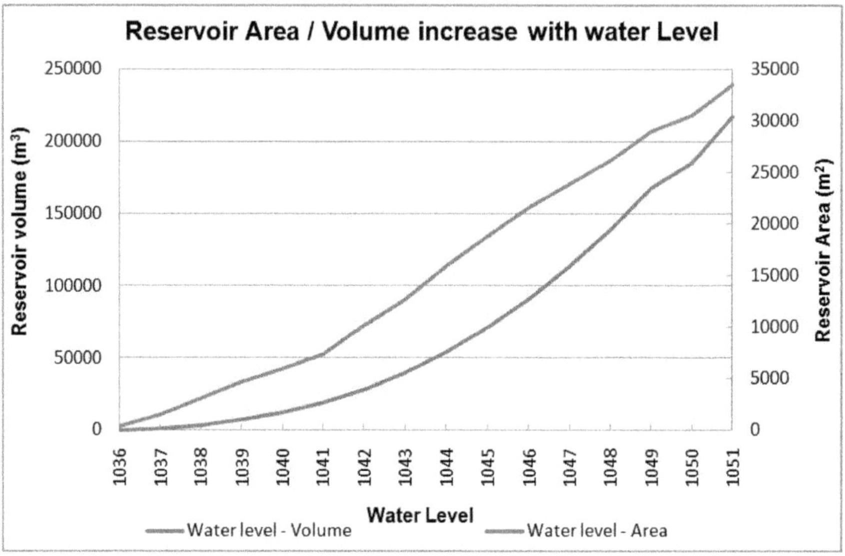

Figure 5. Water level–Surface and Water level–Volume diagrams.

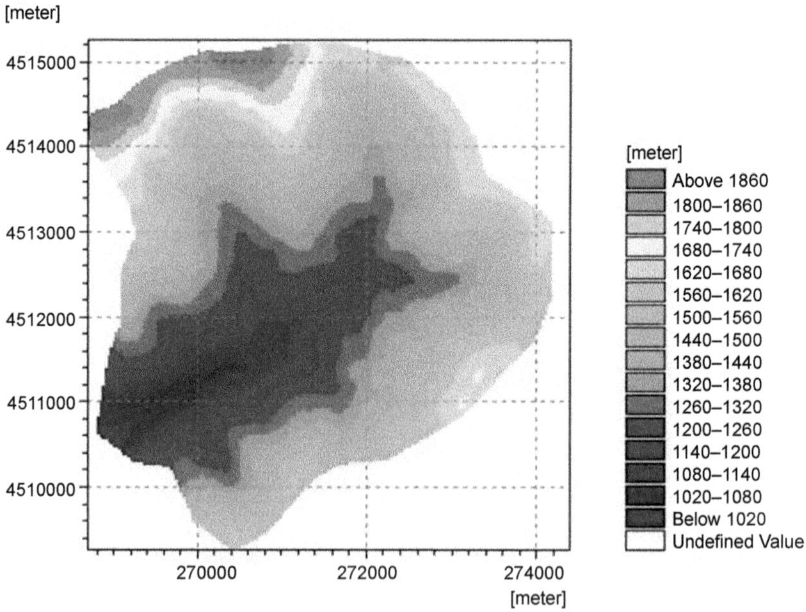

Figure 6. The Ano Mela DEM in MIKE SHE.

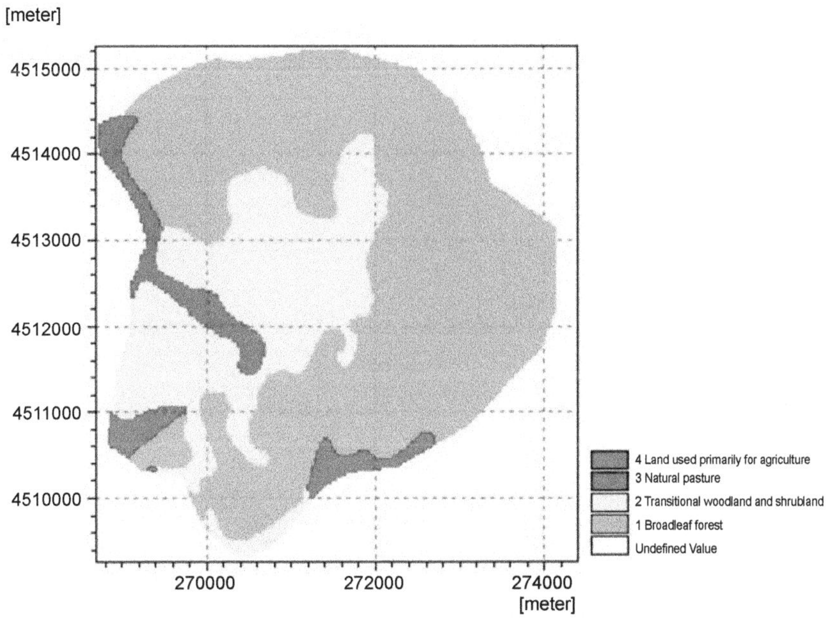

Figure 8. Land Use.

Chapter 13

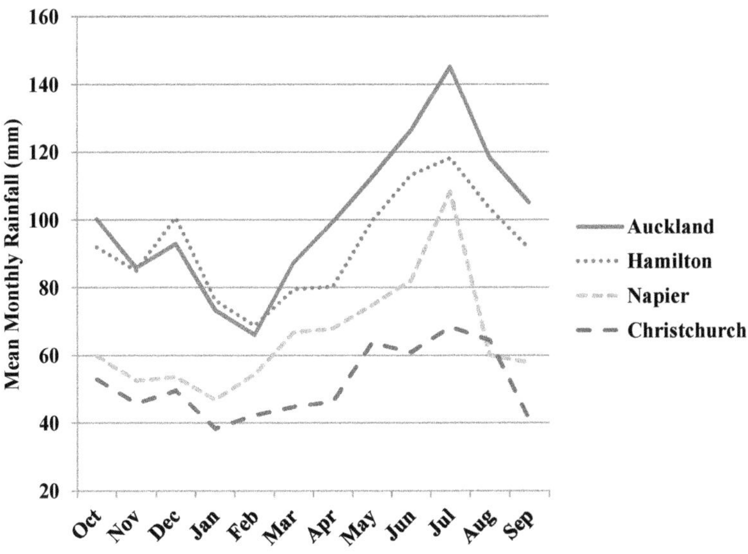

Figure 2. NZ Monthly Rainfall for Selected Cities.

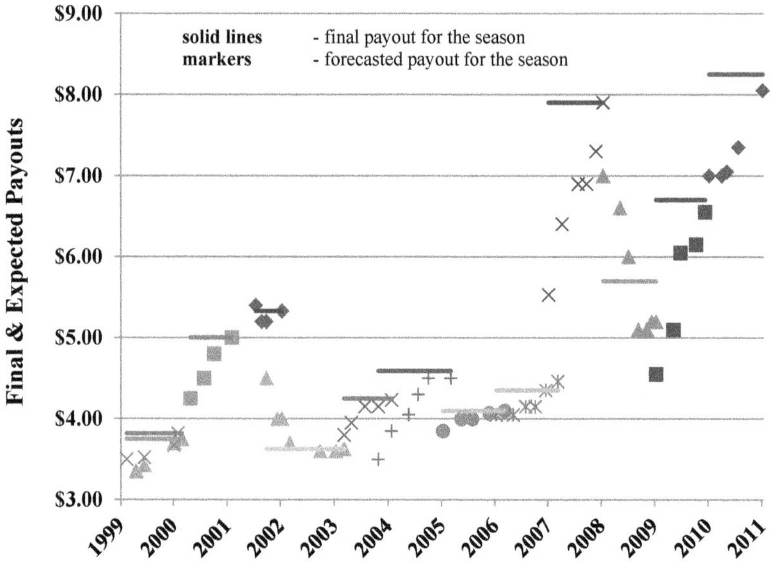

Figure 4. Forecasted vs. Actual Payout.
Source: Fonterra 2000–2012.

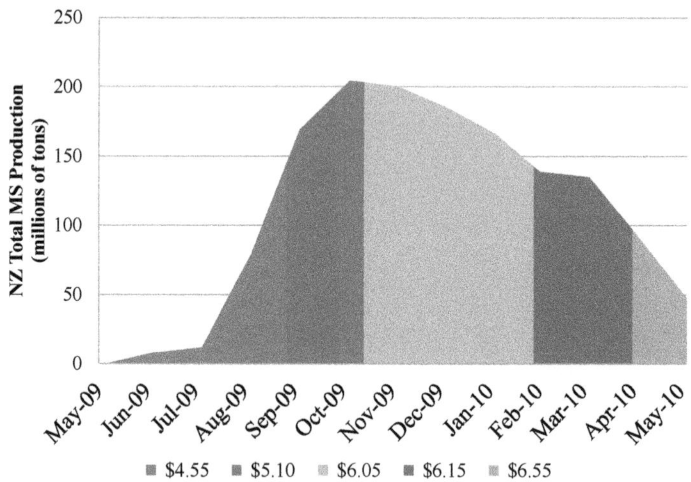

Figure 5. Output-Weighted Forecast Estimation.
Data Sources: DCANZ (2012); Fonterra (2009, 2010).

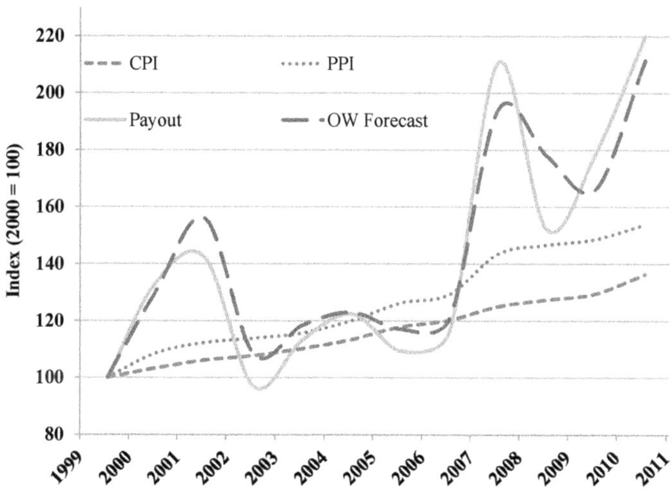

Figure 6. CPI, PPI and Payout Indices (1999=100).

Chapter 14

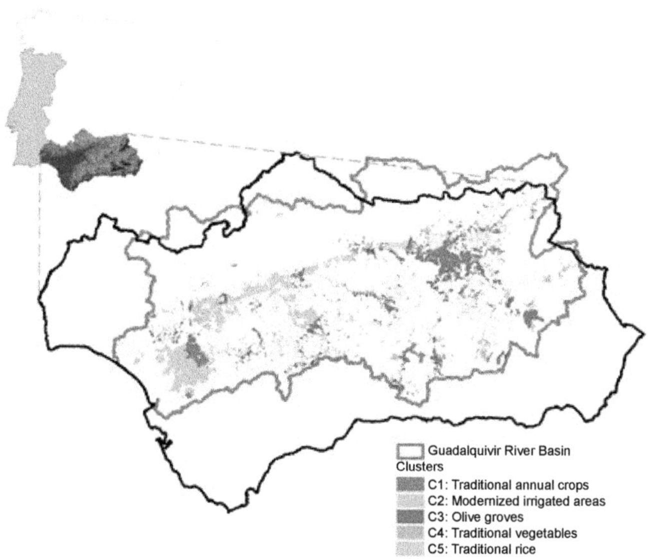

Figure 1. Clusters identified in the GRB.
Source: Gómez-Limón et al. (2012).

For Product Safety Concerns and Information please contact our EU representative GPSR@taylorandfrancis.com
Taylor & Francis Verlag GmbH, Kaufingerstraße 24, 80331 München, Germany

www.ingramcontent.com/pod-product-compliance
Ingram Content Group UK Ltd.
Pitfield, Milton Keynes, MK11 3LW, UK
UKHW021504240926
475702UK00035B/357
* 9 7 8 1 4 8 2 2 3 8 3 9 6 *